Statistics and Computing

Series Editors:
W. Eddy
W. Härdle
S. Sheather
L. Tierney

Statistics and Computing

Härdle/Klinke/Turlach: XploRe: An Interactive Statistical
    Computing Environment
Venables/Ripley: Modern Applied Statistics with S-Plus

W.N. Venables
B.D. Ripley

Modern Applied Statistics with S-Plus

With 124 Figures

Springer-Verlag

New York Berlin Heidelberg London Paris
Tokyo Hong Kong Barcelona Budapest

W.N. Venables
Department of Statistics
University of Adelaide
Adelaide, South Australia 5005
Australia

B.D. Ripley
Professor of Applied Statistics
University of Oxford
1 South Parks Road
Oxford OX1 3TG
England

Series Editors:

W. Eddy	W. Härdle	S. Sheather	L. Tierney
Department of	Institut für Statistik und	Australian Graduate	School of Statistics
Statistics	Ökonometrie	School of Management	University of
Carnegie Mellon	Humboldt-Universität zu	PO Box 1	Minnesota
University	Berlin	Kensington	Vincent Hall
Pittsburgh, PA	Spandauer Str. 1	New South Wales 2033	Minneapolis, MN
15213	D-10178 Berlin	Australia	55455
USA	Germany		USA

Library of Congress Cataloging-in-Publication Data
    Modern applied statistics with S-PLUS / W.N. Venables, B.D.
Ripley.
        p.   cm. -- (Statistics and computing)
    Includes bibliographical references and index.
        ISBN 0-387-94350-1 (New York). -- ISBN 3-540-94350-1 (Berlin)
        1. S-Plus.   2. Statistics--Data processing.   3.Mathematical
    statistics--Data processing.   I. Ripley, Brian D., 1952-
    II. Title.   III. Series.
    QA276.4.V46   1994
    005.369--dc20   94-21589

Printed on acid-free paper.

Production managed by Jim Harbison; manufacturing supervised by Jacqui Ashri.
Photocomposed pages prepared from the authors' PostScript files.
Printed and bound by Hamilton Printing Co., Rensselaer, NY.
Printed in the United States of America.

9 8 7 6 5 4 3 2 (Second corrected printing)

ISBN 0-387-94350-1 Springer-Verlag New York Berlin Heidelberg
ISBN 3-540-94350-1 Springer-Verlag Berlin Heidelberg New York

Preface

Increases in computer power and falling costs have changed dramatically the concept of the 'statistician's calculator'. From 1903 to 1963 this would have been mechanical or electro-mechanical supplemented by a slide rule. In 1973 the HP-35 scientific calculator was available to the wealthy. By 1983 we used BASIC on a microcomputer or a statistical package on a time-sharing central service, but graphical output was cumbersome if available at all. By 1993 a statistician could have a workstation displaying high-resolution graphics with megaflops of computing power and hundreds of megabytes of disc storage. In this era the statistical system S has flourished.

It is perhaps misleading to call S a statistical system. It was developed at AT&T's Bell Laboratories as a flexible environment for data analysis, and in its raw form implements rather few statistical concepts. It does however provide the tools to implement rather elegantly many statistical ideas, and because it can be extended by user-written procedures (both in its own language and in C and FORTRAN), it has been used as the basis of a number of statistics systems, notably S-PLUS from the StatSci division of MathSoft, which has implemented a number of topics (such as time series and survival analysis) using the S language. (The S system is now marketed exclusively by StatSci.)

The user community has contributed a wealth of software within S. This book shows its reader how to extend S, and uses this facility to discuss procedures not implemented in S-PLUS, thereby providing fairly extensive examples of how this might be done.

This is not a text in statistical theory, but does cover modern statistical methodology. Each chapter summarizes the methods discussed, in order to set out the notation and the precise method implemented in S. (It will help if the reader has a basic knowledge of the topic of the chapter, but several chapters have been successfully used for specialized courses in statistical methods.) Our aim is rather to show how we analyse datasets using our 'statistical calculator' S-PLUS. In doing so we aim both to show how S can be used and how the availability of a powerful and graphical system has altered the way we approach data analysis and allows penetrating analyses to be performed routinely. Once calculation became easy, the statistician's energies could be devoted to understanding his or her dataset.

The core S language is not very large, but it is quite different from most other statistics systems. We describe the language in some detail in the early chapters, but these are probably best skimmed at first reading; Chapter 1 contains the most basic ideas, and each of Chapters 2 and 3 are divided into 'basic' and 'advanced' sections. Once the philosophy of the language is grasped, its consistency and logical design will be appreciated.

The chapters on applying S to statistical problems are largely self-contained, although Chapter 6 describes the language used for linear models, and this is used in several later chapters. We expect that most readers will want to pick and choose amongst the later chapters, although they do provide a number of examples of S programming, and so may be of interest to S programmers not interested in the statistical material of the chapter.

This book is intended both for would-be users of S-PLUS (or S) as an introductory guide and for class use. The level of course for which it is suitable differs from country to country, but would generally range from the upper years of an undergraduate course (especially the early chapters) to Masters' level. (For example, almost all the material is covered in the M.Sc. in Applied Statistics at Oxford.) Exercises are provided only for Chapters 2–5, since otherwise these would detract from the best exercise of all, using S to study datasets with which the reader is familiar. Our library provides many datasets, some of which are not used in the text but are there to provide source material for exercises. (We would stress that it is preferable to use examples from the user's own subject area and experience where this is possible.)

Both authors take responsibility for the whole book, but Bill Venables was the lead author for Chapters 1–4, 6, 7 and 9, and Brian Ripley for Chapters 5, 8 and 10–15. (The ordering of authors reflects the ordering of these chapters.)

The datasets and software used in this book are available for electronic distribution from sites in the USA, England and Australia. Details of how to obtain the software are given in Appendix A. They remain the property of the authors or of the original source, but may be freely distributed provided the source is acknowledged. We have tested the software as widely as we are able (including under both Unix and Windows versions of S-PLUS), but it is inevitable that system dependencies will arise. We are unlikely to be in a position to assist with such problems.

The authors may be contacted by electronic mail as

```
venables@stats.adelaide.edu.au
ripley@stats.ox.ac.uk
```

and would appreciate being informed of errors and improvements to the contents of this book.

Acknowledgements:

This book would not be possible without the S environment which has been principally developed by Rick Becker, John Chambers and Allan Wilks, with substantial input from Doug Bates, Bill Cleveland, Trevor Hastie and Daryl Pregibon. The code for survival analysis is the work of Terry Therneau. The S-PLUS code is the work of a much larger team acknowledged in the manuals for that system.

We are grateful to the many people who have read and commented on one or more chapters and who have helped us test the software, as well as to those whose problems have contributed to our understanding and indirectly to examples and exercises. We can not name all, but in particular we would like to thank Adrian Bowman, Michael Carstensen, Sue Clancy, David Cox, Anthony Davison, Peter Diggle, Matthew Eagle, Nils Hjort, Francis Marriott, David Smith, Patty Solomon and StatSci for the provision of a very early copy of S-PLUS 3.2 for final testing of our material.

Bill Venables
Brian Ripley
April 1994

Added at second printing

We have taken the opportunity to incorporate some of the changes planned for S-PLUS 3.3, notably the inclusion of `survival4`. The software has been altered to remove all FORTRAN and to incorporate dynamically-loaded code in the Windows version. Chapter 13 uses corrected versions of the tree-pruning routines which are now supplied, and is the only place where substantial changes occur.

S-PLUS 3.3 will also incorporate classes for various types of matrices and 'trellis' graphics implementing the ideas of Cleveland (1993). We do not yet have access to these (and so have not described them), but believe that they will become preferred methods in due course.

Bill Venables
Brian Ripley
February 1995

Contents

Appendices

Typographical Conventions

Throughout this book S language constructs and commands to the operating system are set in a monospaced typewriter font `like this`.

We often use the prompts `$` for the operating system (it is the standard prompt for the Unix Bourne shell) and `>` for S-PLUS. However, we do *not* use prompts for continuation lines, which are indicated by indentation. One reason for this is that the length of line available to use in a book column is less than that of a standard terminal window, so we have had to break lines which were not broken at the terminal.

Some of the S-PLUS output has been edited. Where complete lines are omitted, these are usually indicated by

    . . . .

in listings; however most *blank* lines have been silently removed. Much of the S-PLUS output was generated with the options settings

```
options(width=65, digits=5)
```

in effect, whereas the defaults are 80 or 87 (depending on the release) and 7. Not all S-PLUS functions consult these settings, so on occasion we have had to manually reduce the precision to more sensible values.

Chapter 1

Introduction

Statistics is fundamentally concerned with the understanding of structures in data. One of the effects of the information-technology era has been to make it much easier to collect extensive datasets with minimal human intervention. Fortunately the same technological advances allow the users of statistics access to much more powerful 'calculators' to manipulate and display data. This book is about the modern developments in applied statistics which have been made possible by the widespread availability of workstations with high-resolution graphics and computational power equal to a mainframe of a few years ago. Workstations need software, and the S system developed at AT&T's Bell Laboratories provides a very flexible and powerful environment in which to implement new statistical ideas. Thus this book provides both an introduction to the use of S and a course in modern statistical methods.

S is an integrated suite of software facilities for data analysis and graphical display. Among other things it offers

- an extensive and coherent collection of tools for statistics and data analysis,

- a language for expressing statistical models and tools for using linear and non-linear statistical models,

- graphical facilities for data analysis and display either at a workstation or as hardcopy,

- an effective object-oriented programming language which can easily be extended by the user community.

The term *environment* is intended to characterize it as a planned and coherent system built around a language and a collection of low-level facilities, rather than the perceived 'package' model of an incremental accretion of very specific, high level and sometimes inflexible tools. Its great strength is that functions implementing new statistical methods can be built on top of the low-level facilities.

Furthermore, most of the environment is open enough that users can explore and if they wish, change the design decisions made by the original implementors. Suppose you do not like the output given by the regression facility (as we have frequently felt about statistics packages). In S you can write your own summary routine, and the system one can be used as a template from which to start. S (but not S-PLUS) is distributed as source code, so sufficiently persistent users can always find out the exact algorithm used, but in most cases this can be deduced by listing the S functions invoked.

We have made extensive use of the ability to extend the environment to implement (or re-implement) statistical ideas within S. All the S functions which are used and our datasets are available in machine-readable form; see Appendix A for details of what is available and how to install it.

The name S arose long ago as a compromise name, in the spirit of the programming language C (also from Bell Laboratories).

We will assume the reader has S-PLUS, an enhanced version of S supplied in binary form by the StatSci division of MathSoft. However most of our examples apply equally well to S; the key differences are outlined in Appendix C. We refer to the system as S unless we are invoking features unique to S-PLUS.

System dependencies

We have tried as far as is practicable to make our descriptions independent of the computing environment and the exact version of S or S-PLUS in use. Clearly some of the details must depend on the environment; we used S-PLUS 3.2 on Sun SparcStations under OpenLook to compute all the examples, but also tested them under S-PLUS 3.2 for Windows.

S and S-PLUS are normally used at a workstation running a graphical user interface (GUI) such as OpenLook, Motif and Microsoft Windows, or an X-terminal running one of the first two of these. We will assume such a graphical environment. It is possible to run S and S-PLUS in an ordinary terminal window and to display graphics on a graphics terminal, but less conveniently. In particular, the ability to cut and paste is very convenient in writing S code; the examples in this book were developed by cut and paste between the actual LaTeX source and a window running an S-PLUS session.

Most versions of S and S-PLUS run under Unix, but versions of S-PLUS are available for PCs running Windows and DOS. These suffer from only one major limitation for using this book; source code in C and FORTRAN (used in a number of our extensions) can only be used if the appropriate compilers are available, which they rarely are. Thus part of our spatial statistics code and a few of the other methods are unavailable to such users. Further, the survival analysis functions we use will only become available in future releases of S-PLUS for Windows; releases 3.1 and 3.2 have a different and less powerful set of routines for survival analysis. (So do the Unix releases, but users can compile the later code.) The DOS version of S-PLUS is now obsolete, and will not be mentioned further.

One system dependency is the mapping of mouse buttons. We refer to button 1, usually the left button, and button 2, the middle button on Unix interfaces (or perhaps both together of two) and the right button under Windows.

Reference manuals

The basic references are Becker, Chambers & Wilks (1988) for the basic environment and Chambers & Hastie (1992) for the statistical modelling, object-oriented programming and data structuring features; these should be supplemented by checking the on-line help pages for changes and corrections. Our aim is not to be comprehensive nor to replace these manuals, but rather to explore much further the use of S and S-PLUS to perform statistical analyses.

1.1   A quick overview of S

Most things done in S are permanent; in particular data, results and functions are all stored in operating system files. These are referred to as *objects*. On most Unix systems these files are stored in the `.Data` sub-directory of the current directory under the same name, so that a variable `stock` is stored (in an internal format) in the file `.Data/stock`. (The Windows version of S-PLUS uses a `_DATA` sub-directory and translates names to conform to the limitations of the DOS file system.)

Variables can be used as scalars, matrices or arrays, and S provides extensive matrix manipulation facilities. Further, objects can be made up of collections (known as *lists*) of such variables, allowing complex objects such as the result of a regression calculation. This means that the result of a statistical procedure can be saved for further analysis in a future session. Typically the calculation is separated from the output of results, so one can perform a regression and then print various summaries and compute residuals and leverage plots from the saved regression object.

Technically S is a function language. Elementary commands consist of either *expressions* or *assignments*. If an expression is given as a command, it is evaluated, printed, and the value is discarded. An assignment evaluates an expression and passes the value to a variable but the result is not printed automatically. An expression can be as simple as `2 + 3` or a complex function call. Assignments are indicated by the *assignment operator* `<-`. For example

```
> 2 + 3
[1] 5
> sqrt(3/4)/(1/3 - 2/pi^2)
[1] 6.6265
> mean(hstart)
[1] 137.99
> m <- mean(hstart); v <- var(hstart)/length(hstart)
> m/sqrt(v)
[1] 32.985
```

Here `>` is the S prompt, and the `[1]` states that the answer is starting at the first element of a vector.

More complex objects will have a short summary printed instead of full details. This is achieved by the object-oriented programming mechanism; complex objects have *classes* assigned to them which determine how they are printed, summarized and plotted.

S can be extended by writing new functions, which then can be used in the same way as built-in functions (and can even replace them). This is very easy; for example to define functions to compute the standard deviation and the two-tailed p–value of a t statistic we can write

```
std.dev <- function(x) sqrt(var(x))
t.test.p <- function(x, mu=0) {
    n <- length(x)
    t <- sqrt(n) * (mean(x) - mu) / std.dev(x)
    2 * (1 - pt(abs(t), n - 1))
}
```

It would be useful to give both the t statistic and its p-value, and the most common way of doing this is by returning a list; for example we could use

```
t.stat <- function(x, mu=0) {
    n <- length(x)
    t <- sqrt(n) * (mean(x) - mu) / std.dev(x)
    list(t = t, p = 2 * (1 - pt(abs(t), n - 1)))
}
z <- rnorm(300, 1, 2)   # generate 300 N(1, 4) variables.
t.stat(z)
$t:
[1] 8.2906
$p:
[1] 3.9968e-15

unlist(t.stat(z, 1))   # test mu=1, compact result
        t       p
 -0.56308 0.5738
```

The first call to t.stat prints the result as a list; the second tests the non-default hypothesis $\mu = 1$ and using unlist prints the result as a numeric vector with named components.

Linear statistical models can be specified by a version of the commonly-used Wilkinson–Rogers (1973) notation, so that

```
time ~ dist + climb
time ~ transplant/year + age + prior.surgery
```

refer to a regression of time on both dist and climb, and of time on year within each transplant group and on age, with a different intercept for each type of prior surgery. This notation has been extended to survival and tree models and to allow smooth non-linear terms.

1.2   Getting started

S makes use of the file system and for each project we strongly recommend that you have a separate working directory to hold the files for that project. We describe our suggested procedure separately for each operating system.

Unix

The suggested procedure for the first occasion on which you use S is:

1. Create a separate directory, say SwS, for this project, which we suppose is 'Statistics with S', and make it your working directory.

   ```
   $ mkdir SwS
   $ cd SwS
   ```

   Copy any data files you need to use with S to this directory.

2. Create a sub-directory of SwS called .Data by

   ```
   $ mkdir .Data
   ```

   This sub-directory is for use by S itself and hence has a Unix 'dot name' to be hidden from casual inspection.

3. Start the system with

   ```
   $ Splus
   ```

   Those used to a terminal environment such as Emacs may prefer to use the -e flag which allows the inbuilt line editor to be used (discussed in Appendix E).

4. At this point S commands may be issued (see later). The prompt is > unless the command is incomplete, when it is +.

5. To quit the S program the command is

   ```
   >  q()
   $
   ```

If you do not create a working directory .Data, S-PLUS will use that sub-directory of your home directory, or create such as directory for you (with a warning). (S will ask if a directory should be created.) To keep projects separate, we strongly recommend that you *do* create a working directory.

For subsequent sessions the procedure is simpler: make SwS the working directory and start the program as before:

```
$ cd SwS
$ Splus
```

issue S commands, terminating with the

```
>  q()
$
```

On the other hand, to start a new project start at step 1.

Windows

This can be a little tricky, as the configuration of S-PLUS can over-ride the intuitive behaviour, and it is very easy for different users inadvertently to share a common working directory. We indicate DOS commands, but the File Manager can also be used.

1. Create a separate directory, say SWS, for this project, and a subdirectory _DATA

   ```
   C:\ MD SWS
   C:\ CD SWS
   C:\SWS MD _DATA
   ```

2. Copy any data files you need to use with S-PLUS to this directory.

3. Within Windows, open the S-Plus for Windows Group from the Program Manager window, and click on the S-Plus for Windows icon. Select Copy from the File menu, and select the S-Plus for Windows Group from the To Group list box. Choose Properties from the File menu, and edit the working directory field to give the project's directory, and the description field to reflect the project.

   Double-click on the project's S-PLUS icon.

4. At this point S commands may be issued (see later). The prompt is > unless the command is incomplete, when it is +. The usual Windows command-line recall and editing is available.

5. To quit the S program the command is

   ```
   > q()
   ```

   or you can exit from the File menu (which will request confirmation).

To keep projects separate, we strongly recommend that you *do* create a working directory. (It is also possible to use the HOME environment variable to select the working directory, but if the default HOME directory contains a _DATA subdirectory, the default is used. As such a directory will be created if any user ever forgets to set the environment variable, this approach is likely to come to grief.)

For subsequent sessions the procedure is much simpler; just launch S-PLUS by double-clicking on the project's icon. On the other hand, to start a new project start at step 1.

1.3 Bailing out

One of the first things we like to know with a new program is how to get out of trouble. S and S-PLUS are generally very tolerant, and can be interrupted by Ctrl-C. (This means hold down the key marked Control or Cntrl and hit the second key.) This will interrupt the current operation, back out gracefully (so, with rare exceptions, it is as if it had not been started) and return to the S prompt.

Sometimes this is not sufficient, for example if a long sequence of operations has been pasted into a terminal window. C-shell users can always suspend the process (with Ctrl-Z) and then kill it using jobs and then kill %*n* for the appropriate *n*. Again, because S only commits changes at the end of an operation, this is almost always harmless.

Sometimes it is necessary to get to the operating system. On a multi-window system we suggest you use another window! If that is not possible, C-shell users can suspend the S process. However, it is also possible to issue commands to the operating system by starting the line with ! , for example

```
> !date
Thu Feb 17 22:23:03 GMT 1994
```

Note: this starts a new copy of the shell or command interpreter. In Windows this starts a new window and deletes it immediately, so use the dos or win3 functions (page 105) instead.

1.4   Getting help with functions and features

There are three separate help systems, depending on the program version. The approach is intuitive, despite the complexity of the explanation needed to cover all the possibilities.

(1) S and S-PLUS under Unix have a help facility similar to the man facility of Unix. This can be invoked from the command line. For example, to get information on the S function var the command is

```
> help(var)
```

which will put up a pager in the terminal window running S to view the help file. If you prefer, a separate help window (which can be left up) can be obtained by

```
> help(var, window=T)
```

A faster alternative (to type) is

```
> ?var
```

The help page can be printed by

```
> help(var, offline=T)
```

provided the system manager has set this facility up when the system was installed.

For a feature specified by special characters and a few other cases (one is "function"), the argument must be enclosed in double or single quotes, making it an entity known in S as a character string. For example two alternative ways of getting help on the list component extraction function, [[, are

```
> help("[[")
> ?"[["
```

(2) Using S-PLUS running under a Unix windowing system there is another way to interact with the help system. For example, under Motif we can use

```
> help.start(gui = "motif")
```

to bring up a window listing the help topics available. Items may be selected interactively from a series of menus, and the selection process causes other windows to appear with the help information. These may be scanned at the screen and then either dismissed or sent to a printer. (Similar facilities are available under Open-Look and SunView by setting gui="openlook" or gui="sunview".) This help system is shut down with help.off and *not* by quitting the help window.

If help or ? are used when this help system is running, the requests are sent to its window.

(3) The Windows version of S-PLUS has a help system accessible from the Help item on the main menu, and this can give a tutorial to its use. It is also possible to use help or ?, which for inbuilt functions send the request to the Windows help system. For user-written functions (for example, our library) the help file is displayed in a pager (default Notepad) window.

1.5 An introductory session

The best way to learn S is by using it. We invite readers to work through the following familiarization session and see what happens. First-time users may not yet understand every detail, but the best plan is to type what you see and observe what happens as a result.

As explained above you should first create a special directory and make it your working directory for the session, then start S-PLUS or S.

The whole session takes most first-time users about an hour at the appropriate leisurely pace. The left column gives commands; the right column gives brief explanations and suggestions.

`help.start(gui="motif")`	Start the help facility.
	This step only applies if you are working in a windowing system in S-PLUS. If not, skip this step and move on. Replace `"motif"` by `"openlook"`, `"athena"` or `"sunview"` as appropriate, or use the Windows help system.
Explore the help windows using the mouse.	
Close the help window	Close it to an icon; do not quit the window.

`motif()`	Or the appropriate graphics device, such as `openlook()` or `win.graph()`.
	Turn on the graphics window. You may need to re-position and re-size to make it convenient to work with both windows.
`library(MASS)`	A command to make available our datasets. Your local advisor can tell you the correct form for your system.

`x <- rnorm(50)` `y <- rnorm(50)`	Generate two pseudo-random normal vectors of x and y coordinates, and assign them to `x` and `y`.
`h <- chull(x, y)`	Find their convex hull in the plane.
`plot(x, y)` `polygon(x[h], y[h], dens=15,` `angle=30)`	Plot the points in the plane, and mark in their convex hull, filling it with lines at the rate of 15 to the inch, at an angle of 30 degrees to the horizontal. The result should be similar to Figure 1.1(a).
`objects()`	List the S objects which are now in the `.Data` directory.
`rm(x,y,h)`	Remove objects no longer needed. (The clean up phase.)
`x <- rnorm(1000)` `y <- rnorm(1000)`	Generate 1000 pairs of normal variates
`hist(c(x, y+2), 25)`	Histogram of mixture of normal distributions. Experiment with the number of bins (25) and the shift (2) of the second component.
`contour(hist2d(x,y,,,8,8))`	2D histogram on an 8×8 grid
`persp(hist2d(x,y,,,8,8))`	Perspective plot.
`image(hist2d(x,y,,,8,8))`	Greyscale plot. You may have to alter the colour scheme in the window (from a menu) to see this at its best.

`x <- seq(1, 20, 0.5)` `x`	Make $x = (1, 1.5, 2, \ldots, 19.5, 20)$ and list it.
`w <- 1 + x/2` `y <- x + w*rnorm(x)`	`w` will be used as a 'weight' vector and to give the standard deviations of the errors.

```
dum <- data.frame(x, y, w)
dum
rm(x, y, w)
```
Make a *data frame* of three columns named x, y and w, and look at it. Remove the original x, y and w.

```
fm <- lm(y ~ x, data=dum)
summary(fm)
```
Fit a simple linear regression of y on x and look at the analysis.

```
fm1 <- lm(y ~ x, data=dum,
  weight=1/w^2)
summary(fm1)
```
Since we know the standard deviations, we can do a weighted regression.

```
lrf <- loess(y ~ x, dum)
```
Fit a smooth regression curve using a modern regression function.

```
attach(dum)
```
Make the columns in the data frame visible as variables.

```
plot(x, y)
```
Make a standard scatterplot. To this plot we will add the three regression lines (or curves) as well as the known true line.

```
lines(spline(x, fitted(lrf)))
```
First add in the local regression curve using a spline interpolation between the calculated points.

```
abline(0, 1, lty=3)
```
Add in the true regression line (intercept 0, slope 1) with a different line type.

```
abline(fm)
```
Add in the unweighted regression line. abline() is able to extract the information it needs from the fitted regression object.

```
abline(fm1, lty=4)
```
Finally add in the weighted regression line, in line type 4. This one should be the most accurate estimate, but may not be, of course. One such outcome is shown in Figure 1.1(b).

You may be able to make a hardcopy of the graphics window by selecting the Print option from the Graph menu.

```
plot(fitted(fm), resid(fm),
  xlab="Fitted Values",
  ylab="Residuals")
```
A standard regression diagnostic plot to check for heteroscedasticity, that is, for unequal variances. The data are generated from a heteroscedastic process, so can you see this from this plot?

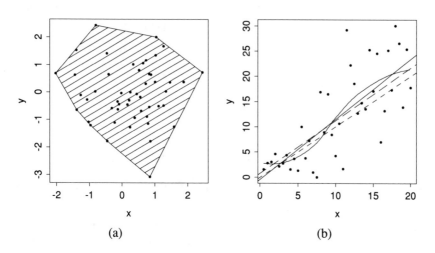

(a)   (b)

Figure 1.1: A convex hull and an heteroscedastic regression plot.

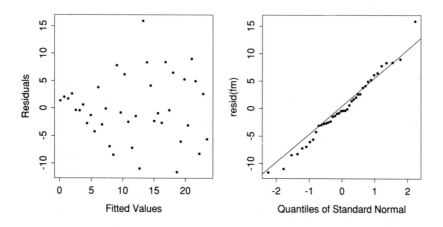

Figure 1.2: Two residual plots for the artificial heteroscedastic regression data.

`qqnorm(resid(fm))` `qqline(resid(fm))`	A normal scores plot to check for skewness, kurtosis and outliers. (Note that the heteroscedasticity may show as apparent non-normality.) Diagnostic plots are shown in Figure 1.2.
`detach()` `rm(fm,fm1,lrf,dum)`	Remove the data frame from the search list and clean up again.

We look next at a set of data on record times of Scottish hill races against distance and total height climbed.

`hills`	List the data
`pairs(hills)`	Show all the pairwise scatterplots (Figure 1.3).
`brush(as.matrix(hills))`	Try highlighting points and see how they are linked in the scatterplots (Figure 1.4). Also try rotating the points in 3D.
`attach(hills)`	Make columns available by name.

`plot(dist, time)`
`identify(dist, time,`
`  row.names(hills))`
	Use mouse button 1 to identify outlying points, and button 2 to quit. Their row numbers are returned.

`abline(lm(time ~ dist))`	Show least squares regression line.
`abline(ltsreg(dist, time),` `  lty=3)`	Fit a very resistant line. See Figure 1.5.
`detach()`	Clean up again.

We can explore further the effect of outliers on a linear regression by designing our own examples interactively. Try this several times.

`plot(c(0,1), c(0,1), type="n")` `xy <- locator(type="p")`	Make our own dataset by clicking with button 1, then with button 2 to finish.

`abline(lm(y ~ x, xy), col=4)`
`abline(rreg(xy$x, xy$y),`
`    lty=3, col=3)`
`abline(ltsreg(xy$x, xy$y),`
`    lty=2, col=2)`
	Fit least-squares and robust regression lines, and a resistant line. (Omit the `col=` on a mono display.) Repeat to try the effect of outliers, both vertically and horizontally.

`rm(xy)`	Clean up again.

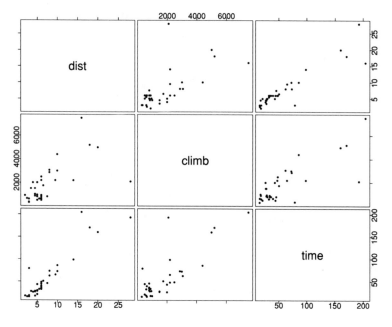

Figure 1.3: Pairs plot for data on Scottish hill races.

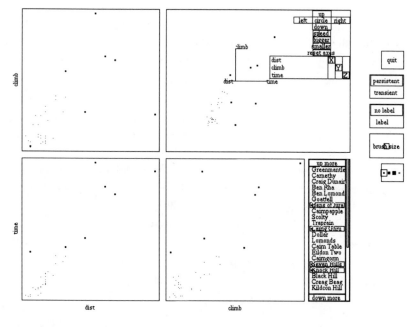

Figure 1.4: Screendump of a `brush` plot of dataset `hills` (from **openlook**).

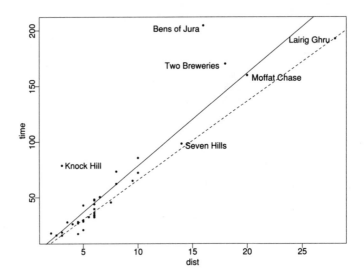

Figure 1.5: Annotated plot of time vs distance for `hills` with regression line and resistant line (dashed).

We now look at data from the 1879 experiment of Michelson to measure the speed of light. There are five experiments (column `Expt`) and each has 20 runs (column `Run`) and `Speed` is the recorded speed of light, in km/sec, less 299000. (The currently accepted value on this scale is 734.5).

`attach(michelson)`	Make the columns visible by name.
`search()`	The *search list* is a sequence of places, either directories or data frames, where S looks for objects required for calculations.

`plot(Expt, Speed,`
  `main="Speed of Light Data",`
  `xlab="Experiment No.")`
	Compare the five experiments with simple boxplots. The result is shown in Figure 1.6.

`fm <- aov(Speed ~ Run + Expt)` `summary(fm)`	Analyse as a randomized block design, with *runs* and *experiments* as factors.

```
          Df Sum of Sq Mean Sq F Value    Pr(F)
Run       19    113344    5965  1.1053 0.36321
Expt       4     94514   23629  4.3781 0.00307
Residuals 76    410166    5397
```

`fm0 <- update(fm, . ~ .-Run)` `anova(fm0, fm)`	Fit the sub-model omitting the nonsense factor, *runs*, and compare using a formal analysis of variance.

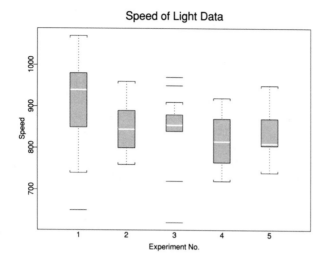

Figure 1.6: Boxplots for the speed of light data.

```
Analysis of Variance Table
Response: Speed

          Terms Resid. Df    RSS Test Df Sum of Sq F Value    Pr(F)
  1        Expt        95 523510
  2 Run + Expt         76 410166 +Run 19    113344  1.1053 0.36321
```

detach()	Clean up before moving on.
rm(fm, fm0)	

---

type in the listing in Figure 1.7

butterfly	Note how naming the function object
butterfly()	butterfly without the parentheses gives the function definition. Adding the parentheses causes it to be evaluated, which should give you a picture of a rather mathematical looking butterfly in your graphics window, Figure 1.8.
usa()	An S map, which can be used for geographical display, Figure 1.8.
q()	Quit the S program

```
butterfly <- function()
{
    theta <- seq(0, 24 * pi, len = 2000)
    radius <- exp(cos(theta)) - 2 * cos(4 * theta) +
                              sin(theta/12)^5
    plot(radius * sin(theta), - radius * cos(theta),
        type = "l", axes = F, xlab = "", ylab = "")
}
```

Figure 1.7: The source of the butterfly function. The indentation is unimportant, but be careful where lines are broken.

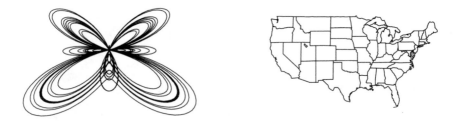

Figure 1.8: A mathematical butterfly and an S map

1.6   What next?

We hope that you now have a flavour of S and are inspired to delve more deeply. We suggest that you read Chapter 2, perhaps cursorily, and the first three sections of Chapter 3. Thereafter, tackle the statistical topics that are of interest to you. Chapters 5 to 15 are fairly independent, and contain cross-references where they do interact. Chapters 7, 8 and 9 build on Chapter 6, especially its first two sections.

Chapter 4 comes early, because it is about S not about statistics, but will be most useful to advanced users and to those writing extensions to S. Many users are surprised to find themselves in this category; the philosophy of S encourages writing re-usable functions, friends get to hear about them, ...

Chapter 2

The S Language

S is a language for the manipulation of objects. It aims to be both an interactive language (like, for example, a Unix shell language) as well as a complete programming language with some convenient object-oriented features. In this chapter we shall be concerned with the interactive language, and hence certain language constructs used mainly in programming will be postponed to Chapter 4.

This chapter divides into two parts. The first six sections give an informal overview of the language which will suffice at first reading. Later sections are much more formal, and useful for answering "why did it do that?" questions. Appendix B gives brief synopses of commonly used S functions, with references to their description in the main text.

2.1   A concise description of S objects

In this section we discuss the most important types of S objects and the simplest ways to manipulate them,

Naming conventions

Standard S names for objects are made up from the upper and lower case roman letters, the digits, 0–9, in any non-initial position and also the period, ' . ', which behaves as a letter except in names such as .37 where it acts as a decimal point. Some important conventions make use of a period in object names and users will become aware of these in due course. For example 'three dots' : ... is a reserved identifier, only used in the context of defining functions (see pages 25 and 95). Other reserved identifiers include break, for, function, if, in, next, repeat, return and while.

Non-standard names are allowed using any collection of characters, but special steps have to be taken to use them. We suggest that they are avoided, and have never found need of them ourselves.

Note that S is *case sensitive*, so Alfred and alfred are distinct S names and that the underscore, _ , is *not* available as a letter in the S standard name alphabet. (The period is commonly used to separate words in names.)

Avoid using system names for your own objects; in particular avoid c, s, t, C, F, T, diff, range and tree. At best these generate warning messages, at worst incorrect results.

Language layout

Commands to S are either expressions or assignments. Commands are separated by either a semi-colon, ; , or a newline. The # symbol marks the rest of the line as comments.

The S prompt is > unless the command is syntactically incomplete, when the prompt changes to +.[1] The only way to extend a command over more than one line is to ensure that at the end of each line *it is* syntactically incomplete.

In this book we will often omit both prompts and indicate continuation by simple indenting. When we do include prompts it is usually to separate input from output within a display.

An expression command is evaluated, printed, and the value assigned to a object named .Last.value. For example

```
> 1 - pi + exp(1.7)
[1] 3.332355
```

(The next command expression overwrites this .Last.value.) This rule allows any object to be printed by giving its name (but it is not then assigned to .Last.value). Note that pi is the value of π. Giving the name of an object will normally print it or a short summary; this can be done explicitly using the function print, and summary will often give a full description.

An assignment command evaluates an expression and passes the value to a variable but the result is not printed. The recommended assignment symbol is the combination, <- , so an assignment in S looks like

```
a <- 6
```

which gives the object a the value 6. To improve readability of your code we strongly recommend that you put at least one space before and after binary operators, especially the assignment symbol. (We regard the use of _ for assignments as unreadable, but it is allowed.)

Almost always an assignment can be used itself as an expression with value equal to the value assigned. Thus multiple assignments are available:

```
b <- a <- 6
```

which are evaluated from right to left, and here gives both a and b the value 6.

If you forget to assign an expression, do remember that the result is in .Last.value. This is also useful if the result is unexpectedly not printed; just print(.Last.value).

---

[1] These prompts can be altered: see Section 2.11.

Vectors and matrices

Objects in S are vectors, functions or lists. Note that there are no scalars; vectors of length one are used instead. Vectors consist of numeric or logical values or character strings (and may not mix them). Normally it is unnecessary to be aware if the numeric values are integer, real or even complex, and whether they are stored to single or double precision; S handles all the possibilities in a unified way.

The simplest way to create a vector is to specify its elements by the function c (for concatenate):

```
> mydata <- c(2.9, 3.4, 3.4, 3.7, 3.7, 2.8, 2.8, 2.5, 2.4, 2.4)
> colours <- c("red", "green", "blue", "white", "black")
> x1 <- 25:30
> x1
[1] 25 26 27 28 29 30
> mydata[7]
[1] 2.8
> colours[3]
[1] "blue"
```

and the colon expression specifies a range of integers. The [1] indicates that the printout of the vector starts at element one. Individual elements of a vector are accessed by specifying them by number in square brackets, as in these examples. Character strings may be entered with either double or single quotes (in matching pairs), but will always be printed with double quotes.

Logical values are represented as T and F:

```
> mydata > 3
 [1] F T T T T F F F F F
```

and may also be entered as TRUE and FALSE. Logical operators apply element-by-element to vectors, as do arithmetical expressions.

The elements of a vector can also be named and accessed by name:

```
> names(mydata) <- c('c','j','b','e','i','h','g','d','f','a')
> mydata
  c   j   b   e   i   h   g   d   f   a
2.9 3.4 3.4 3.7 3.7 2.8 2.8 2.5 2.4 2.4
> names(mydata)
 [1] "c" "j" "b" "e" "i" "h" "g" "d" "f" "a"
> mydata["e"]
  e
3.7
```

More generally, several elements of a vector can be picked out by giving a vector of element names or numbers or a logical vector:

```
> letters[1:5]
[1] "a" "b" "c" "d" "e"
> mydata[letters[1:5]]
  a   b   c   d   e
2.4 3.4 2.9 2.5 3.7
> mydata[mydata > 3]
  j   b   e   i
3.4 3.4 3.7 3.7
```

`letters` is an S vector containing the 26 lowercase English letters. We will refer to taking *subsets* of vectors (although to mathematicians they are subsequences).

Elements of a vector can be <u>omitted</u> by giving negative indices, for example,

```
> mydata[-c(3:5)]
  c   j   h   g   d   f   a
2.9 3.4 2.8 2.8 2.5 2.4 2.4
```

but in this case the names can not be used.

There are many special uses of vectors. For example, they can appear to be matrices, arrays or factors. This is handled by giving the vectors *attributes*. All objects have two attributes, their `mode` and their `length`:

```
> mode(mydata)
[1] "numeric"
> mode(letters)
[1] "character"
> mode(sin)
[1] "function"
> length(mydata)
[1] 10
> length(letters)
[1] 26
> length(sin)
[1] 2
```

(The length of a function is one plus the number of arguments, so not very useful.) Some objects have names, and these are the `names` attribute. (For functions the names are those of the arguments, plus an empty one.)

Giving a vector a `dim` attribute causes it to treated as a matrix or array, for example,

```
> dim(mydata) <- c(2,5)
> mydata
     [,1] [,2] [,3] [,4] [,5]
[1,]  2.9  3.4  3.7  2.8  2.4
[2,]  3.4  3.7  2.8  2.5  2.4
attr(, "names"):
 [1] "c" "j" "b" "e" "i" "h" "g" "d" "f" "a"
> dim(mydata) <- NULL
```

Notice how the names are retained as an attribute, and how the matrix is filled down columns rather than across rows. The final assignment removes the `dim` attribute and restores `mydata` to a named vector. A simpler way to create a matrix from a vector is to use `matrix`:

```
> matrix(mydata, 2, 5)
      [,1] [,2] [,3] [,4] [,5]
[1,]   2.9  3.4  3.7  2.8  2.4
[2,]   3.4  3.7  2.8  2.5  2.4
```

when the names are discarded. The function `matrix` can also fill matrices by row,

```
> matrix(mydata, 2, 5, byrow=T)
      [,1] [,2] [,3] [,4] [,5]
[1,]   2.9  3.4  3.4  3.7  3.7
[2,]   2.8  2.8  2.5  2.4  2.4
```

As these displays suggest, matrix elements can be accessed as `mat[m,n]`, and whole rows and columns by `mat[m,]` and `mat[,n]` respectively.

Arrays are multi-way extensions of matrices, formed by giving a `dim` attribute of length three or more. They can be accessed by `A[r,s,t]` and so on. Sections 2.7 and 2.8 discuss matrices and arrays in more detail.

Lists

A list is used to collect together items of different types. For example, an employee record might be created by

```
Empl <- list(employee="Anna", spouse="Fred", children=3,
             child.ages=c(4,7,9))
```

As this example shows, the elements of a list do not have to be of the same length. The components of a list are always *numbered* and may always be referred to as such. Thus `Empl` is a list of length 4, and the individual components may be referred to as `Empl[[1]]`, `Empl[[2]]`, `Empl[[3]]` and `Empl[[4]]`. Further, since `Empl[[4]]` is a vector, `Empl[[4]][1]` is its first entry. However, it is much more convenient to refer to the components by name, in the form

```
> Empl$employee
[1] "Anna"
> Empl$child.ages[2]
[1] 7
```

The names of components may be abbreviated down to the minimum number of letters needed to identify them uniquely. Thus `Empl$employee` may be minimally specified as `Empl$e` since it is the only component whose name begins with the letter 'e', but `Empl$children` must be specified as at least `Empl$childr` because of the presence of another component called `Empl$child.ages`.

In many respects lists are like vectors. For example, the vector of component names is in fact simply a `names` attribute of the list like any other object and may be handled as such. For example to change the component names of `Empl` to `a`, `b`, `c` and `d` we can use

```
> names(Empl) <- letters[1:4]
> Empl[3:4]
$c:
[1] 3

$d:
[1] 4 7 9
```

where we select *components* as if this were a vector (and not `[[3:4]]` as might have been expected).

The concatenate function, `c`, can also be used to concatenate lists, so

```
Empl <- c(Empl, service=8)
```

would add a component for years of service.

The function `unlist` converts a list to a vector:

```
> unlist(Empl)
 employee spouse children child.ages1 child.ages2 child.ages3
  "Anna"    "Fred" "3"         "4"         "7"         "9"
> unlist(Empl, use.names=F)
[1] "Anna" "Fred" "3"     "4"     "7"     "9"
```

which can be useful for a compact print-out (as here). (Mixed types will all be converted to character, giving a character vector.)

Factors

A factor is a special type of vector, normally used to hold a categorical variable, for example

```
> citizen <- factor(c("UK","US","No","Oz","UK","US","US"))
> citizen
[1] UK US No Oz UK US US
```

Although this is entered as a character vector, it is printed without quotes. Appearances here are deceptive, and a special `print` method is used. Internally the factor is stored as a set of codes, and an attribute giving the *levels*:

```
> print.default(citizen)
[1] 3 4 1 2 3 4 4
attr(, "levels"):
[1] "No" "Oz" "UK" "US"
attr(, "class"):
[1] "factor"
> codes(citizen)
[1] 3 4 1 2 3 4 4
```

Why might we want to use this rather strange form? Using a factor signals to many of the statistical functions that this is a categorical variable (rather than just a list of labels), and so it is treated specially. As just one example, using a factor signals to the function `tree` that a classification rather than regression tree is required (see Chapter 13).

Notice that the levels are sorted into alphabetical order, and the codes assigned accordingly. Some of the statistical functions give the first level a special status, so it may be necessary to specify the levels explicitly:

```
> citizen <- factor(c("UK","US","No","Oz","UK","US","US"),
      levels = c("US", "Fr", "No", "Oz", "UK"))
> citizen
[1] UK US No Oz UK US US
Levels:
[1] "US" "Fr" "No" "Oz" "UK"
```

Note that levels which do not occur can be specified, it which case the levels *are* printed. This often occurs when subsetting factors.

Sometimes the levels of a categorical variable are naturally ordered, as in

```
> income <- ordered(c("Mid","Hi","Lo","Mid","Lo","Hi","Lo"))
> income
[1] Mid Hi  Lo  Mid Lo  Hi  Lo

  Hi < Lo < Mid
```

Again the effect of alphabetic ordering is not what is required, and we need to set the levels explicitly, or use an assignment to `ordered`:

```
> inc <- ordered(c("Mid","Hi","Lo","Mid","Lo","Hi","Lo"),
      levels=c("Lo", "Mid", "Hi"))
> inc
[1] Mid Hi  Lo  Mid Lo  Hi  Lo

  Lo < Mid < Hi
> ordered(income) <- c("Lo", "Mid", "Hi")
> income
[1] Mid Hi  Lo  Mid Lo  Lo  Hi  Lo

  Lo < Mid < Hi
```

Functions may treat ordered factors in an even more special way (they are also factors).

Note that whereas assigning to `ordered` permutes the current levels, assigning to `levels` merely changes the names of the levels and leaves the codes unchanged.

Data frames

A data frame is the type of object normally used in S to store a data matrix. It should be thought of as a list of variables of the same length, but possibly of different types (numeric, character or logical). Consider our data frame `painters`:

```
> painters
            Composition Drawing Colour Expression School
   Da Udine          10       8     16          3      A
   Da Vinci          15      16      4         14      A
 Del Piombo           8      13     16          7      A
 Del Sarto           12      16      9          8      A
 Fr. Penni            0      15      8          0      A
   ....
```

which has four numerical variables and one character variable. Since it is a data frame, it is printed in a special way. The components are printed as columns (rather than as rows as vector components of lists are) and there is a set of names, the `row.names`, common to all variables.

```
> row.names(painters)
[1] "Da Udine"       "Da Vinci"        "Del Piombo"
[4] "Del Sarto"      "Fr. Penni"       "Guilio Romano"
[7] "Michelangelo"   "Perino del Vaga" "Perugino"
    ....
```

Further, neither the row names nor the character variable appear in quotes.

Data frames can be accessed in ways similar to matrices. In particular, they can be addressed by row and column indices:

```
> painters[1:5, c(2, 4)]
            Drawing Expression
   Da Udine       8          3
   Da Vinci      16         14
 Del Piombo      13          7
 Del Sarto       16          8
 Fr. Penni       15          0
```

Variables which satisfy suitable restrictions (having the same length, and the same names, if any) can be collected into a data frame by the function `data.frame`, which resembles `list`:

```
mydat <- data.frame(y=mpg, x1=dist, x2=weather, x3=dry)
```

although data frames are most commonly created by reading a file (see `read.table` on page 33).

There is a side effect of `data.frame` which needs to be considered; all character and logical columns are converted to factors unless their names are included in `I()`, so for example

```
mydat <- data.frame(y=mpg, x1=dist, x2=I(weather), x3=I(dry))
```

preserves `weather` and `dry` as character vectors.

Compatible data frames can be joined by `cbind`, which adds columns of the same length (and names, if any), and `rbind`, which adds rows from a data frame with compatible columns.

It is also possible to include matrices and lists within data frames. If a matrix is supplied to `data.frame` it is as if its columns were supplied individually; suitable labels are concocted. If a list is supplied, it is treated as if its components had been supplied individually.

Coercion

There are a series of functions named `as.xxx` which convert to the specified type in the best way possible. For example, `as.matrix` will convert a numerical data frame to a numerical matrix, and a data frame with any character or factor columns to a character matrix. The function `as.character` is often useful to generate names and other labels.

Functions `is.xxx` test if their argument is of the required type. Note that these do not always behave as one might guess; for example `is.vector(mydata)` will be *false* as this tests for a 'pure' vector without any attributes such as names. Similarly, `as.vector` has the (sometimes useful) side effect of discarding all attributes.

2.2   Calling conventions for functions

Functions may have their arguments *specified* or *unspecified* when the function is defined. (We saw how to write simple functions on page 4.)

When the arguments are unspecified there may be an arbitrary number of them and the order in which they occur is unimportant. They are shown as ... when the function is defined or printed. Examples of functions with unspecified arguments include the concatenation function, `c(...)`, and the parallel maximum and minimum functions `pmax(...)` and `pmin(...)`.

Where the arguments are specified there are two conventions for supplying values for the arguments when the function is called:

1. arguments may be specified in the same order in which they occur in the function definition, in which case the values are supplied in order, and

2. arguments may be specified as `name=value`, when the order in which the arguments appear is irrelevant. The name may be abbreviated providing it partially matches just one named argument.

It is important to note that these two conventions may be mixed. A call to a function may begin with specifying the arguments in positional form but specify some later arguments in the named form. After the first named argument the remaining arguments must be specified in the named form. For example,

```
polygon(x1, y1, border=F, density=10)
polygon(x=x1, y=y1, border=F, density=10)
```

are equivalent calls.

Functions with named arguments also have the option of specifying *default values* for those arguments, in which case if a value is not specified when the function is called the default value is used. For example, the function `polygon` has an argument list defined as

```
polygon(x, y, density=-1, angle=45, border=T, col=par("col"))
```

so that our previous calls can also be specified as

```
polygon(x1, y1, 10, ,F)
```

and in all cases the default value of `angle` (45°) is used. Using the positional form and omitting values, as in this last example, is rather prone to error, so the named form is preferred except for the first couple of arguments.

Some functions (for example `paste`) have both unspecified and specified arguments, in which case the specified arguments must be named and given after all unspecified arguments.

The argument names and any default values for an S function can be found from the on-line help, by printing the function itself or succinctly using the `args` function. For example,

```
> args(hist)
function(x, nclass, breaks, plot = TRUE,
        probability = FALSE, ..., xlab = deparse(substitute(x)))
    NULL
```

shows the arguments, their order and those default values which are specified for the histogram function. (The return value from `args` always ends with `NULL`.) Note that even when no default value is specified the argument itself may not need to be specified. If no value is given for `nclass` or `breaks` when the `hist` function is called, default values are calculated within the function. Unspecified arguments are passed on to a plotting function called from within `hist`.

Functions are considered in much greater detail in Chapter 4.

2.3   Arithmetical expressions

We have seen that a basic unit in S is a vector. Arithmetical operations are performed on vectors, element by element. The standard operators + - * / ^ are available, where ^ is the power (or exponentiation) operator (giving x^y).

Vectors may be empty. The expression `numeric(0)` is both the expression to create an empty numeric vector and the way it is represented when printed. It has length zero. It may be described as "a vector such that if there were any elements in it, they would be numbers"!

Vectors can be complex, and almost all the rules for arithmetical expression apply equally to complex quantities. A complex number is entered in the form `3.1 + 2.7i`, with no space before the `i`. Function `Re` and `Im` return the real and imaginary parts. Note that complex arithmetic is not used unless explicitly requested, so `sqrt(x)` for `x < 0` is an error unless `x` is complex (use `as.complex(x)` if this is desired).

The recycling rule

The expression `2 + y` is a syntactically natural way to add `2` to each element of the vector `y`, but `2` is a vector of length 1 and `y` may be a vector of any length. A convention is needed to handle vectors occurring in the same expression but not all of the same length. The value of the expression is a vector with the same length as that of the longest vector occurring in the expression. Shorter vectors are *recycled* as often as need be, perhaps fractionally, until they match the length of the longest vector. In particular a single number is repeated the appropriate number of times. Hence

```
x <- c(10.4, 5.6, 3.1, 6.4, 21.7)
y <- c(x, 0, x)
v <- 2*x + y + 1
```

generates a new vector `v` of length 11 constructed by

1. repeating the number `2` five times to match the length of the vector `x` and multiplying element by element, and

2. adding together, element by element, `2*x` repeated 2.2 times, `y` as it stands, and `1` repeated eleven times.

Some functions

Some examples of standard functions follow.

1. There are several functions to convert to integers; `round` will normally be preferred, and rounds to the nearest integer. (It can also round to any number of digits in the form `round(x, 3)`. Using a negative number rounds to a power of ten, so that `round(x,-3)` rounds to thousands.) Each of `trunc`, `floor` and `ceiling` round in a fixed direction, towards zero, down and up respectively.

2. Other arithmetical operators are `%/%` for integer divide and `%%` for modulo reduction.[2]

3. The common functions are available, including `abs`, `log`, `log10`, `sqrt`, `exp`, `sin`, `cos`, `tan`, `acos`, `asin`, `atan`, `cosh`, `sinh` and `tanh` with their usual meanings. Note that the value of each of these is a vector of the same length as their argument. The function `log` has a second argument, the base of the logarithms, defaulting to e.

   Less common functions are `gamma` and `lgamma` ($\log_e \Gamma(x)$).

4. There are functions `sum` and `prod` to form the sum and product of a whole vector, as well as cumulative versions `cumsum` and `cumprod`.

---

[2] The result of `e1 %/% e2` is `floor(e1/e2)` if `e2!=0` and `0` if `e2==0`. The result of `e1 %% e2` is `e1-floor(e1/e2)*e2` if `e2!=0` and `e1` otherwise (see Knuth, 1968, §1.2.4). Thus `%/%` and `%%` always satisfy `e1==(e1%/%e2)*e2+e1%%e2`.

5. The functions `max(x)` and `min(x)` select the largest and smallest elements of a vector `x`. The functions `cummax` and `cummin` give cumulative maxima and minima.

6. The functions `pmax(x1, x2, ...)` and `pmin(x1, x2, ...)` take an arbitrary number of vector arguments and return the element-by-element maximum or minimum values respectively. Thus the result is a vector of length that of the longest argument and the recycling rule is used for shorter arguments. For example

   ```
   xtrunc <- pmax(0, pmin(1,x))
   ```

   is a vector like `x` but with negative elements replaced by `0` and elements larger than 1 replaced by `1`.

7. The function `range(x)` returns `c(min(x), max(x))`. If `range`, `max` or `min` is given several arguments these are first concatenated into a single vector.

8. Two useful statistical functions are `mean(x)` which calculates the sample mean, which is the same as `sum(x)/length(x)`, and `var(x)` which gives the sample variance, namely `sum((x-mean(x))^2)/(length(x)-1)`.[3]

9. `sort` returns a vector of the same size as `x` with the elements arranged in increasing order. See page 44 for further details. The function `rev` reverses a vector.

Operator precedence

The formal precedence of operators is given in Table 2.1. However, as usual it is better to use parentheses to group expressions rather than rely on remembering these rules. They can be found on-line from `help(Syntax)`.

Generating regular sequences

There are several ways in S to generate sequences of numbers. For example, `1:30` is the vector `c(1,2, ...,29,30)`. The colon operator has a high precedence within an expression, so `2*1:15` is the vector `c(2,4,6, ...,28,30)`. Put `n <- 10` and compare the sequences `1:n-1` and `1:(n-1)`.

The construction `30:1` may be used to generate a sequence in reverse order.

The function `seq` is a more general facility for generating sequences. It has five arguments, only some of which may be specified in any one call. The first two arguments, named `from` and `to`, if given specify the beginning and end of the sequence, and if these are the only two arguments the result is the same as the colon operator. That is `seq(2,10)` and `seq(from=2,to=10)`, give the same vector as `2:10`.

---

[3] If the argument to `var` is an $n \times p$ matrix the value is a $p \times p$ sample covariance matrix obtained by regarding the rows as sample vectors.

Table 2.1: Precedence of operators, from highest to lowest.

`$`	for list extraction
`[, [[`	vector and list element extraction
`^`	exponentiation
`-`	unary minus
`:`	sequence generation
`%%, %/%, %*%`	and other special operators %...%
`* /`	multiply and divide
`+ -`	addition and subtraction
`< > <= >= == !=`	comparison operators
`!`	logical negation
`&	&& ||`	logical operators
`~`	formula
`<<-`	assignment within a function (see page 108)
`<-, _, ->`	assignment

The third and fourth arguments to `seq` are named by and `length`, and specify a step size and a length for the sequence. If by is not given, the default by=1 is used. For example,

```
s3 <- seq(-5, 5, by=0.2)
s4 <- seq(length=51, from=-5, by=0.2)
```

generate in both s3 and s4 the vector $(-5.0, -4.8, -4.6, \ldots, 4.6, 4.8, 5.0)$.

The fifth argument is named `along` and has a vector as its value. If it is the only argument given it creates a sequence 1, 2, ..., length(*vector*), or the empty sequence if the value is empty. (This makes `seq(along=x)` preferable to `1:length(x)` in most circumstances.) If specified rather than to or `length` its length determines the length of the result.

A related function is `rep` which can be used to repeat an object in various ways. The simplest form is

```
s5 <- rep(x, times=5)
```

which will put five copies of x end-to-end in s5.

A `times=v` argument may specify a vector of the same length as the first argument, x. In this case the elements of v must be non-negative integers, and the result is a vector obtained by repeating each element in x a number of times as specified by the corresponding element of v. Some examples will make the process clear:

```
x <- 1:4        # puts c(1,2,3,4)           into x
i <- rep(2, 4)  # puts c(2,2,2,2)           into i
y <- rep(x, 2)  # puts c(1,2,3,4,1,2,3,4)   into y
z <- rep(x, i)  # puts c(1,1,2,2,3,3,4,4)   into z
w <- rep(x, x)  # puts c(1,2,2,3,3,3,4,4,4,4) into w
```

As a more useful example, consider a two-way experimental layout with 4 row classes, 3 column classes and 2 observations in each of the 12 cells. The observations themselves are held in a vector y of length 24 with column classes stacked above each other, and row classes in sequence within each column class. Our problem is to generate two indicator vectors of length 24 which will give the row and column class, respectively, of each observation. Since the 3 column classes are the first, middle and last 8 observations each, the column indicator is easy. The row indicator requires three calls to rep:

```
> colc <- rep(1:3,rep(8,3));   colc
> [1] 1 1 1 1 1 1 1 1 2 2 2 2 2 2 2 2 3 3 3 3 3 3 3 3
> rowc <- rep(rep(1:4,rep(2,4)),  3); rowc
> [1] 1 1 2 2 3 3 4 4 1 1 2 2 3 3 4 4 1 1 2 2 3 3 4 4
```

These can also be generated arithmetically using the ceiling function:

```
> 1 + (ceiling(1:24/8) - 1) %% 3 -> colc; colc
> [1] 1 1 1 1 1 1 1 1 2 2 2 2 2 2 2 2 3 3 3 3 3 3 3 3
> 1 + (ceiling(1:24/2) - 1) %% 4 -> rowc; rowc
> [1] 1 1 2 2 3 3 4 4 1 1 2 2 3 3 4 4 1 1 2 2 3 3 4 4
```

In general the expression 1 + (ceiling(1:n/r) - 1) %% m generates a sequence of length n consisting of the numbers 1, 2, ..., m each repeated r times, similar to the GLIM function %gl.

Logical expressions

Logical vectors are most often generated by *conditions*. The logical operators are <, <=, >, >= (which have self-evident meanings), == for exact equality and != for exact inequality. If c1 and c2 are logical expressions, c1 & c2 is their intersection ('and'), c1 | c2 is their union ('or') and !c1 is the negation of c1.

Logical vectors may be used in ordinary arithmetic. They are *coerced* into numeric vectors, F becoming 0 and T becoming 1. For example, assuming the value or values in sd are positive

```
No.extremes <- sum(y > ybar + 3*sd | y < ybar - 3*sd)
```

would count the number of elements in y that were further than 3*sd from ybar on either side. The right hand side can be expressed more concisely as sum(abs(y-ybar) > 3*sd).

The functions any and all are useful to collapse a logical vector. The function all.equal provides a way to test for equality up to a tolerance if appropriate.

The missing value marker, NA

Not all the elements of a vector may be known. When an element or value is "not available" or a "missing value', a place within a vector may be reserved for it by assigning the special value NA.

In general any operation on an NA becomes an NA. The motivation for this rule is simply that if the specification of an operation is incomplete, the result cannot be known and hence is not available.

The function is.na(x) gives a logical vector of the same length as x with values which are true if and only if the corresponding element in x is NA.

```
ind <- is.na(z)
```

Notice that the logical expression x == NA is not equivalent to is.na(x). Since NA is not really a value but a marker for a quantity that is not available, the first expression is incomplete whereas the second is not. Thus x == NA is a vector of the same length as x *all* of whose values are NA irrespective of the elements of x.

S functions differ considerably in their ability to handle missing values. Many omit all rows of a data matrix containing a missing value, but the na.action argument to statistical functions may allow other possibilities. The default is usually na.fail, to stop, but na.omit omits all rows containing a missing value. The function na.gam.replace replaces missing numerical values in a data frame by the mean of the non-missing values in the column, and adds a level NA to factors and ordered factors (resulting in an unordered factor).

Missing values are output as NA, and can be input as NA or by ensuring that a value is missing (for example, a field is blank).

Elementary matrix operations

We have seen that a matrix is just a data vector with a dim attribute specifying a double index. However, S contains many operators and functions for matrices; for example t(X) is the transpose function. The functions nrow(A) and ncol(A) give the number of rows and columns in the matrix A.

The operator %*% is used for matrix multiplication. Vectors which occur in matrix multiplications are if possible promoted either to row or column vectors, whichever is multiplicatively coherent. (This may be ambiguous, as we shall see.) Note carefully that if A and B are square matrices of the same size, then

```
A * B
```

is the matrix of element by element products whereas

```
A %*% B
```

is the matrix product. If x is a vector, then

```
x %*% A %*% x
```

is a quadratic form $x^T A x$, where x is the column vector and T denotes transpose.

Note that x %*% x seems to be ambiguous, as it could mean either $x^T x$ or xx^T. A more precise definition of %*% is that of an inner product rather than a matrix product, so in this case $x^T x$ is the result. (For xx^T use x %o% x; see page 46.)

Matrices can be built up from other vectors and matrices by the functions cbind and rbind. Informally cbind forms matrices by binding together vectors or matrices column-wise, and rbind binds row-wise. The arguments to cbind must be either vectors of any length, or matrices with the same column size, that is the same number of rows. The result is a matrix with the concatenated arguments forming the columns. If some of the arguments to cbind are vectors they may be shorter than the column size of any matrices present, in which case they are cyclically extended to match the matrix column size (or the length of the longest vector if no matrices are given). The function rbind performs the corresponding operation for rows. In this case any vector arguments, possibly cyclically extended, are taken as rows. Note that we have already seen cbind and rbind operating on data frames.

Further matrix operations are discussed in sections 2.7 and 2.8.

2.4   Reading data

Large data objects will usually be read as values from external files rather than entered during an S session at the keyboard. The S input facilities are simple and their requirements are fairly strict and rather inflexible. There is a clear presumption by the designers of S that you will be able to modify your input files using other tools, such as file editors and the Unix utilities sed and awk, to fit in with the requirements of S. There are some tools within S that can cater for non-standard situations, as we discuss below.

If variables are to be held in data frames, as we strongly suggest they should be, an entire data frame can be read directly with the read.table function. There is also a more general input function, scan, that is useful in special circumstances.

The read.table function

In order to be read into a data frame, an external file should have a standard form:

1. The first line of the file should have a *name* for each variable in the data frame.
2. Each additional line of the file has as its first item a *row label* and the values for each variable, separated by spaces, tabs or both. Character strings containing blanks must be contained within quotes, which are otherwise optional.

The first few lines of a file to be read as a data frame are shown in Figure 2.1. By default numeric items (except row labels) are read as numeric variables and

	Price	Floor	Area	Rooms	Age	Cent.heat
01	52.00	111.0	830	5	6.2	no
02	54.75	128.0	710	5	7.5	no
03	57.50	101.0	1000	5	4.2	no
04	57.50	131.0	690	6	8.8	no
05	59.75	93.0	900	5	1.9	yes
	....					

Figure 2.1: Input file form with names and row labels.

non-numeric variables, such as Cent.heat in the example, as factors. (This can be changed if necessary via the argument as.is.)

The function read.table is used to read in the data frame

```
HousePrice <- read.table("houses.dat")
```

Often it is convenient to omit the row labels and use the default, the row number. In this case the file should omit the row label column (as in Figure 2.2), and we must specify header=T in the call to read.table:

```
HousePrice <- read.table("houses1.dat", header=T)
```

Price	Floor	Area	Rooms	Age	Cent.heat
52.00	111.0	830	5	6.2	no
54.75	128.0	710	5	7.5	no
57.50	101.0	1000	5	4.2	no
57.50	131.0	690	6	8.8	no
59.75	93.0	900	5	1.9	yes
....					

Figure 2.2: Input file form without row labels.

Note that data frames *must* have both row and column labels so read.table supplies default values where necessary. The default labels can be surprisingly unhelpful, so we recommend always supplying at least column labels.

Data file names

Unix users can use any legal file name for data files. So can Windows users, but they have to take care how they specify them, as backslashes (\) within names have to be doubled inside S, for example as

```
"c:\\mywork\\splus\\sws\\file.dat"
```

This can also be specified with slashes as

```
"c:/mywork/splus/sws/file.dat"
```

which can be easier to use, especially for users conversant with Unix. The filename
clipboard may be used in Windows to refer to the Windows clipboard for input
or output (but the file length is limited to 32Kb).

The function scan

The simplest use of the scan function is to input a single vector from the keyboard:

```
> x <- scan()
1: 23.4 45.6 77.8 12.9
5: 20 10 11
8: 33
9:
> x
[1] 23.4 45.6 77.8 12.9 20.0 10.0 11.0 33.0
```

The default arguments specify that input is to come from the keyboard. Data items
are entered as the prompt changes to n: where n is the index of the next item to
be read. Reading is terminated by an empty input line (only from the keyboard)
or by ^D.

Data may be read as a vector from an external file using scan with the file
name as argument. Suppose, for example, the file mat.dat contains

12	5	4	3
5	17	2	1
6	4	19	0
4	5	1	21

where each line is intended to become a row of an S matrix M. To read the matrix
we use

```
M <- matrix(scan("mat.dat"), ncol=4, byrow=T)
```

Note that it is not necessary to say how many rows the matrix has since this is
deduced from the number of columns and the total number of items read. The
argument byrow=T indicates that the incoming vector is to fill the matrix by rows
rather than by columns.

Another common way to use scan is similar to read.table but with more
flexibility. Suppose the data vectors are of equal length and are to be read in
parallel from a data file input.dat. Suppose that there are three vectors, the
first of mode character and the remaining two of mode numeric. The first step is
to use scan to read in the three vectors as a list, as follows

```
indat <- scan("input.dat", list(id="", x=0, y=0))
```

The second argument is a dummy list structure that establishes the mode of the
three vectors to be read. If we wish to access the variables separately they may
either be re-assigned to variables:

```
label <- indat$id; x <- indat$x; y <- indat$y
```

or the list itself may be attached to the search list.

As a final example, suppose we are to read in a large file with 50 numeric variables, and each case occupies 5 lines of the file `big.dat`. We are willing to to label the variables X1, X2, ..., X50. The first step is to construct the template list:[4]

```
inlist <- as.list(rep(0,50))   # coercion to list
names(inlist) <- paste("X", 1:50, sep="")
```

To read the file we must also specify that each case will occupy more than one line of the data file:

```
datlist <- scan("big.dat", what=inlist, multi.line=T)
```

This is only possible using the `multi.line=T` argument with `scan`; multi-line files cannot yet be read with `read.table`.

There is a function `count.fields` that will count the number of fields on each line of a file, which can be useful in locating faulty lines in a large file.

Reading non-standard data files

Occasionally it is helpful to use a different field separator from the default, which is any consecutive sequence of blanks, tabs or newlines (known as "whitespace"). For example, some spreadsheet programs' output files are written with fields separated by commas. The argument `sep=","` of `read.table` and `scan` will allow this. (In this case empty and blank fields will be read as missing values `NA`.) Using `sep="\t"` specifies that the fields are to be separated by a single tab character, thus allowing strings to be read that themselves contain blanks, and `sep="\n"` allows only the newline as a field separator, so every line will be read as a single field.

If input data are to be read from a file where data items occupy specific columns without guaranteed whitespace separators between items, the first option to consider is to modify the file so that there is separating whitespace. This makes the data easier to read and simpler to check directly from the data file if necessary. If it is necessary to use a file with fixed-width fields there is an argument, `widths`, to `scan` that allows fixed width input by specifying an integer vector of field widths. There is also a function `make.fields` that can be used to convert a file with fixed width, non-separated, input fields into a file with separated fields. It should be noted that the `widths` argument of `scan` uses the `make.fields` function itself and a temporary file, so its use with very large files may pose a problem.

Another solution to handling complex or non-standard input is to read in each line of the file as a character string vector and extract the data items later. For example, suppose we wish to read 40 variables from each line of a file, `v40.dat`. Each variable occupies two columns, making 80 columns of data in all. Missing values are denoted by a blank entry. We may use

---

[4] `paste` is discussed on page 39.

```
chdata <- scan("v40.dat", what="", sep="\n")
```

Specifying `sep="\n"` ensures that data items are separated only by the newline, and `what=""` establishes that the data to be read in are to be of mode `character` and so `chdata` is a character vector with each complete line of the data file as its elements.

To extract each column in numeric form we need to use a simple loop

```
dat <- list()
for(i in 1:40)
        dat[[i]] <- as.numeric(substring(chdata, 2*i-1, 2*i))
```

Coercion to numeric of a fully blank string results in a missing value, as required. To make the data more easily accessible it is a good idea to convert it to a data frame:

```
dat <- as.data.frame(dat)
```

2.5   Finding S objects

It is important to understand where S keeps its objects and where it looks for objects on which to operate. The objects that S creates at the interactive level during a session it (usually) stores permanently, as *files* in the .Data sub-directory of your working directory. (Subdirectory _DATA in the Windows versions.) On the other hand, objects created at a higher level, such as within a function, are kept in what is known as a *local frame* which is only temporary and such objects are deleted when the function is finished.[5]

Each permanent object is held as a separate file of the same name[6] and so may be manipulated by the usual operating system commands such as, in Unix, rm, cp and mv, although this is rarely needed.[7] What is important to note is that when you restart your S session at a later time, objects created in previous sessions are still available.

This explains why we recommend that you should use separate working directories for different jobs. For some users common names for objects are single letter names like x, y and so on, and if two projects share the same .Data subdirectory their objects easily become mixed up. However, data frames provide another convenient way of encapsulating and partitioning collections of related objects within the same .Data directory.

When S looks for an object, it searches through a sequence of places known as the *search list*. Usually the first entry in the search list is the .Data subdirectory of the current working directory. The names of the places currently making up the search list are given by invoking the function

---

[5] For precise details see page 107.

[6] Except perhaps in the Windows versions of S-PLUS where the allowable file names have length restrictions.

[7] One use is to transfer objects as binary files to a compatible computer.

```
search()
```

To get a list of objects currently held in the first place on the search list, use the command

```
objects()
```

The names of the objects held in any place in the search list can be displayed by giving the `objects` function an argument. For example,

```
objects(2)
```

lists the contents of the entity at position 2 of the search list. It is also possible to list selectively, using the `pattern` argument of `objects`. This restricts the listing to objects matching the pattern (in the sense of `grep` regular expressions) For example, with our library `MASS` attached at position 8,

```
> objects(8, pattern=".*lda")
[1] "lda"          "predict.lda"
> objects(8, pattern="lda$")
[1] "lda"          "predict.lda"
```

both list all objects whose names end with `lda`. As these examples show, the wildcard matching in `grep` is not that of the operating systems; `.` matches any character (use `\.` to match `.`) and `.*` matches zero or more occurrences of any character, that is any character string.

Conversely the function `find(`*object*`)` discovers where an object appears on the search list, perhaps more than once. For example (on one of our systems under S-PLUS 3.1):

```
> search()
[1] ".Data"
[2] "/usr/local/newsplus/splus/.Functions"
[3] "/usr/local/newsplus/stat/.Functions"
[4] "/usr/local/newsplus/s/.Functions"
[5] "/usr/local/newsplus/s/.Datasets"
[6] "/usr/local/newsplus/stat/.Datasets"
[7] "/usr/local/newsplus/splus/.Datasets"
> find(detach)
[1] "/usr/local/newsplus/splus/.Functions"
[2] "/usr/local/newsplus/s/.Functions"
```

The object named `detach` occurs on two places in the search list; there is a S-PLUS version as well as an S version. Since the S-PLUS version occurs before the S version on the search list, this is the one that will be found and used.

The 'places' on the search list can be of two types.[8] As well as directories of S files, they can also be lists or list-like, which include data frames. In the S literature any entity that can be placed on the search list is sometimes referred to as a *dictionary*, or as a *database*. The directory at position 1 (normally `.Data`) is

---

[8] In truth there are further types of compiled and user databases which we have never encountered.

called the *working directory*. There is a specialized use of a directory, a library, discussed in Appendix D.

If several different objects with the same name occur on the search list all but the first will be masked and normally unreachable. However the function get may be used to select an object from any given position on the search list. For example,

```
s.detach <- get("detach", 4)
```

copies the object called detach from position 4 on the search list to an object called s.detach in the working directory.[9] To see if your objects are masking system functions and datasets, use the command masked().

Extra directories, lists or data frames can be added to this list with the attach function and removed with the detach function. Examples were included in the introductory session. Normally a new entity is attached at position 2, and detach() removes the entity at position 2, normally the result of the last attach. All the higher-numbered databases are moved up or down accordingly. Note that lists can be attached by attach(alist) but must be detached by detach("alist"). If a list is attached, a copy is used, so any subsequent changes to the original list will not be reflected in the attached copy. When a list is detached the copy is normally discarded, but if any changes have been made to that database it will be saved unless the argument save=F is set. The name used for the saved list is of the form .Save.alist.2 (and is reported by detach) unless save is a character string when the database will be saved under that name.[10]

To remove objects permanently the function rm is used:

```
rm(x, y, z, ink, junk)
```

whose arguments are just the names of the objects to be discarded. The function remove can be used to remove objects with non-standard names.

A considerable degree of *caching* of databases is performed, so their view may differ inside and outside the S session. To avoid this, use the synchronize function. With no argument, it writes out objects to the current working directory. With a numerical vector argument, it re-reads the specified directories on the search list, which can be necessary if some other process has altered the database.

2.6   Character vector operations

Take care to understand character vectors. Unlike say C, they are vectors of character strings, not of characters, and most operations are performed string-by-string on the vector(s).

---

[9] Note that the object name as the argument to get must be given in quotes. With find the argument may also be given in quotes, and must be if it is a nonstandard name.

[10] This description differs from the help page, which is incorrect.

Note that " " is a legal character string with no characters in it, known as the empty string. This should be contrasted with `character(0)` which is empty character vector, that is, "the empty vector, but if there were any elements in it they would be strings". As vectors, " " has length 1 and `character(0)` has length 0.

Character vectors may created by assignments and may be concatenated by the `c` function. They may also be used in logical expressions, such as `"a" < "bill"`, in which case lexicographic ordering applies using the ASCII collating sequence.

There are several functions for operating on character vectors. The function `nchar(text)` gives (as a vector) the number of characters in each element of its character vector argument. The function `paste` takes an arbitrary number of arguments, coerces them to strings or character vectors if necessary and joins them together, element by element, as character vectors. For example

```
> paste(c("X","Y"), 1:4)
[1] "X 1" "Y 2" "X 3" "Y 4"
```

Any short arguments are re-cycled in the usual way. By default the joined elements are separated by a blank; this may be changed by using the argument, `sep=string`, often the empty string:

```
> paste(c("X","Y"), 1:4, sep="")
[1] "X1" "Y2" "X3" "Y4"
```

Another argument, `collapse`, allows the result vector to be concatenated into a single long string. It specifies another character string, and if NULL, the default, no such global concatenation takes place. Otherwise the value is placed between each element string in the final result. For example

```
> paste(c("X","Y"), 1:4, sep="", collapse=" + ")
[1] "X1 + Y2 + X3 + Y4"
```

Substrings of the strings of a character vector may be extracted (element-by-element) using the `substring` function. It has three arguments,

```
substring(text, first, last = 1000000)
```

where `text` is the character vector, `first` is a vector of first character positions to be selected and `last` is a vector of character positions for the last character to be selected. If `first` or `last` are shorter vectors than `text` they are re-cycled in the usual way. For example, the dataset `state.name` is a character vector of length 50 containing the names of the states of the United States of America in alphabetic order. To extract the first four letters in the names of the last seven states:

```
> substring(state.name[44:50], 1, 4)
[1] "Utah" "Verm" "Virg" "Wash" "West" "Wisc" "Wyom"
```

Note the use of the index vector, `[44:50]`, to select the last seven states.

The function `abbreviate` provides a more general mechanism for generating abbreviations. In this example it gives

```
> as.vector(abbreviate(state.name[44:50]))
[1] "Utah" "Vrmn" "Vrgn" "Wshn" "WsVr" "Wscn" "Wymn"
> as.vector(abbreviate(state.name[44:50], use.classes=F))
[1] "Utah" "Verm" "Virg" "Wash" "WVir" "Wisc" "Wyom"
```

We used `as.vector` to suppress the names attribute of the vector, which contains the full names!

☞   This is the end of the informal part of Chapter 2. You might like to take a rest, start skimming, or move on to Chapter 3 at this point.

2.7   Indexing vectors, matrices and arrays

We have already seen how subsets of the elements of a vector (or an expression evaluating to a vector) may be selected by appending to the name of the vector an *index vector* in square brackets. We now consider indexing more formally. For vector objects, index vectors can be any of four distinct types:

1. **A logical vector.** The index vector must be of the same length as the vector from which elements are to be selected. Values corresponding to T in the index vector are selected and those corresponding to F omitted. For example,

   ```
   y <- x[!is.na(x)]
   ```

   creates an object y which will contain the non-missing values of x, in the same order as they originally occurred. Note that if x has any missing values, y will be shorter than x. Another example is

   ```
   z <- (x+y)[!is.na(x) & x > 0]
   ```

   which creates an object z and places in it the values of the vector x+y for which the corresponding value in x was positive (and non-missing).

2. **A vector of positive integers.** In this case the values in the index vector must lie in the the the set { 1, 2, ..., length(x) }. The corresponding elements of the vector are selected and concatenated, in that order, in the result. The index vector can be of any length and the result is of the same length as the index vector. For example x[6] is the sixth component of x and x[1:10] selects the first 10 elements of x (assuming $\text{length}(x) \geqslant 10$, otherwise there will be an error). For another example we use the dataset `letters`, a character vector of length 26 containing the lowercase letters:

   ```
   > letters[1:3]
   [1] "a" "b" "c"
   > letters[1:3][c(1:3,3:1)]
   [1] "a" "b" "c" "c" "b" "a"
   ```

3. A vector of negative integers. This specifies the values to be *excluded* rather than included. Thus

```
> y <- x[-(1:5)]
```

drops the first five elements of `x`.

4. A vector of character strings. This possibility only applies where an object has a `names` attribute to identify its components. In that case a subvector of the names vector may be used in the same way as the positive integers in case **2.** For example,

```
> fruit <- c(5, 10, 1, 20)
> names(fruit) <- c("orange", "banana", "apple", "peach")
> lunch <- fruit[c("apple","orange")]
> lunch
 apple orange
     1      5
```

results in `fruit` having the vector value `c(1,5)`, retaining the `names` attribute `c("apple","orange")`.

A vector with an index expression attached can also appear on the receiving end of an assignment, in which case the assignment operation is performed only on those elements of the vector implied by the index. For example,

```
x[is.na(x)] <- 0
```

replaces any missing values in `x` by zeros and

```
y[y < 0] <- -y[y < 0]
```

has the same effect as

```
y <- abs(y)
```

The case of a zero index falls outside the rules given above. A zero index in a vector of an expression being assigned passes nothing, and a zero index in a vector to which something is being assigned accepts nothing. For example

```
> a <- 1:4
> a[0]
numeric(0)
> a[0] <- 10
> a
[1] 1 2 3 4
```

Another case to be considered is if the absolute value of an index falls outside the range 1, ..., `length(x)`. In an expression this is an error. In an assignment using a positive index greater than `length(x)` extends the vector, assigning missing values to any gap, and a negative value less than `-length(x)` is ignored.

The functions `replace` and `append` are convenience functions; `replace(x, pos, values)` returns a copy of `x` with `x[pos] <- values` without affecting the original. The function `append(x, values, after=length(x))` appends `values` to `x` in the specified place, and returns a copy without changing `x`.

Array indices

An *array* can be considered as a multiply indexed collection of data entries. Any array with just two indices is called a *matrix*, and this special case is perhaps the most important.

A *dimension vector* is a vector of positive integers of length at least 2. If its length is k then the array is k-dimensional, or as we prefer to say, k-indexed. The values in the dimension vector give the upper limits for each of the k indices. Index ranges for S objects always start at 1 (unlike those of C which start at 0). Note that singly subscripted arrays do not exist other than as vectors.

A vector can be used by S as an array only if it has a dimension vector as its dim attribute. Suppose, for example, a is a vector of 150 elements. Either of the assignments

```
a <- array(a, dim=c(3,5,10)) # make a a 3x5x10 array
dim(a) <- c(3,5,10)          # alternative direct form
```

gives it the dim attribute that allows it to be treated as a $3 \times 5 \times 10$ array. The elements of a may now be referred to either with one index, as before, as in a[!is.na(z)] *or* with three, comma separated indices, as in a[2,1,5].

To create a matrix the function array may be used, or the simpler function matrix. For example, to create a 10×10 matrix of zeros:

```
Zmat <- matrix(0, nrow=10, ncol=10)
```

Note that a vector used to define an array or matrix is recycled if necessary, so the 0 here is repeated 100 times.

It is important to know how the two indexing conventions correspond; which element of the vector is a[2,1,5] ? S arrays use 'column major order' (as used by FORTRAN but not C). This means the first index moves fastest, and the last slowest. So the correspondence is

a[1]	$\longleftrightarrow$	a[1,1,1],	a[5] $\longleftrightarrow$ a[2,2,1],	
a[2]	$\longleftrightarrow$	a[2,1,1],	... $\longleftrightarrow$ ...,	
a[3]	$\longleftrightarrow$	a[3,1,1],	a[149] $\longleftrightarrow$ a[3,5,9],	
a[4]	$\longleftrightarrow$	a[1,2,1],	a[150] $\longleftrightarrow$ a[3,5,10]	

A formal definition of column major order can be given by saying if the dimension vector is $(d_1, d_2, \ldots, d_m)$ then the element with multiple index $i_1, i_2, \ldots, i_m$ corresponds to the element with single index

$$i_1 + \sum_{j=2}^{m} \left\{ (i_j - 1) \prod_{k=1}^{j-1} d_k \right\}$$

The function matrix has an additional argument byrow which if set to true allows a matrix to be generated from a vector in row-major order.

Each of the dimensions can be given a set of names, just as for a vector. The names are stored in the `dimnames` attribute which is a list of (possibly `NULL`) vectors of character strings. For example we can name the first two dimensions of a by

```
dimnames(a) <- list(letters[1:3],
                    c("i", "ii", "iii", "iv", "v"), NULL)
```

A k-indexed array may be used with each index in any of the four forms allowed for vector indexing. There are two additional possibilities:

5. **Any array index position may be empty.**   In this case the index range implied is the entire range allowed for that index.

6. **An array may be indexed by a matrix.**   In this case if the array is k-indexed the index matrix must be an $m \times k$ matrix with integer entries and each row of the index matrix is used as an index vector specifying one element of the array. Thus the matrix specifies m elements of the array to be extracted or replaced.

So if a is a $3 \times 5 \times 10$ array, then a[1:2,,] is the $2 \times 5 \times 10$ array obtained by omitting the last level of the first index. The same sub-array could be specified in this case by a[-3,,] .

To give a simple example of a matrix index, consider extracting the diagonal elements of a square matrix X, that is, X[1,1], X[2,2], ..., X[n,n] in a vector, say Xd, and then zeroing the diagonal.[11]

```
n <- dim(X)[1]         # number of rows of X
i <- matrix(1:n, n, 2) # recycling rule
Xd <- X[i]             # extract diagonals
X[i] <- 0              # replace each by zero.
```

Note that by default a[2,,] is not a $1 \times 5 \times 10$ array but a 5×10 array. Also a[2,,1] and Xd are vectors and not arrays. In general if any index range reduces to a single value the corresponding element of the dimension vector is removed in the result. This default convention is sometimes helpful and sometimes not. To override it a named argument drop=F can be given in the array reference:

```
sua <- a[2,,]          # a   5x10 matrix
sub <- a[2,,, drop=F]  # a 1x5x10 array
```

Note that the drop convention also applies to columns (but not rows) of data frames, so if subscripting leaves just one column, a vector is returned.

However, this convention does not apply to matrix operations. Thus

```
X <- matrix(xvals, 10, 3)
X[,1]
X %*% c(1,3,5)
```

---

[11] In fact this example is unnecessary since there is a function `diag` for both purposes.

result in a ten-element vector and a 10×1 matrix respectively, which appears inconsistent. (One can argue for either convention, as reducing $1 \times n$ matrices to vectors is often undesirable.) The function `drop` forces dropping, so `drop(X %*% c(1,3,5))` returns a vector and `X[,1, drop=F]` returns a matrix. (Function writers have often overlooked these rules, which can result in puzzling or incorrect behaviour when just one observation or variable meets some criterion.)

Array arithmetic

Arrays may be used in ordinary arithmetic expressions and the result is an array formed by element-by-element operations on the data vector. The `dim` attributes of operands generally need to be the same, and this becomes the dimension vector of the result. So if `A`, `B` and `C` are all arrays of the same dimensions

```
D <- 2*A*B + C + 1
```

makes `D` a similar array with data vector the result of the evident element by element operations. However the precise rule concerning mixed array and vector calculations has to be considered a little more carefully. From experience we have found the following to be a reliable guide, although it is to our knowledge undocumented and hence liable to change.

- The expression is scanned from left to right.
- Any short vector operands are extended by recycling their values until they match the size of any other operands.
- As long as short vectors and arrays, only, are encountered, the arrays must all have the same `dim` attribute or an error results.
- Any vector operand longer than some previous array immediately converts the calculation to one in which all operands are coerced to vectors. A diagnostic message is issued if the size of the long vector is not a multiple of the (common) size of all previous arrays.
- If array structures are present and no error or coercion to vector has been precipitated, the result is an array structure with the common `dim` attribute of its array operands.

Sorting

The S function `sort` at its simplest takes one vector argument and returns a vector of sorted values. The vector to be sorted may be numeric or character, and if there is a names attribute the correspondence of names is preserved. As a simple example consider sorting `mydata`:

```
> mydata
  c   j   b   e   i   h   g   d   f   a
2.9 3.4 3.4 3.7 3.7 2.8 2.8 2.5 2.4 2.4
> sort(mydata)
  f   a   d   h   g   c   j   b   e   i
2.4 2.4 2.5 2.8 2.8 2.9 3.4 3.4 3.7 3.7
```

Note that the ordering of tied values is preserved.

The second argument to `sort` allows partial sorting. The argument specifies an index or set of indices which represent the order statistics guaranteed to be correct in the result. The values between these reference indices will all be intervening values, but may not be in sorted order. For example, consider finding the median of a large sample from the $N(0, 1)$ distribution:

```
> x <- rnorm(100001)
> sort(x, partial=50001)[50001]
[1] 0.0028992
> median(x)      # as a check
[1] 0.0028992
```

This is most often used to find quantiles, for which the function `quantile` is provided.

A more flexible sorting tool is `sort.list`,[12] which produces an index vector which will arrange its argument in increasing order. Thus `x[sort.list(x)]` returns the same value as `sort(x)` and `x[sort.list(-x)]` is the vector x arranged in *decreasing* order. Suppose y is a vector and X a matrix with `length(y)` rows. We wish to arrange y in increasing order and re-arrange the rows of X so that the original correspondence with the values in y is maintained. We can use

```
i <- sort.list(y); y <- y[i]; X <- X[i,]
```

A further sorting function is `order`. It takes an arbitrary number of arguments and returns the index vector which would arrange the first in increasing order, with ties broken by the second, and so on. For example, to arrange employees by age, by salary within age, and by employment number within salary, and list them together with their number of dependents, we might use:

```
m <- order(age, salary, No)
cbind(age[m], salary[m], No[m], depndts[m])
```

All three functions have an argument `na.last` that determines the handling of missing values. With `na.last=NA` (the default for `sort`) missing values are deleted; with `na.last=T` (the default for `sort.list` and `order`) they are put last, and with `na.last=F` they are put first.

2.8   Matrix operations

The standard matrix operators and indexing of matrices have already been covered in Sections 2.3 and 2.7. In this section we cover more specialized calculations involving matrices. Many (but not all) of these will work with numerical data frames, which can always be coerced by `as.matrix(dataframe)` (see page 51).

The function `crossprod` forms "crossproducts", meaning that

---

[12] The use of "list" is unfortunate here since it has nothing to do with S list objects.

```
Xdashy <- crossprod(X, y)
```

calculates $X^T y$. This matrix could be calculated as `t(X) %*% y` but using
`crossprod` is more transparent and efficient. If the second argument is omitted
it is taken to be the same as the first. Thus `crossprod(X)` calculates the matrix
$X^T X$.

An important operation on arrays is the *outer product*. If `a` and `b` are two
numeric arrays, their outer product is an array whose dimension vector is obtained
by concatenating their two dimension vectors, (order is important), and whose
data vector is obtained by forming all possible products of elements of the data
vector of `a` with those of `b`. The outer product is formed by the operator `%o%`:

```
ab <- a %o% b
```

and by the function `outer`:

```
ab <- outer(a, b, "*")
ab <- outer(a, b)          # as "*" is the default.
```

The multiplication function may be replaced by an arbitrary function of two
variables. For example if we wished to evaluate the function

$$f(x, y) = \frac{\cos(y)}{1 + x^2}$$

over a regular grid of values with x– and y–coordinates defined by the S vectors
`x` and `y` respectively, we could use:

```
f <- function(x, y) cos(y)/(1 + x^2) # define the function
z <- outer(x, y, f)                  # use it.
```

The function `diag` either creates a diagonal matrix from a vector argument, or
extracts as a vector the diagonal of a matrix argument. Used on the assignment side
of an expression it allows the diagonal of a matrix to be replaced. For example, to
form a covariance matrix in multinomial fitting we could use

```
> p <- dbinom(0:4, size=4, prob=0.1)  # an example prob vector
> CC <- -(p %o% p)
> diag(CC) <- p*(1-p)
> CC
           [,1]         [,2]         [,3]         [,4]         [,5]
[1,]   0.22563279 -0.19131876 -0.03188646 -0.00236196 -6.561e-05
[2,]  -0.19131876  0.20656944 -0.01417176 -0.00104976 -2.916e-05
[3,]  -0.03188646 -0.01417176  0.04623804 -0.00017496 -4.860e-06
[4,]  -0.00236196 -0.00104976 -0.00017496  0.00358704 -3.600e-07
[5,]  -0.00006561 -0.00002916 -0.00000486 -0.00000036  9.999e-05
```

In addition `diag(n)` for a positive integer `n` generates an $n \times n$ identity matrix.
The function `vecnorm` gives the p-norm of a vector, by default for $p = 2$.

The functions `apply` and `sweep`

The function `apply` allows functions to operate on an array using sections successively. For example, consider the dataset `iris` which is a $50 \times 4 \times 3$ array of four observations on 50 specimens of each of three species. Suppose we want the means for each variable by species; we can use `apply`.

The arguments of `apply` are

1. The name of the array, `X`.

2. An integer vector, `MARGIN`, giving the indices defining the sections of the array to which the function is to be separately applied. It is helpful to note that if the function applied has a scalar result, the result of `apply` is an array with `dim(X)[MARGIN]` as its dimension vector.

3. The function, or the name of the function, `FUN`, to be applied separately to each section.

4. Any additional arguments needed by the function as it is applied to each section.

Thus we need to use

```
> apply(iris, c(2,3), mean)
          Setosa Versicolor Virginica
Sepal L.  5.006      5.936     6.588
Sepal W.  3.428      2.770     2.974
Petal L.  1.462      4.260     5.552
Petal W.  0.246      1.326     2.026
> apply(iris, c(2,3), mean, trim=0.1)
          Setosa Versicolor Virginica
Sepal L. 5.0025     5.9375    6.5725
Sepal W. 3.4150     2.7800    2.9625
Petal L. 1.4600     4.2925    5.5100
Petal W. 0.2375     1.3250    2.0325
```

where we also show how arguments can be passed to the function, in this case to give a trimmed mean. If we want the overall means we can use

```
> apply(iris, 2, mean)
 Sepal L. Sepal W. Petal L. Petal W.
   5.8433   3.0573    3.758   1.1993
```

Note how dimensions have been dropped to give a vector. If the result of `FUN` is itself a vector of length `d`, say, then the result of `apply` is an array with dimension vector `c(d, dim(X)[MARGIN])`, with single-element dimensions dropped. Note that matrix results are reduced to vectors; if we ask for the covariance matrix for each species,

```
> ir.var <- apply(iris, 3, var)
> dim(ir.var) <- c(4,4,3) # expect a warning
> dimnames(ir.var) <- list(dimnames(iris)[[2]],
        dimnames(iris)[[2]], dimnames(iris)[[3]])
```

we get a 16×3 matrix. We can add back the dimensions, but lose the `dimnames`.

The function `apply` is often useful to replace explicit loops. Earlier versions of the function used loops internally, but from S-PLUS 3.2 they are replaced by a call to an internal function. Note too that for *linear* calculations it is rather inefficient. We can form the means by matrix multiplication:

```
> ir1 <- matrix(iris, 50)
> matrix(rep(1/50, 50) %*% ir1, 4,
        dimnames=list(dimnames(iris)[[2]], dimnames(iris)[[3]]))
          Setosa Versicolor Virginica
Sepal L.  5.006     5.936     6.588
Sepal W.  3.428     2.770     2.974
Petal L.  1.462     4.260     5.552
Petal W.  0.246     1.326     2.026
```

which will be very much faster on larger examples, but is much less transparent.

The function `aperm` is often useful with array/matrix arithmetic of this sort. It permutes the indices, so that `aperm(iris, c(2,3,1))` is a $4 \times 3 \times 50$ array.

Function `sweep`

The functions `apply` and `sweep` are often used together. For example, having found the means of the `iris` data, we may want to remove them by subtraction or perhaps division. We can use sweep in each case:

```
ir.means <- apply(iris, c(2,3), mean)
sweep(iris, c(2,3), ir.means)
log(sweep(iris, c(2,3), ir.means, "/"))
```

Of course, we could have subtracted the log means in the second case.

Functions operating on matrices

The standard operations of linear algebra are either available as functions or can easily be programmed, making S a flexible matrix manipulation language (if rather slower than specialized matrix languages).

The function `solve` inverts matrices and solves systems of linear equations; `solve(A)` inverts A and `solve(A, b)` solves A `%*%` x = b. (If the system is over-determined, the least-squares fit is found, but matrices of less than full rank give an error.)

The function `chol` returns the Choleski decomposition $A = U^T U$ of a non-negative definite symmetric matrix. (Note that this is an unusual convention; more commonly the lower-triangular form $A = LL^T$ with $L = U^T$ is used.) Function `backsolve` solves upper triangular systems of matrices, and is often used in conjunction with `chol`.

Eigenvalues and eigenvectors

The function `eigen` calculates the eigenvalues and eigenvectors of a square matrix. The result is a list of two components, `values` and `vectors`. If we only need the eigenvalues we can use:

```
eigen(Sm, only.values=T)$values
```

Real symmetric matrices are known to have real eigenvalues, and the calculation for this case can be much simpler and more stable. A further named argument, `symmetric`, may be used to specify whether or not a matrix is (to be regarded as) symmetric. The default value is `T` if the matrix exactly equals its transpose, otherwise `F`.

Singular value decomposition

An $n \times p$ matrix X has a *singular value decomposition* (SVD) of the form

$$X = U\Lambda V^T$$

where U and V are $n \times \min(n, p)$ and $p \times \min(n, p)$ matrices of orthonormal columns, and Λ is a diagonal matrix. Conventionally the diagonal elements of Λ are ordered in decreasing order; the number of non-zero elements is the rank of X. A proof of its existence can be found in Golub & Van Loan (1989), and a discussion of the *statistical* value of SVDs in Thisted (1988).

The function `svd` takes a matrix argument, `M`, and calculates the singular value decomposition. The components of the result are `u` and `v`, the orthonormal matrices and `d`, a vector of singular values. If either U or V is not required its calculation can be avoided by the argument `nu=0` or `nv=0`.

The QR decomposition

A faster decomposition than the SVD is the QR decomposition, defined as

$$M = QR$$

where Q is a matrix of orthonormal columns and R is an upper triangular matrix. (See Golub & Van Loan (1989, §5.2)). The function `qr(M)` implements the algorithm detailed in Golub & Van Loan, and hence returns the result in a somewhat inconvenient form to use directly. The result is a list that can be used by other tools. For example

```
M.qr <- qr(M)            # QR decomposition
Q <- qr.Q(M.qr)          # Extract the Q matrix
R <- qr.R(M.qr)          # Extract the R matrix
y.res <- qr.resid(M.qr, y)  # Project onto error space
```

The last command finds the residual vector after projecting the vector `y` onto the column space of `M`. Other tools that use the result of `qr` include `qr.fitted` for fitted values and `qr.coef` for regression coefficients.

Determinant and trace

No function is provided for finding the determinant of a square matrix M. There are several ways to write determinant functions. Often it is known in advance that a determinant will be non-negative or the sign is not needed, in which case methods to calculate the absolute value of the determinant suffice.

Since M is square, the matrices U and V of the SVD are orthogonal matrices. Thus the absolute value of the determinant is the product of the singular values. Hence the function

```
absdet <- function(M) prod(svd(M, nu=0, nv=0)$d)
```

calculates the absolute value of the determinant of M.

Another possibility is to use the QR decomposition, which is faster but numerically slightly less stable. On looking carefully at the output of the qr function the appropriate calculation can be seen to be

```
absdet1 <- function(M) abs(prod(diag(qr(M)$qr)))
```

The sign is reversed for each column calculated, so the (signed) determinant is given by[13]

```
det1 <- function(M) prod(diag(qr(M)$qr)) *
        ifelse(nrow(M)%%2, 1, -1)
```

Note however, that this depends on the details of the QR algorithm used (Householder reflections) and is therefore subject to change.

If the sign is unknown and important, the determinant may be calculated as the product of the eigenvalues. However these will in general be complex and the result may have complex roundoff error even though the exact result is known to be real. A simple function to perform the calculation is

```
det <- function(M) Re(prod(eigen(M, only.values=T)$values))
```

As a further simple example the trace of a square matrix may be written as a function

```
tr <- function(M) sum(diag(M))
```

Converting data frames to and from matrices

The coercion function as.data.frame takes a matrix argument (of mode numerical, character or logical) and produces a data frame from it using the columns as variables. If the matrix has a dimnames attribute it is retained in the names and row names of the data frame. If not, names are constructed in a default manner with the variable (column) names incorporating the original matrix name.

There are two functions which may be used for converting a data frame into a matrix, and they behave slightly differently.

---

[13] ifelse is discussed on page 86.

1. `as.matrix(dataframe)` produces a matrix from the argument data frame of mode numeric if all constituent variables in the data frame are numeric, and of mode character otherwise. So any factor will cause the entire matrix result to be of mode character. Note that it is not possible to have mixed mode matrices.

2. `data.matrix(dataframe)` produces a numeric matrix from the argument data frame in all circumstances. Any factors among the variables are first coerced to numeric, and the levels information of the factors is retained in the attribute `column.levels` of the outcome.

2.9   Functions operating on factors and lists

Consider a (hypothetical) sample survey of 30 professional people from the states and territories of Australia[14] given by:

```
state <- c("tas", "sa", "qld", "nsw", "nsw", "nt",  "wa",
           "wa", "qld", "vic", "nsw", "vic", "qld", "qld",
           "sa",  "tas", "sa", "nt",  "wa",  "vic", "qld",
           "nsw", "nsw", "wa", "sa",  "act", "nsw", "vic",
           "vic", "act")
self.emply <- c(T,T,T,T,F,T,F,T,F,T,T,T,F,T,F,
                T,F,T,T,F,F,T,F,T,T,F,T,T,T,T)
incomes <- c(60, 49, 40, 61, 64, 60, 59, 54, 62, 69,
             70, 42, 56, 61, 61, 61, 58, 51, 48, 65,
             49, 49, 41, 48, 52, 46, 59, 46, 58, 43)
profession.survey <- data.frame(state, self.emply, incomes)
rm(state, self.emply, incomes)
```

where we note that character and logical vectors become factors in a data frame. The `table` function shows the frequencies of each factor:

```
> attach(profession.survey)
> table(state)
 act nsw nt qld sa tas vic wa
   2   6  2   5  4   2   5  4
> table(self.emply)
 FALSE TRUE
    10   20
> table(self.emply, state)
       act nsw nt qld sa tas vic wa
 FALSE   1   2  0   3  2   0   1  1
 TRUE    1   4  2   2  2   2   4  3
```

---

[14] There are eight states and territories in Australia, namely the Australian Capital Territory, New South Wales, the Northern Territory, Queensland, South Australia, Tasmania, Victoria and Western Australia.

Note that if more than one factor is given to `table`, a multi-way frequency table is produced. The related function `tabulate` (which is called by `table`) operates on a non-negative numeric vector and returns a vector of frequencies of each value taken.

The combination of a vector and a labelling factor is an example of what is called a *ragged array*, since the subclass sizes are usually irregular. (When the subclass sizes are all the same the indexing may be done implicitly and more efficiently using arrays.)

To continue the previous example, we now consider the incomes. If the first argument to `plot` is a factor, the method function used, `plot.factor`, produces boxplots:

```
plot(state, incomes)
```

The result is shown in Figure 2.3. To calculate the mean income for each state

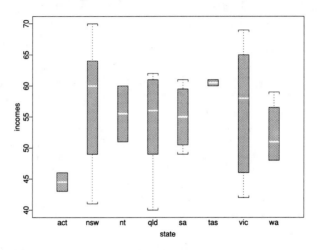

Figure 2.3: Boxplots of incomes by state.

from this sample we now use the function `tapply`:

```
> tapply(incomes, state, mean)
  act    nsw    nt  qld sa  tas vic    wa
 44.5 57.333 55.5 53.6 55 60.5  56 52.25
```

giving a vector with the components labelled by the levels of `state`. The function `tapply` is used to apply a function, here `mean`, to each subset of `incomes` defined by the levels of the second argument, `state`. The result is a vector of the same length as the `levels` attribute of the factor, with the `levels` attribute of the factor giving the `names` attribute of the result.

To take the example a little further, suppose we needed to calculate the standard errors of the state income means. To do this we write an S function to calculate the standard error for any given vector:

```
> std.err <- function(x) sqrt(var(x)/length(x))
> tapply(incomes, state, std.err)
 act    nsw  nt   qld     sa tas   vic     wa
 1.5 4.3102 4.5 4.1061 2.7386 0.5 5.244 2.6575
```

Finding the usual 95% confidence limits (without assuming equal variances) for the state mean incomes is now an easy exercise. (The function `tapply` could be used with the `length` function to find the sample sizes, and the `qt` function to find the percentage points of the appropriate t-distributions; see Chapter 5.)

The function `tapply` can be used to handle more complicated indexing of a vector by multiple categories, by specifying a list of indices. For example, we find

```
> tapply(incomes, self.emply, mean)
 FALSE  TRUE
  56.1 54.05
> tapply(incomes, list(self.emply, state), mean)
        act   nsw   nt    qld   sa   tas   vic wa
FALSE   46 52.50   NA 55.667 59.5   NA 65.00 59
 TRUE   43 59.75 55.5 50.500 50.5 60.5 53.75 50
```

which copes correctly with the empty cells in the table.

The function `split` takes as arguments a data vector and a vector defining groups within it. The value is a list of vectors, one component for each group. For example

```
> split(incomes, self.emply)
$FALSE:
  5  7  9 13 15 17 20 21 23 26
 64 59 62 56 61 58 65 49 41 46
$TRUE:
  1  2  3  4  6  8 10 11 12 14 16 18 19 22 24 25 27 28 29 30
 60 49 40 61 60 54 69 70 42 61 61 51 48 49 48 52 59 46 58 43
```

Note how the group variable values are used as the `names` attribute.

The function `lapply` and `sapply` are the analogues of `tapply` and `apply` for lists or vectors. In a sense `lapply` is the primitive function, since in S-PLUS 3.2 it is called by all the others. It takes two arguments, a list and a function, and applies the function to each component of the list, to return a list. (Optional arguments to the function can as with `apply` and `tapply` be given as further named arguments.) The function `sapply` has the same call but may take a vector or a list as its first argument. The value is usually simplified to a vector or matrix. An alternative way to produce summaries by employment category is

```
> sapply(split(incomes, self.emply), mean)
 FALSE  TRUE
  56.1 54.05
> inc.summ <- function(x) c(mean(x), std.err(x))
> sapply(split(incomes, self.emply), inc.summ)
        FALSE    TRUE
[1,] 56.1000 54.0500
[2,]  2.5667  1.9201
```

and the second form produces a neater result than the alternative

```
> tapply(incomes, self.emply, inc.summ)
$FALSE:
[1] 56.1000  2.5667
$TRUE:
[1] 54.0500  1.9201
```

Note that although `sapply` is usually applied to lists it may be applied to any S object, for example numeric vectors.

2.10   Input/Output facilities

In this section we cover a miscellany of topics connected with input and output.

Writing data to a file

A character representation of any S vector may be written on an output device (including the session window) by the `write` function

```
write(x, file="outdata")
```

where the session window is specified by `file=""`. Little is allowed by way of format control, but the number of columns can be specified by the argument `ncolumns` (default 5 for numeric vectors, 1 for character vectors). This is sometimes useful to print out vectors or matrices in a fixed layout. For data frames, `write.table` is provided from S-PLUS 3.2:

```
> write.table(dataframe, file="", sep=",")
> write.table(painters, "", sep="\t", dimnames.write="col")
Composition     Drawing Colour  Expression      School
10        8         16       3        A
15        16        4        14       A
8         13        16       7        A
12        16        9        8        A
     ....
```

Omitting `dimnames.write="col"` writes out both row and column labels (if any) and setting it to F omits both.

We have found the following simple function useful to print out matrices or data frames:

```
write.matrix <- function(x, file="", sep=" ")
{
      x <- as.matrix(x)
      p <- ncol(x)
      cat(dimnames(x)[[2]],format(t(x)), file=file,
            sep=c(rep(sep, p-1), "\n"))
}
```

This produces a neatly formatted layout, and starts with the column labels (if any) ready for reading in by `read.table` with `header=T`.

Executing commands from, or diverting output to, a file

If commands are stored on an external file, say `commands.q` in the current directory, they may be executed at any time in an S session with the command

```
source("commands.q")
```

It is often useful to use `options(echo=T)` before a source file, to have the commands echoed. (If used in the file itself, it takes effect *after* the file is completed.) Similarly

```
sink("record.lis")
```

will divert all subsequent output from the session window to an external file `record.lis`. The command

```
sink()
```

restores output to the window once more.

Two functions are useful for externally manipulating or transmitting S objects. The function `dump` takes as its first argument a character vector giving the names of S objects to be written on an external file, by default the file `dumpdata`. These are written in *assignment form* so that executing the file `dumpdata` with the `source` command will re-create the objects.

```
dump(c("a", "x", "ink"), file="outdata") # dump objects
    ....
source("outdata") # re-create a, x and ink
```

The dumped objects are in a form which is fairly easy to read and to edit, but reading such a file with `source` can be slow, particularly for large numeric objects.

The function `data.dump` does a similar job to `dump` but the dumped objects may only be re-created using the companion function `data.restore` to read the file, and this operation is relatively fast. These two functions are intended for transmission of S objects between remote or incompatible computers, and unlike `dump` are guaranteed to preserve the storage mode (integer, single or double precision) of S objects.

General printing

The function `cat` is similar to `paste` with argument `collapse=""` in that it coerces its arguments to character strings and concatenates them. However instead of returning a character string result it prints out the result in the session window or optionally on an external file. For example to print out today's date on our system:

```
> d <- date()
> cat("Today's date is:", substring(d,1,10),
            substring(d,25,28), "\n")
Today's date is: Sun Mar  6 1994
```

which is needed occasionally for dating output from within a function. Note that an explicit newline ("\n") is needed.

Other arguments to `cat` allow the output to be broken into lines of specified length, and optionally labelled:

```
> cat(1,2,3,4,5,6, fill=8, labels=letters)
a 1 2
c 3 4
e 5 6
```

and `fill=T` fills to the current output width.

The function `format` provides the most general way to prepare data for output. It coerces data to character strings in a common format, which can then be used by `cat`. For example, the print function `print.summary.lm` for summaries of linear regressions contains

```
cat("\nCoefficients:\n")
print(format(round(x$coef, digits = digits)), quote = F)
cat("\nResidual standard error:",
    format(signif(x$sigma, digits)), "on", rdf,
    "degrees of freedom\n")
cat("Multiple R-Squared:", format(signif(x$r.squared, digits)),
    "\n")
cat("F-statistic:", format(signif(x$fstatistic[1], digits)),
    "on", x$fstatistic[2], "and", x$fstatistic[3],
    "degrees of freedom, the p-value is", format(signif(1 -
    pf(x$fstatistic[1], x$fstatistic[2], x$fstatistic[3]),
    digits)), "\n")
```

Note the use of `signif` and `round` to specify the accuracy required. (For `round` the number of digits are specified, whereas for `signif` it is the number of significant digits.) To see the effect of `format` *vs* `write`, consider:

```
> write(iris[,1,1], "", 15)
5.1 4.9 4.7 4.6 5 5.4 4.6 5 4.4 4.9 5.4 4.8 4.8 4.3 5.8
5.7 5.4 5.1 5.7 5.1 5.4 5.1 4.6 5.1 4.8 5 5 5.2 5.2 4.7
4.8 5.4 5.2 5.5 4.9 5 5.5 4.9 4.4 5.1 5 4.5 4.4 5 5.1
4.8 5.1 4.6 5.3 5
> cat(format(iris[,1,1]), fill=60)
5.1 4.9 4.7 4.6 5.0 5.4 4.6 5.0 4.4 4.9 5.4 4.8 4.8 4.3 5.8
5.7 5.4 5.1 5.7 5.1 5.4 5.1 4.6 5.1 4.8 5.0 5.0 5.2 5.2 4.7
4.8 5.4 5.2 5.5 4.9 5.0 5.5 4.9 4.4 5.1 5.0 4.5 4.4 5.0 5.1
4.8 5.1 4.6 5.3 5.0
```

There is a tendency to output values such as 0.6870000000000001, even after rounding to (here) 3 digits. (Not from `print`, but from `write`, `cat`, `paste`, `as.character` and so on.) Use `format` to avoid this.

By default the accuracy of printed and converted values is controlled by the `options` parameter `digits`, which defaults to 7. Prior to S-PLUS 3.2, `format` did not consult this value.

2.11   Customizing your S environment

The S environment can be customized in many ways, down to replacing system functions by your own versions. For multi-user systems, all that is normally desirable is to use the `options` command to change the defaults of some variables, and if it is appropriate to change these for every session of a project, to use `.First` to set them (see the next subsection).

The function `options` accesses or changes the dataset `.Options`, which can also be manipulated directly. Its exact contents will differ between operating systems and S-PLUS releases, but one example is

```
> unlist(options())
        echo prompt continue width length         keep check digits
     "FALSE" "> "    "+ "     "80" "48"   "function" "0"    "7"
         memory object.size audit.size        error   show
     "2147483647" "5000000"   "500000"    "dump.calls" "TRUE"
     compact scrap free warn editor expressions reference
     "100000" "500" "1"  "0"  "vi"   "256"      "1"
     contrasts.factor contrasts.ordered  ts.eps  pager
     "contr.helmert"   "contr.poly"      "1e-05" "less"
```

(The function `unlist` has converted all entries to character strings.) Other options (such as `gui`) are by default unset. Calling `options` with no argument or a character vector argument returns a list of the current settings of all options, or those specified. Calling it with one or more `name=value` pairs resets the values of component `name`, or sets it if it was unset. For example

```
> options("width")
$width:
[1] 80
> options("width"=65)
> options(c("length", "width"))
$length:
[1] 48
$width:
[1] 65
```

The meaning of the more commonly used options are given in Table 2.2.

There is a similar command, `ps.options`, to customize the actions of the `postscript` graphics driver in the Unix versions of S-PLUS.

Session startup and finishing functions

If the `.Data` subdirectory of the working directory contains a function `.First` this function is silently executed at the start of any S session. This allows some automatic customization of the session which may be particular to that working directory. A typical `.First` function might include commands such as

Table 2.2: Commonly used options to customize the S environment.

width	The page width, in characters. Not always respected.
length	The page length, in lines. Used to split up listings of large objects, repeating column headings. Set to a very large value to suppress this.
digits	number of significant digits to use in printing. Set this to 17 for full precision.
gui	The preferred graphical user interface style.
echo	Logical for whether expressions are echoed before being evaluated. Useful when reading commands from a file.
prompt	The primary prompt.
continue	The command continuation prompt.
editor	The default text editor for ed and fix.
error	Function called to handle errors. See Section 4.3.
warn	The level of strictness in handling warnings. The default, 0, collects them together; 1 reports them immediately and 2 makes any warning an error condition.
memory	The maximum memory (in bytes) that can be allocated.
object.size	The maximum size (in bytes) of any object.

```
.First <- function()
{
  options(prompt = "> ", continue = "+   ", digits = 5,
     length = 99999, gui = "motif", editor="vi")
  ps.options(paper = "a4", font = 3, pointsize = 10,
     horizontal = F)
  library(MASS, first=T)
}
```

If it were known that the project would always use the same windowing system it might be appropriate to open a graphics window and also a help window with statements such as:

```
motif("-geometry 600x500-0+0")
help.start()
```

If there is a function .Last it is executed when the session is terminated. A typical .Last function on a Unix system where disc space was always in short supply might be:

```
.Last <- function()
{
   unix("rm -f ps.out.*.ps") # remove unwanted PostScript files
   cat("Adios.\n")
}
```

2.12   History and audit trails

Unless specifically disabled, S keeps a compact record of all commands in a file
.Audit in the current working directory. This can be accessed in two ways. The
history command will retrieve commands (by default the last 10) and allow
them to be edited and re-submitted, and again does so for the last command.
Both can search for commands matching a pattern, their first argument. For
example,

```
history(max=50, rev=F, evaluate=F)
```

will list the last up to 50 commands in chronological order, and

```
again("lda", ed=T)
```

allows the last command containing the string "lda" to be edited and then submit-
ted.

The command Splus AUDIT used from the operating system prompt allows
a much more detailed investigation of the audit trail. It has a cryptic command
language detailed in its help page. For details of the power of this facility, see
Becker & Chambers (1988).

The audit trail can grow large. The option audit.size, default 0.5Mb, if
reached triggers a warning at the beginning of a session. Run

```
Splus TRUNC_AUDIT n
```

at any time (outside S) to truncate the file to about n bytes (default 100000).

2.13   Exercises

The exercises in Chapter 4 also exercise the material of this chapter, but are rather
harder.

2.1.   How would you find the index(es) of specified values within a vector? For
example, where is the hill race (in hills) with a climb of 2100 feet?

2.2.   The column ftv in data frame birthwts counts the number of visits.
Reduce this to a factor with levels 0, 1 and '2 or more'. [Hint: manipulate the
levels, or investigate functions cut and merge.levels.]

2.3.   Write a simple function to compute the median absolute deviation (used in
robust statistics; see Chapter 8) median$[x - \mu]$ with default μ the sample median.
S-PLUS users should compare their answer with the function mad.

2.4.   Use tapply or sapply to give the 95% confidence intervals for the mean
income of professionals by Australian state based on the data in Section 2.9.

2.5.   For a matrix X, we want to find for each row which row of a matrix M
is nearest in Euclidean distance. Use apply to find out. (You may assume that
there is a unique nearest row.)

2.6.   Suppose x is an object with named components and out is a character
string vector. How would you make a new object obtained from x by *excluding*
any components whose names are in out ?

Chapter 3

Graphical Output

S-PLUS provides comprehensive graphics facilities, from simple facilities for producing common diagnostic plots by plot (*object*) to fine control over publication-quality graphs. In consequence, the number of graphics parameters is huge. In this chapter, we build up the complexity gradually. Most readers will not need to go beyond the first 3 sections, and indeed the material later in this chapter is not used elsewhere in this book. However, we *have* needed to make use of it, especially in matching existing graphical styles.

Some graphical ideas are best explored in their statistical context, so that, for example, histograms are covered in Chapter 5, survival curves in Chapter 11, Chernoff's faces and biplots in Chapter 12 and time-series graphics in Chapter 14. Table 3.1 gives an overview of the high-level graphics commands with page references; many of the call sequences can be found in Appendix B.

The handling of graphical devices differs considerably between S-PLUS and S; throughout this chapter we assume that S-PLUS is being used.

There are many books on graphical design. Cleveland (1993) discusses most of the methods of this chapter and detailed design choices (such as aspect ratios of plots and the presence of grids) which can affect the perception of graphical displays. As these are to some extent a matter of personal preference and this is also a guide to S-PLUS, we have kept to the default choices.

3.1 Graphics devices

Before any plotting commands can be used, a graphics device must be opened to receive graphical output. Most commonly this is a window on the screen of a workstation or a plotter or printer. A list of supported devices on the current hardware with some indication of their capabilities is available from the online help system by

```
?Devices
```

(Note the capital letter.) A number of graphics terminals are supported, but as these are now unlikely to be encountered they are not considered here.

A graphics device is opened by giving the command in Table 3.2, possibly with parameters giving the size and position of the window, for example

Table 3.1: High level plotting functions. Page references are given to the most complete description in the text: see also Appendix B.

	page	
abline	66	Adds lines to the current plot in slope-intercept form.
barplot	65	Bar graphs.
biplot	311	Represent rows and columns of a data matrix.
brush spin	302	Dynamic graphics.
contour	69	Contour plot.
coplot	74	Plot one variable against another for given ranges or levels of one or two other variables.
dotchart		Produce a dot chart.
eqscplot	67	Plots with geometrically equal scales (our library).
faces	308	Chernoff's faces plot of multivariate data.
frame	71	Advance to next figure region.
hist	126	Histograms using barplot.
hist2d	141	Two-dimensional histogram calculations.
identify locator	72	Interact with an existing plot.
image	69	High density image plot functions.
image.legend		
interaction.plot		Interaction plot for a two-factor experiment.
legend	73	Add a legend to the current plot.
matplot	82	Multiple plots specified by the columns of a matrix.
mtext	73	Add text in the margins.
pairs	301	All pair-wise plots between multiple variables.
par	76	Set or ask about graphics parameters.
persp perspp	69	Three dimensional perspective plot functions.
persp.setup		
pie		Produce a pie diagram.
plot		Generic plotting function.
polygon		Add a polygon to the present plot, possibly filled.
points lines	66	Adds points or lines to the current plot.
qqplot qqnorm	122	Quantile-quantile and normal Q-Q plots.
scatter.smooth	250	Scatterplot with a smooth curve.
segments arrows	82	Draw line segments or arrows on the current plot.
stars	308	Star plots of multivariate data.
symbols		Draw variable sized symbols on a plot.
text	66	Add text symbols to the current plot.
title	72	Add title(s).

Table 3.2: Some of the graphical devices available in S-PLUS.

`motif`	X11-windows systems.
`openlook`	X11-windows systems.
`iris4d`	Silicon Graphics Iris workstations.
`suntools`	Sun workstations under SunView.
`win.graph`	Windows version of S-PLUS.
`postscript`	PostScript printers.
`hplj`	Hewlett-Packard LaserJet printers.
`hpgl`	Hewlett-Packard HP-GL plotters.
`win.printer`	Use Windows printing.
`pic`	A `troff` picture processing language.
`fig`	For use with the `xfig` drawing package.[1]

```
motif("-geometry 600x400-0+0")
```

opens a small graphics window initially positioned in the top right hand corner of the screen. (Note that `motif` may be used on any X11 display, not just one running the Motif window manager, although `openlook` is preferred for OpenLook window managers.) Many screen devices support the argument `ask`, which if true will request permission to clear the screen and start the next plot. (This can be set later by `par(ask=T)` or using `dev.ask`.)

Graphics devices may be closed using

```
graphics.off()
```

This will close all the graphics devices currently open and perform any further wrap-up actions that need to be taken, such as sending trailer information to printer devices. Note that quitting with `q()` will automatically perform a `graphics.off` operation.

It is possible to have several graphical devices open at once. By default the most recently opened one is used, but `dev.set` can be used to change the current device (by number). The function `dev.list` lists currently active devices, and `dev.off` closes the current device, or one specified by number. There are also commands `dev.cur`, `dev.next` and `dev.prev` which return the number of the current, next or previous device on the list.

The `motif` and `openlook` devices have a Copy option on their Graph menu which allows a (smaller) copy of the current plot to the copied to a new window, perhaps for comparison with latter plots. (The copy window can be dismissed by the Destroy item from its menu.) There is also a `dev.copy` function which copies the current plot to the specified device (by default the next device on the list).

Many of the graphics devices on windowing systems have menus of choices, for example to make hardcopies and to alter the colour scheme in use. (Under

---

[1] Available from `statlib`; see Appendix D.

Windows the colour scheme is selected for all graphics windows from the **Options** menu on the main menu bar.)

Graphical hardcopy

There are several ways to produce a hardcopy of a plot on a suitable printer or plotter. The `motif` and `openlook` devices have a `Print` item on their `Graph` menu which will send a full-page copy of the current plot to the default printer. A little more control (the orientation and the print command) is available from the `Printing` item on the `Options` or `Properties` menu.

Users of the **Windows** version of S-PLUS can use the **Print** option on the **File** menu. This prints the current (normally topmost) window, so bring the desired graphics window to the top first, or make it current via its control menu.

The function `printgraph` will copy the current plot to a POSTSCRIPT or LASERJET printer, and allows the size and orientation to be selected, as well as paper size and printer resolution where appropriate.

The function `dev.print` will copy the current plot to a printer device (default `postscript`) and allow size, orientation, pointsize of text and so on to be set.

Finally, it is normally possible to open an appropriate printer device and repeat the plot commands, although this does preclude interacting with the plot on-screen.

A warning: none of these methods appear to work correctly for `brush` and `spin` plots, except printing directly from a `suntools` window. The plots in this book were produced by screen dumps using **X11** facilities. **Windows** users can copy to their clipboard and then paste into other applications.

Hardcopy to a file

It is very useful to be able to copy a plot to a file for inclusion in a paper or book (as here). Since each of the hardcopy methods allows the printer output to be re-directed to a file, there are many possibilities. If the plot will be rescaled subsequently, the simplest way is to edit the print command in the `Printing` item of a `motif` or `openlook` device to be

```
cat > plot_file_name <
```

or to make a Bourne-shell command file `rmv` by

```
$ cat > rmv
mv $2 $1
^D
$ chmod +x rmv
```

place this in your path and use `rmv plot_file_name`. (This avoids leaving around temporary files with names like `ps.out.0001.ps`.) **Windows** users can use the **Print** option on the **File** menu to select printing to a file and should also check the **Setup** box to include the header in the file (at least for POSTSCRIPT printers.). The graphics window can then be printed in the usual way.

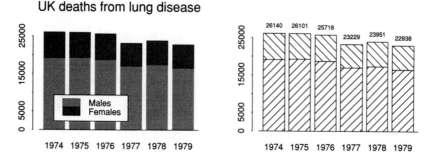

Figure 3.1: Two different styles of barchart showing the annual UK deaths from certain lung diseases. In each case the lower block is for males, the upper block for females.

POSTSCRIPT users will probably want Encapsulated PostScript (EPSF) format files. These are produced automatically by the procedure in the last paragraph, and also by setting `onefile=F` as an argument to the `postscript` device under Unix. Note that these are *not* EPSI files and include neither a preview image nor the PICT resource that MACINTOSH EPSF files require (but can often be used as a TEXT file on a MACINTOSH).

3.2   Basic plotting functions

The function `plot` is a generic function, which when applied to many S objects will give a plot. (A list of plot methods is given in Table 4.4 on page 111.) An example is shown in Figure 1.2. (However, in many cases the plot does not seem appropriate to us.) Many of the plots appropriate to univariate data such as boxplots and histograms are considered in Chapter 5.

Barcharts

The function to display barcharts is `barplot`. This has many options (described in the online help), but some simple uses are shown in Figure 3.1. (Many of the details are covered in Section 3.3.)

```
lung <- aggregate(ts.union(mdeaths, fdeaths), 1)
# Splus 3.1 users replace ts.union by tsmatrix
barplot(t(lung), names=dimnames(lung)[[1]],
    main="UK deaths from lung disease")
legend(locator(1), c("Males", "Females"), fill=c(2,3))
loc <- barplot(t(lung), names=dimnames(lung)[[1]], style="old",
    dbangle=c(45,135), density=10)
total <- apply(lung,1,sum)
text(loc, total + par("cxy")[2], total, cex=0.7)
```

Line and scatterplots

The default plot function takes arguments x and y, vectors of the same length, or a matrix with two columns, or a list (or dataframe) with components x and y and produces a simple scatterplot. The axes, scales, titles and plotting symbols are all chosen automatically, but can be overridden with additional graphical parameters which can be included as named arguments in the call. The most commonly used ones are:

`type="c"`	Type of plot desired. Values for c are:
	p for points only (the default),
	l for lines only,
	b for both points and lines, (the lines miss the points),
	s, S for step functions (s specifies the level of the step at the left end, S at the right end),
	o for overlaid points and lines,
	h for high density vertical line plotting, and
	n for no plotting (but axes are still found and set).
`axes=L`	If F all axes are suppressed. Default T, axes are automatically constructed.
`xlab="string"` `ylab="string"`	Give labels for the x– and/or y–axes (default: the names, including suffices, of the x and y coordinate vectors).
`sub="string"` `main="string"`	sub specifies a title to appear under the x–axis label and main a title for the top of the plot in larger letters. (default: both empty).
`xlim=c(lo ,hi)` `ylim=c(lo, hi)`	Approximate minimum and maximum values for x– and/or y–axis settings. These values are automatically rounded to make them 'pretty' for axis labelling.

The functions `points`, `lines`, `text` and `abline` can be used to add to a plot, possibly one created with `type="n"`. Brief summaries are:

`points(x,y,...)`	Add points to an existing plot (possibly using a different plotting character). The plotting character is set by pch= and the size of the character by cex= or mkh=.
`lines(x,y,...)`	Add lines to an existing plot. The line type is set by lty= and width by lwd=. The type options may be used.
`text(x,y, labels,...)`	Add text to a plot at points given by x,y. labels is an integer or character vector; labels[i] is plotted at point (x[i],y[i]). The default is seq(along=x). The character size is set by cex=.
`abline(a,b,...)` `abline(h=c,...)` `abline(v=c,...)` `abline(`*lmobject*`,...)`	Draw a line in intercept and slope form, (a,b), across an existing plot. h=c may be used to specify y–coordinates for the heights of horizontal lines to go across a plot, and v=c similarly for the x–coordinates for vertical lines. The coefficients of a suitable *lmobject* are used.

These are the most commonly used graphics functions; we have seen examples of their use in Chapter 1, and will see many more. (There is also a function symbols described in Appendix B that we do not use in this book.) The plotting characters available for plot and points can be characters of the form pch="o" or numbered from 0 to 18, which uses the marks shown in Figure 3.2.

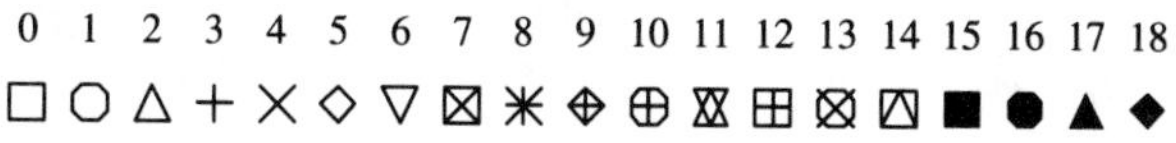

0 1 2 3 4 5 6 7 8 9 10 11 12 13 14 15 16 17 18

□ ○ △ + × ◇ ▽ ⊠ ✳ ✧ ⊕ ⊠ ⊞ ⊠ ⧄ ■ ● ▲ ◆

Figure 3.2: Plotting symbols or marks, specified by pch=n .

Size of text and symbols

Confusingly, the size of plotting characters is selected in one of two very different ways. For plotting characters (by pch="o") or text (by text), the parameter cex (for 'character expansion') is used. This defaults to the global setting (which defaults to 1), and rescales the character by that factor. For a mark set by pch=n, the size is controlled by the mkh parameter which gives the height of the symbol *in inches*. (This will be clear for printers; for screen devices the default plot region is about 8in × 6in and this is not changed by re-sizing the window.) However if mkh=0 the size is also controlled by cex and the default size of each symbol is approximately that of 0. Care is needed in changing cex on a call to plot, as this will also change the size of the axis labels. It is better to use, for example,

```
plot(x, y, type="n")              # axes only
points(x, y, pch=4, mkh=0, cex=0.7)   # add the points
```

Equally scaled plots

There are many plots, for example in multivariate analysis, which represent distances in the plane and for which it is essential to have a scaling of the axes which is geometrically accurate. This can be done in many ways, but most easily by our function eqscplot which behaves as the default plot function but shrinks the scale on one axis until geometrical accuracy is attained.

Multivariate plots

The plots we have seen so far deal with one or two variables. To view more we have several possibilities. A *scatterplot matrix* or pairs plot shows a matrix of scatterplots for each pair of variables, as we saw in Figure 1.3, which was produced by pairs(hills). It is also possible to use a function as the panel argument which will produce the content of each scatterplot. For example, for the dataset on fertility and socio-economic factors on Swiss provinces in about 1888 we can use

```
swiss.df <- data.frame(Fertility=swiss.fertility, swiss.x)
points.lines <- function(x, y, ...)
{
  points(x, y, ...)
```

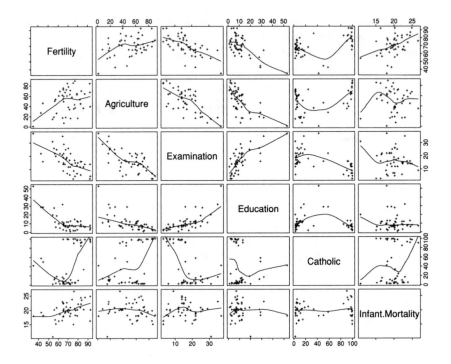

Figure 3.3: An enhanced pairs plot for the Swiss fertility data.

```
      lines(loess.smooth(x, y), ...)
}
pairs(swiss.df, panel=points.lines, pch=3, mkh=0.03, cex=0.4)
```

to produce Figure 3.3. This has a smooth curve fitted robustly by `loess` (see page 250) added to each panel. We will see other panel functions to identify groups of points by colour or symbol in Chapter 12.

Dynamic graphics

The S-PLUS function `brush` allow interaction with the (lower half) of a scatter-plot matrix. Examples are shown in Figure 1.4 on page 13 and Figure 12.1 on page 302. As it is much easier to understand these by using them, we suggest you try

```
brush(as.matrix(hills))
```

and experiment.

Points can be highlighted (marked with a symbol) by moving the brush (a rect-angular window) over them with button 1 held down. When a point is highlighted, it is shown highlighted in all the displays. Highlighting is removed by brushing with button 2 held down. It is also possible to add or remove points by clicking with button 1 in the scrolling list of row names.

One of four possible (device-dependent) marking symbols can be selected by clicking button 1 on the appropriate one in the display box on the right. The marking is by default persistent, but this can be changed to 'transient' in which only points under the brush are labelled (and button 1 is held down). It is also possible to select marking by row label as well as symbol.

The brush size can be altered under Unix by picking up a corner of the brush in the brush size box with the mouse button 1 and dragging to the required size. Under Windows, move the brush to the background of the main brush window, hold down the left mouse button and drag the brush to the required size.

The plot produced by brush will also show a three-dimensional plot (unless spin=F), and this can be produced on its own by spin. Clicking with mouse button 1 will select three of the variables for the x, y and z axes. The plot can be spun in several directions and re-sized by clicking in the appropriate box. The speed box contains a vertical line or slider indicating the current position.

Plots from brush and spin can only be terminated by clicking with the mouse button 1 on the quit box or button.

Obviously brush and spin are only available on suitable screen devices, including motif, openlook, suntools and win.graph.

Plotting surfaces

The functions contour, persp and image allow the display of a function defined on a two-dimensional regular grid. We will see examples of their use in Chapters 5, 8, 10 and 15. We anticipate an example from Chapter 15 of plotting a smooth topographic surface for Figure 3.4. On Unix library(spatial) is needed.

```
topo.loess <- loess(z ~ x * y, topo, degree=2, span = 0.25)
topo.mar <- list(x = seq(0, 6.5, 0.2), y=seq(0, 6.5, 0.2))
topo.lop <- predict(topo.loess, expand.grid(topo.mar))
par(pty="s")  # square plots
contour(topo.mar$x, topo.mar$y, topo.lop, xlab="", ylab="",
    levels = seq(700,1000,25), cex=0.7)
points(topo)
image(topo.mar$x, topo.mar$y, topo.lop, cex=0.7)
points(topo)
par(pty="m")
topo.p1 <- persp(topo.mar$x, topo.mar$y, topo.lop, cex=0.4)
points(perspp(topo$x, topo$y, topo$z, topo.p1))
topo.p2 <- persp(topo.mar$x, topo.mar$y, topo.lop, cex=0.4,
    eye=c(3, 80, 2000))
points(perspp(topo$x, topo$y, topo$z, topo.p2))
```

This generates values of the surface on a regular 33×33 grid generated by expand.grid, and shows a contour plot, a greylevel plot and perspective plots from two viewpoints. (The eye position is in user (x, y, z) coordinates, and was chosen by experiment.)

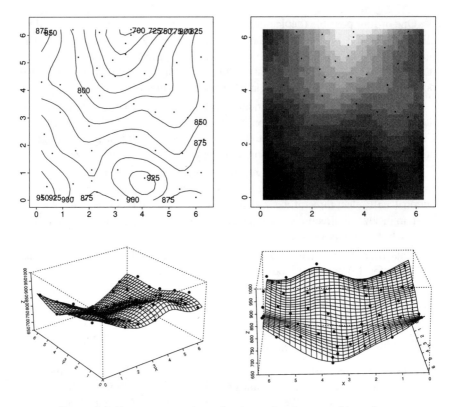

Figure 3.4: Four representations of an interpolated topographic surface.

The first three lines generate vectors of (sorted) x and y coordinates, and a matrix of surface values over the rectangular grid determined by those vectors. Each of the three functions displays data in this form; `contour` will choose contour levels, or (as here) they can be specified. Calling `contour` with argument `add=T` allows further contour levels to be added, or contours added to a plot generated some other way (such as an image plot).

The function `perspp` maps coordinates from (x, y, z)–space to their (x, y) positions on a perspective plot.

3.3 Enhancing plots

In this section we cover a number of ways that are commonly used to enhance plots, without reaching the level of detail of Section 3.5.

Some plots (such as Figure 1.1) have been square, whereas others are rectangular. Which is selected by the graphics parameter `pty`. Setting `par(pty="s")` selects a square plotting region, whereas `par(pty="m")` selects a maximally sized (and therefore usually rectangular) region.

Multiple figures on one plot

We have already seen several examples of plotting two or more figures on a single plot, apart from scatterplot matrices. The graphics parameters `mfrow` and `mfcol` subdivide the plot into an array of figure regions. They differ in the order in which the regions are filled. Thus

```
par(mfrow=c(2,3))
par(mfcol=c(2,3))
```

both select a 2×3 array of figures, but with the first they are filled along rows, and with the second along columns. A new figure region is selected for each new plot, and figure regions can be skipped by using `frame`.

All but two of the multi-figure plots in this book were produced with `mfrow`. Most of the side-by-side plots were produced with `par(mfrow=c(2,2))`, but using only the first two figure regions.

The `split.screen` function provides an alternative and more flexible way of generating multiple displays on a graphics device. An initial call such as

```
split.screen(figs=c(3,2))
```

subdivides the current device surface into a 3×2 array of *screens*. The screens created in this example are numbered 1 to 6, *by rows*, and the original device surface is known as screen 0. The current screen is then screen 1 in the upper left corner, and plotting output will fill the screen as it would a figure. Unlike multi-figure displays, the next plot will use the same screen unless another is specified using the `screen` function. For example the command

```
screen(3)
```

causes screen 3 to become the next current screen.

On screen devices the function `prompt.screen` may be used to define a screen layout interactively. The command

```
split.screen(prompt.screen())
```

allows the user to define a screen layout by clicking mouse button 1 on diagonally opposite corners. In our experience this requires a steady hand, although there is a `delta` argument to `prompt.screen` that can be used to help in aligning screen edges. (This was used for Figure 3.4.) Alternatively, if the `figs` argument to `split.screen` is specified as an N by 4 matrix, this divides the plot into N screens (possibly overlapping) whose corners are specified by giving (xl, xu, yl, yu) as the row of the matrix (where the whole region is $(0, 1, 0, 1)$).

The `split.screen` function may be used to subdivide the current screen recursively, thus leading to non-regular arrangements. In this case the screen numbering sequence continues from where it had reached.

Split-screen mode is terminated by a call to `close.screen(all=T)`; individual screens can be shut by `close.screen(n)`.

Use of the `fig` parameter to `par` provides an even more flexible way to subdivide a plot; see Section 3.5.

Adding information

The basic plots produced by `plot` often need additional information added to give context, particularly if they are not going to be used with a caption. We have already seen the use of `xlab`,`ylab`, `main` and `sub` with scatterplots. These arguments can all be used with the function `title` to add titles to existing plots. The first argument is `main`, so

```
title("A Useful Plot?")
```

adds a main title to the current plot.

Further points and lines are added by the `points` and `lines` functions. We have seen how plot symbols can be selected with `pch=`. The line type is selected by `lty=`. This is device-specific, but usually includes solid lines (1) and a variety of dotted, dashed and dash-dot lines. Line width is selected by `lwd=`, with standard width being 1, and the effect being device-dependent. On colour devices it is also possible to select a colour number by parameter `col=`, again device-specific. (Colours are often rendered as greylevels on printer devices.)

The `type="s"`, `"S"` argument to `lines` allows 'staircase' lines. It is better to use `stepfun` with dashed lines, as the dash sequence is not restarted at each change of angle (at least on the `postscript` device).

Identifying points interactively

The function `identify` has a similar calling sequence to `text`. The first two arguments give the $x-$ and $y-$coordinates of points on a plot and the third argument gives a vector of labels for each point. (The first two arguments may be replaced by a single list argument with two of its components named x and y, or by a two-column matrix.) The labels may be a character string vector or a numeric vector (which is coerced to character). Then clicking with mouse button 1 near a point on the plot causes its label to be plotted; labelling all points or clicking anywhere in the plot with button 2 terminates the process. (The precise position of the click determines the label position, in particular to left or right of the point.) We saw an example in Figure 1.5 on page 14. The function returns a vector of index numbers of the points which were labelled.

In Chapter 1 we used the `locator` function to add new points to a plot. This function is most often used in the form `locator(1)` to return the (x, y) coordinates of a single button click to place a label or legend, but can also be used to return the coordinates of a series of points, terminated by clicking with mouse button 2.

Adding further axes and grids

It is sometimes useful to add further axis scales to a plot, as in Figure 9.1 on page 224 which has scales for both kilograms and pounds. This is done by the function `axis`. There we used

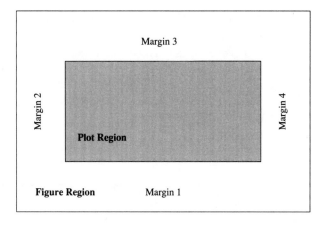

Figure 3.5: Anatomy of a graphics figure.

```
attach(wtloss)
# alter margin 4; others are default
par(mar=c(5.1, 4.1, 4.1, 4.1))
plot(Days, Weight, type="p",ylab="Weight (kg)")
Wt.lbs <- pretty(range(Weight*2.205))
axis(side=4, at=Wt.lbs/2.205, lab=Wt.lbs, srt=90)
mtext("Weight (lb)", side=4, line=3)
detach()
```

This adds an axis on side 4 (labelled clockwise from the bottom; see Figure 3.5) with labels rotated by 90° (srt=90) and then uses mtext to add a label 'underneath' that axis. Other parameters are explained in Section 3.5. Please read the on-line documentation very carefully to determine which graphics parameters are used in which circumstances. (For example, why did we need srt=90 for axis but not mtext ?)

Grids can be added by using axis with long tick marks, setting parameter tck=1 (yes, obviously). For example, a dotted grid is created by

```
axis(1, tck=1, lty=2)
axis(2, tck=1, lty=2)
```

and the location of the grid lines can be specified using at= .

Adding legends

Legends are added by the function legend. Since it can label many types of variation such as line type and width, plot symbol, colour, and fill type, its description is very complex. All calls are of the form

```
legend(x, y, legend, ...)
```

where x and y give either the upper left corner of the legend box or both upper left and lower right corners. These are often most conveniently specified on-screen by using locator(1) or locator(2). Argument legend is a character vector giving the labels for each variation. The remaining arguments are vectors of the same length as legend giving the appropriate coding for each variation, by lty=, lwd=, pch=, col=, fill=, angle= and density=.

By default the legend is contained in a box; the drawing of this box can be suppressed by argument bty="n".

Mathematics in labels

Users frequently wish to include the odd subscript, superscript and mathematical symbol in labels. There is no general solution, but for the postscript driver Alan Zaslavsky's package postscriptfonts adds these features. It is available from statlib: see Appendix D.

3.4   Conditioning plots

Conditioning plots are also known as *coplots*, and produced by the function coplot. Two variables are plotted against each other in a series of plots with the values of a third variable restricted to a series of overlapping ranges. This allows us to see how a relationship between two variables changes as a third variable varies. The variables are specified by an S formula of the type y ~ x | z, which implies "plot y against x for a series of conditioning intervals determined by z". The coplot function has a panel argument to specify what information is to appear in the panels and an optional data argument. Coplots are discussed in detail by Cleveland (1993).

Suppose we wished to examine the relationship between Fertility and Education in the Swiss fertility data (examined above as a pairs plot) as the variable Catholic ranges from predominantly non-Catholic to mainly Catholic provinces. We add a smooth fit to each panel. (Function points.lines is defined on page 67.)

```
coplot(Fertility ~ Education | Catholic, data=swiss.df,
       panel=points.lines)
```

The result is shown in Figure 3.6. Fertility generally falls as education rises and rises as the proportion of Catholics in the population rises. Note that the level of education is lower in predominantly Catholic provinces.

Another argument to coplot is given.values which specifies the conditioning intervals. The ancillary function co.intervals can be used to construct suitable ranges. For example,

```
attach(swiss.df)
coplot( Fertility ~ Education | Catholic, data=swiss.df,
    panel=points.lines,
    given.values=co.intervals(Catholic, number=4, overlap=0.25))
```

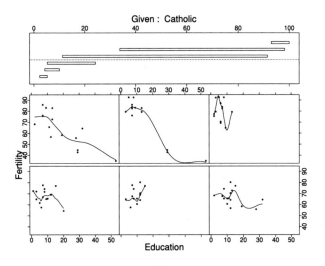

Figure 3.6: The Swiss fertility data. A conditioning plot of fertility against education given percentage of Catholics. The sections are plotted left to right, bottom row first.

specifies that there should be 4 intervals, and each pair of adjacent intervals should have about 25% of their points in common. (The default is 6 intervals with an overlap of about 50%.) If the conditioning variable is a factor then the levels of the factor determine the points in each panel with no overlap.

Conditioning plots may also have two simultaneous conditioning variables. These are specified by a formula such as `y ~ x | z1 * z2`. The conditioning variables are split into subranges as before, and the panels correspond to each pair of subranges. In this case the `given.values` argument, if given, must specify a list of two components, specifying the collection of subranges for each of the two conditioning variables. Let us look again at fertility against education, but conditioning on both Catholic and agriculture simultaneously. Since the sample is small it seems prudent to limit the number of panels to 6 in all.

```
C.ranges <- co.intervals(Catholic, number=2, overlap=0)
A.ranges <- co.intervals(Agriculture, number=3, overlap=0.25)
coplot(Fertility ~ Education | Catholic * Agriculture,
       data=swiss.df, panel=points.lines,
       given.values=list(C.ranges, A.ranges))
```

The result is shown in Figure 3.7. In general the fertility rises with the proportion of Catholics and agriculture and falls with education. There is no convincing evidence of substantial interaction.

A dynamic effect similar to conditioning can be obtained by brushing (with a large brush) over a range of values of one variable.

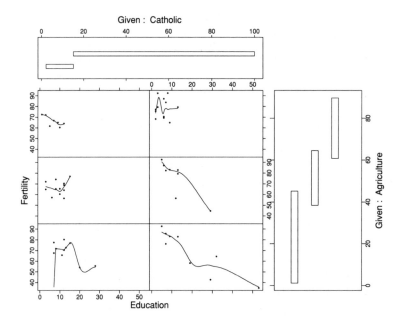

Figure 3.7: The Swiss fertility data. A conditioning plot of fertility on education given Catholic and agriculture.

3.5   Fine control of graphics

The graphics process is controlled by *graphics parameters*, which are set for each graphics device. Each time a new device is opened these parameters for that device are reset to their default values. Graphics parameters may be set, or their current values queried, using the par function. If the arguments to par are of the name=value form the graphics parameter name is set to value, if possible, and other graphics parameters may be reset to ensure consistency. The value returned is a list giving the previous parameter settings. Instead of supplying the arguments as name=value pairs, par may also be given a single list argument with named components.

   If the arguments to par are quoted character strings, "name", the current value of graphics parameter name is returned. If more than one quoted string is supplied the value is an list of the requested parameter values, with named components. The call par() with no arguments returns a list of all the graphics parameters.

   Some of the many graphics parameters are given in Tables 3.3 and 3.4. They have short names, usually of three characters.

The figure region and layout parameters

When a device is opened it makes available a rectangular surface on which one or more plots may appear. Each plot occupies a rectangular section of the device

Table 3.3: Some graphics layout parameters with example settings.

`din, fin, pin`	Absolute device size, figure size and plot region size in inches. `fin=c(6,4)`
`fig`	Define the figure region as a fraction of the device surface. `fig=c(0,0.5,0,1)`
`font`	Small positive integer determining a text font for characters and hence an interline spacing. `font=3`
`mar, mai`	The margin sizes in relative and absolute units respectively. `mar=c(3,3,1,1)+0.1`
`mex`	Number of text lines per interline spacing. `mex=0.7`
`mfg`	Defines a position within a specified multi-figure display. `mfg=c(2,2,3,2)`
`mfrow, mfcol`	Defines a multi-figure display. `mfrow=c(2,2)`
`new`	Logical value indicating whether the current figure has been used or not. `new=T`
`oma, omi, omd`	Defines outer margins in relative or absolute units, or by defining the size of the array of figures as a fraction of the device surface. `oma=c(0,0,4,0)`
`plt`	Define the plot region as a fraction of the figure region. `plt=c(0.1,0.9,0.1,0.9)`
`pty`	Plot type, or shape of plotting region, `"s"` or `"m"`
`uin`	Returns inches per user coordinate for x and y.
`usr`	Limits for the plot region in user coordinates. `usr=c(0.5, 1.5, 0.75, 10.25)`

surface called a *figure*. A figure consists of a rectangular *plot region* surrounded by a *margin* on each side. The margins or sides are numbered 1 to 4, clockwise starting from the bottom. The plot region and margins together make up the *figure region*, as in Figure 3.5. The device surface, figure region and plot region have their vertical sides parallel and hence their horizontal sides also parallel.

The size and position of figure and plot regions on a device surface are controlled by *layout parameters*, most of which are listed in Table 3.3. Lengths may be set in either absolute or relative units. Absolute lengths are in *inches*. Relative lengths are in *text lines*.

Margin sizes are set using `mar` for relative units or `mai` for inches. These are 4-component vectors giving the sizes of the lower, left, upper and right margins in the appropriate units. Changing one causes a consistent change in the other; changing `mex` will change `mai` but not `mar`.

Positions may be specified in relative units using the unit square as a coordinate system for which some enclosing region, such as the device surface or the figure region, is the unit square. The `fig` parameter is a vector of length 4 specifying the current figure as a fraction of the device surface. The first two components give the lower and upper x–limits and the second two give the y–limits. Thus to put a point plot in the left-hand side of the display and a Q-Q plot on the right-hand side we could use:

```
postscript(file="twoplot.ps")   # open a postscript device
par(fig=c(0, 2/3, 0, 1))        # set a figure on the left
plot(x,y)                       # point plot
par(fig=c(2/3, 1, 0, 1))        # set a figure on the right
qqnorm(resid(obj))              # diagnostic plot
dev.off()
```

The left hand figure occupies $2/3$ of the device surface and the right hand figure $1/3$. For regular arrays of figures it is simpler to use mfrow or split.screen.

Positions in the plot region may also be specified in absolute *user coordinates*. Initially user coordinates and relative coordinates coincide, but any high level plotting function changes the user coordinates so that the $x-$ and y–coordinates range from their minimum to maximum values as given by the plot axes. The graphics parameter usr is a vector of length 4 giving the lower and upper $x-$ and y–limits for the user coordinate system. Initially its setting is usr=c(0,1,0,1). Consider another simple example:

```
> motif()               # open a device
> par("usr")            # usr coordinates
[1] 0 1 0 1
> x <- 1:20
> y <- x + rnorm(x)     # generate some data
> plot(x, y)            # produce a scatterplot
> par("usr")            # user coordinates now match the plot
[1]   0.2400 20.7600  1.2146 21.9235
```

Any attempt to plot outside the user coordinate limits causes a warning message unless the general parameter xpd is set to T.

Figure 3.8 shows some of the layout parameters for a multi-figure layout. If the number of rows or columns in a multi-figure display is more than 2, the parameters cex and mex are both set to 0.5. This may produce characters that are too small and some resetting may be appropriate. (On the other hand, for a 2×2 layout the characters will usually be too large.)

Such an array of figures may occupy the entire device surface, or it may have *outer margins*, which are useful for annotations that refer to the entire array. Outer margins are set with the parameter oma (in text lines) or omi (in inches). Alternatively omd may be used to set the region containing the array of figures in a similar way to which fig is used to set one figure. This implicitly determines the outer margins as the complementary region. In contrast to what happens with the margin parameters mar and mai, a change to mex will leave the outer margin size, omi, constant but adjust the number of text lines, oma.

Text may be put in the outer margins by using mtext with parameter outer=T.

Table 3.4: Some of the more commonly used general and high level graphics parameters with example settings.

Text:

adj	Text justification. 0=left justify, 1=right justify, 0.5=centre.
cex	Character expansion. cex=2
csi	Height of font (inches). csi=0.11
font	Font number: device-dependent.
srt	String rotation in degrees. srt=90
cin cxy	Character width and height in inches and usr coordinates (for information, not settable).

Symbols:

col	Colour for symbol, line or region. col=2
lty	Line type: solid, dashed, dotted, etc. lty=2
lwd	Line width, usually as a multiple of default width. lwd=2
mkh	Mark height (inches). mkh=0.05
pch	Plotting character or mark. pch="*" or pch=4 for marks. (See page 67.)

Axes:

bty	Box type, as "o", "l", "7", "c", "n".
exp	Notation for exponential labels. exp=1
lab	Tick marks and labels. lab=c(3,7,4)
las	Label orientation. 0=parallel to axis, 1=horizontal, 2=vertical.
log	Controls log axis scales. log="y"
mgp	Axis location. mgp=c(3,1,0)
tck	Tick mark length as signed fraction of the plot region dimension. tck=-0.01
xaxp yaxp	Tick mark limits and frequency. xaxp=c(2, 10, 4)
xaxs yaxs	Style of axis limits. xaxs="i"
xaxt yaxt	Axis type. "n" (null), "s" (standard), "t" (time) or "l" (log)

High Level:

ask	Prompt before going on to next plot? ask=F
axes	Print axes? axes=F
main	Main title. main="Figure 1"
sub	Subtitle. sub="23-Jun-1994"
type	Type of plot. type="n"
xlab ylab	Axis labels. ylab="Speed in km/sec"
xlim ylim	Axis limits. xlim=c(0,25)
xpd	May points or lines go outside the plot region? xpd=T

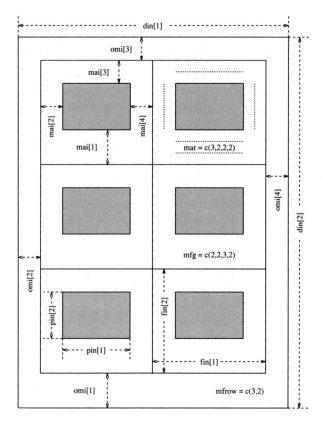

Figure 3.8: An outline of a 3×2 multi-figure display with outer margins showing some graphics parameters. The current figure is at position $(2, 2)$ and the display is being filled by rows. In this figure " fin[1] " is used as a shorthand for par("fin")[1] , and so on.

Common axes for figures

There are at least two ways to ensure that several plots share a common axis or axes.

1. Use the same xlim or ylim (or both) setting on each plot and ensure that the parameters governing the way axes are formed, such as lab, las, xaxs and allies, do not change.
2. Set up the desired axis system with the first plot and then set the low level parameter xaxs="d" , yaxs="d" or both as appropriate. This ensures that the axis or axes are not changed by further high level plot commands on the same device.

An example: A Q-Q normal plot with envelope

In Chapter 5 we recommend assessing distributional form by quantile-quantile plots. A simple way to do this is to plot the sorted values against quantile approxi-

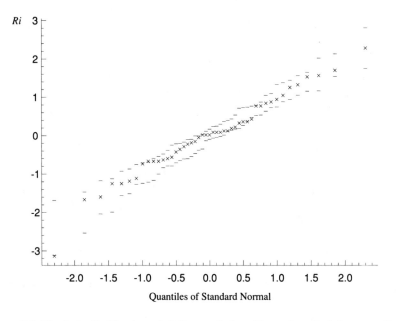

Figure 3.9: The Swiss fertility data. A Q-Q normal plot with envelope for infant mortality.

mations to the expected normal order statistics and draw a line through the 25% and 75% percentiles to guide the eye, performed for the variable `Infant.Mortality` of the Swiss provinces data (studied in the `pairs` plot) by

```
attach(swiss.df)
qqnorm(Infant.Mortality)
qqline(Infant.Mortality)
```

The reader should check the result and compare it with the style of Figure 3.9.

Another suggestion to assess departures is to compare the sample Q-Q plot with the envelope obtained from a number of other Q-Q plots from generated normal samples. This is discussed in Atkinson (1985, § 4.2) and is based on an idea of Ripley (see Ripley 1981, Chapter 8). The idea is simple. We generate a number of other samples of the same size from a normal distribution and scale all samples to mean 0 and variance 1 to remove dependence on location and scale parameters. Each sample is then sorted. For each order statistic the maximum and minimum values for the generated samples form the upper and lower envelopes. The envelopes are plotted on the Q-Q plot of the scaled original sample and form a guide to what constitutes serious deviations from the expected behaviour under normality. Following Atkinson our calculation uses 19 generated normal samples.

We begin by calculating the envelope and the x–points for the Q-Q plot.

```
samp <- cbind(Infant.Mortality, matrix(rnorm(47*19), 47, 19))
samp <- apply(scale(samp), 2, sort)
rs <- samp[,1]
```

```
xs <- qqnorm(rs, plot=F)$x
env <- t(apply(samp[,-1], 1, range))
```

As an exercise in building a plot with specific requirements we will now present the envelope and Q-Q plot in a style very similar to Atkinson's. To ensure that the Q-Q plot has a y-axis large enough to take the envelope we could calculate the y-limits as before, or alternatively use a matrix plot with type=n for the envelope at this stage. The axes are also suppressed for the present:

```
matplot(xs, cbind(rs,env), type="pnn", pch=4, mkh=0.06, axes=F)
```

The argument setting type="pnn" specifies that the first column (rs) is to produce a point plot and the remaining two (env) no plot at all, but the axes will allow for them. Setting pch=4 specifies a 'cross' style plotting symbol (see Figure 3.2) similar to Atkinson's, and mkh=0.06 establishes a suitable size for the plotting symbol.

Atkinson uses small horizontal bars to represent the envelope. We can now calculate a half length for these bars so that they do not overlap and do not extend beyond the plot region. Then we can add the envelope bars using segments:

```
xyul <- par("usr")
smidge <- min(diff(c(xyul[1], xs, xyul[2])))/2
segments(xs-smidge, env[,1], xs+smidge, env[,1])
segments(xs-smidge, env[,2], xs+smidge, env[,2])
```

Atkinson's axis style differs from the default S style in several ways. There are many more tick intervals; the ticks are inside the plot region rather than outside; there are more labelled ticks, and the labelled ticks are longer than the unlabelled. From experience ticks along the x-axis at 0.1 intervals with labelled ticks at 0.5 intervals seems about right but this is usually too close on the y-axis. The axes require four calls to the axis function:

```
xul <- trunc(10*xyul[1:2])/10
axis(1, at=seq(xul[1], xul[2], by=0.1), labels=F, tck=0.01)
xi <- trunc(xyul[1:2])
axis(1, at=seq(xi[1], xi[2], by=0.5), tck=0.02)
yul <- trunc(5*xyul[3:4])/5
axis(2, at=seq(yul[1], yul[2], by=0.2), labels=F, tck=0.01)
yi <- trunc(xyul[3:4])
axis(2, at=yi[1]:yi[2], tck=0.02)
```

Finally we add the L–box, put the x-axis title at the centre and the y-axis title at the top:

```
box(bty="l")                # lower case "L"
ps.options()$fonts
mtext("Quantiles of Standard Normal", side=1, line=2.5, font=3)
mtext("Ri", side=2, line=2, at=yul[2], font=10)
```

where fonts 3 and 10 are Times-Roman and Times-Italic on the device used (postscript under Unix), found from the list given by ps.options.

The final plot is shown in Figure 3.9.

3.6 Exercises

3.1. The data frame survey contains the results of a survey of 237 first-year Statistics students at Adelaide University. For a graphical summary of all the variables, use plot(survey). Note that this produces a dotchart for factor variables, and a normal scores plot for the numeric variables.

One component of this data frame, Exer, is a factor object containing the responses to a question asking how often the students exercised. Produce a barchart of these responses. Use table and pie to create a pie chart of the responses. Do you like this better than the bar plot? Which is more informative? Which gives a better picture of exercise habits of students? The pie function takes an argument names which can be used to put labels on each pie slice. Redraw the pie chart with labels. Alternatively, you could add a legend to identify the slices.

You might like to try the same things with the Smoke variable, which records responses to the question "How often do you smoke?" Note that table and levels ignore missing values; if you wish to include non-respondents in your chart use summary to generate the values, and names on the summary object to generate the labels.

3.2. Make a plot of petal width *vs* petal length of the iris data for a partially-sighted audience, identifying the three species. You will need to double the annotation size, thicken the lines, and change the layout to allow larger margins for the larger annotation.

3.3. Plot $\sin(x)$ against x, using 200 values of x between $-\pi$ and π, but do not plot any axes yet (use parameter axes=F in the call to plot.) Add a y axis passing through the origin using the 'extended' style and horizontal labels. Add an x axis with tick-marks from $-\pi$ to π in increments of $\pi/4$, twice the usual length.

3.4. Cleveland (1993) recommends that the aspect ratio of line plots is chosen so that lines are 'banked' at $45°$. By this he means that the peak slopes should be around $\pm45°$ over much of the plot. Write a function to achieve this for a time-series plot, and try it out on the sunspots dataset.

Chapter 4

Programming in S

The S language is both an interactive language and a language for adding new functions to the S system. It is a complete programming language with control structures, recursion and a useful variety of data types. The S environment provides many functions to handle standard operations, but most users need occasionally to write new functions. This chapter is concerned with designing, writing, testing and correcting your own S functions.

Much of the system software is itself written in the S language. This code may be inspected, providing a wealth of examples. A user who understands S programming can often clarify points of detail from the code itself in cases where the documentation is unclear.

Writing clear, correct programs in any programming language always presents a challenge and although it is an easy, high level language S is no exception. The first three sections of this chapter are relevant to the casual writer of functions; the rest of the chapter is important for those intending to make extensions to S, especially if these are to be used by others.

4.1 Control structures

Control structures are the commands that make decisions or execute loops. These statements are most often used in programs rather than interactively.

Conditional execution of statements

Conditional execution uses either the `if` statement or the `switch` function. The `if` statement has the form

```
if (condition) statement1 else statement2
```

First `condition` is evaluated. If the result is T (or non-zero) the value of the `if` statement is that of `statement1`, otherwise that of `statement2`. The `else` part is optional and omitting it is equivalent to using "`else NULL`". If `condition` yields a vector result only the first component is used and by default a warning is issued. The `if` function can be extended over several lines, and the statements may be compound statements enclosed in braces { }.

Two additional logical operators, `&&` and `||`, are useful with `if` statements. Unlike `&` and `|`, which operate on vectors, these operate on logical expressions, and only evaluate the right-hand expression if the left-hand one is true. This conditional evaluation property can be used as a safety feature, as in the following code:

```
if (is.numeric(x) && min(x) > 0)  sx <- sqrt(x)
else  stop("x must be numeric and all components positive")
```

The expression `min(x) > 0` is invalid for non-numeric `x`.

The functions `any(e)` and `all(e)` are often useful in defining conditions. They evaluate respectively the logical 'or' and 'and' of all components of their argument. For example the usual test for exact symmetry of a matrix is `all(X == t(X))`; a less strict version would be `all(abs(X-t(X)) < eps)` where `eps` is some appropriate error tolerance. Alternatively, we could use `all.equal(X, t(X))`, which uses a tolerance.

The `if` statement should be distinguished from the `ifelse` function, which is its vector counterpart. Its form is

```
ifelse(test, true.value, false.value)
```

In this case `test` is evaluated and coerced to logical if necessary, and both values are evaluated. In those component positions where the value is `T` the corresponding component of `true.value` is the result and elsewhere it is that of `false.value`. Since `ifelse` operates on vectors, it is fast and should be used if possible.

Note that `ifelse` can generate `NA` warning messages even in cases where the result contains none. In an assignment such as

```
y.logy <- ifelse(y <= 0, 0, y*log(y))
```

since all three arguments to `ifelse` are evaluated, warning messages will be generated if `y` has any negative components even though the final result will have no `NA` components. Two other ways of producing the desired result for *frequency* vectors `y` are

```
y.logy <- y * log(y + (y==0))
y.logy <- y * log(pmax(1, y))  # alternative
```

The `switch` function provides a graceful alternative to a sequence of branching `if` statements. For example if you wished to allow the user a choice of tests for equality of variances the conditional statements

```
result <- if (test == "Levene")  levene(y, f)
          else
              if (test == "Cochran") cochran(y, f)
              else bartlett(y, f)
```

would allow three possibilities, Levene's, Cochran's or Bartlett's test (assuming the three functions were available). This sequence of checks can be replaced by the construction:

```
result <-  switch(test,
               Levene = levene(y, f),
               Cochran = cochran(y, f),
               bartlett(y, f)
           )
```

If the first argument to `switch` evaluates to a character string the value of
the expression is that of the named argument whose name matches, otherwise
the value is that of a final unnamed argument, if any. If the actual argument for
the matching string is missing, `switch` uses the next available argument. For
example one could use

```
result <- switch(test,
    Levene =, levene =, "Levene's test" = levene(y, f),
    Cochran =, cochran =, "Cochran's test" = cochran(y, f),
    Bartlett =, bartlett =, "Bartlett's test" =,
    bartlett(y, f))
```

to allow alternative possibilities for specifying the test to be used, still retaining
Bartlett's test as the "catch all". Note that non-standard names may be used if
quoted.

Abbreviated names are not accommodated; the argument name must match
`test` exactly. We can allow abbreviations using `pmatch`:

```
result <- switch(pmatch(test, c("Levene","levene", "Cochran",
    "cochran", "Bartlett","bartlett"), nomatch = ""),
    "1" =, "2" = levene(y, f),
    "3" =, "4" = cochran(y, f),
    bartlett(y, f))
```

The first argument to `switch` may also evaluate to a number, which is coerced
to an integer. Argument names are then ignored and the appropriate numbered
argument among those remaining, if there is one, is selected. There is no default
final argument; if the number is outside the range 1 to `nargs()-1` the result is
`NULL`. If a selected argument position is present but the argument itself is vacant
there is no "drop through" convention. In this case `switch` returns no result,
which can result in a puzzling error.

Loops: the `for`, `while` and `repeat` statements

A `for` loop allows a statement to be iterated as a variable assumes values in a
specified sequence. The statement has the form

```
for(variable in sequence) statement
```

where `in` is a keyword, `variable` is the loop variable and `sequence` is the
vector of values it assumes as the loop proceeds. (This is often of the form `1:10`
or `seq(along=x)`.) The `statement` part will often be a grouped statement and
hence enclosed within braces, `{ }`.

The `while` and `repeat` loops do not make use of a loop variable. Their
forms are

```
while (condition) statement
```

and

```
repeat statement
```

In both cases the commands in the body of the loop are repeated. For a `while` loop the normal exit occurs when `condition` becomes F; the `repeat` statement continues indefinitely unless exited by a `break` statement.

The `next` statement within the body of a `for`, `while` or `repeat` loop causes a jump to the beginning of the next iteration. The `break` statement causes an immediate exit from the loop.

There are also `For` loops, which are similar to `for` loops but 'unroll' the loop and use a separate S process to perform the calculations. Their main use is to tune the time/memory tradeoffs in large calculations, and it has many arguments (see the help page) which can be used in this tuning. The basic usage is of the form

```
For(i = 1:250; res[i] <- myfun(sample(x, replace=T)) )
```

which would run `myfun` on 250 bootstrap samples (see Section 5.6).

A single-parameter maximum-likelihood example

For a simple example with a statistical context we will estimate the parameter λ of the zero-truncated Poisson distribution by maximum likelihood. We will use artificial data but the same problem sometimes occurs in practice.

The probability distribution is specified by

$$\Pr(Y = y) = \frac{e^{-\lambda}\lambda^y}{(1 - e^{-\lambda})\, y!} \qquad y = 1, 2, \ldots$$

and corresponds to observing only non-zero values of a Poisson count. The mean is

$$E(Y) = \frac{\lambda}{1 - e^{-\lambda}}$$

The maximum likelihood estimate $\hat{\lambda}$ is found by equating the sample mean to its expectation:

$$\bar{y} = \frac{\hat{\lambda}}{1 - e^{-\hat{\lambda}}}$$

If this equation is written as $\hat{\lambda} = \bar{y}\,(1 - e^{-\hat{\lambda}})$, Newton's method leads to the iteration scheme

$$\hat{\lambda}_{m+1} = \hat{\lambda}_m - \frac{\hat{\lambda}_m - \bar{y}\,(1 - e^{-\hat{\lambda}_m})}{1 - \bar{y}\,e^{-\hat{\lambda}_m}}$$

which we will now implement. This is a natural situation in which a `while` loop might be used in S. First we generate our artificial sample from a distribution with $\lambda = 1$.

```
> yp <- rpois(50, lam=1)    # full Poisson sample of size 50
> table(yp)
  0  1  2 3 5
 21 12 14 2 1
> y <- yp[yp > 0]            # truncate the zeros; n=29
```

We will have a termination condition based both on convergence of the process and an iteration count limit just for safety. An obvious starting value is $\hat{\lambda}_0 = \bar{y}$.

```
> ybar <- mean(y); ybar
> [1] 1.7586
> lam <- ybar
> it <- 0                    # iteration count
> del <- 1
> while (abs(del) > 0.0001 && (it <- it + 1) < 10) {
    del <- (lam - ybar*(1 - exp(-lam)))/(1 - ybar*exp(-lam))
    lam <- lam - del
    cat(it, lam, "\n")
    }
1 1.32394312696735
2 1.26142504977282
3 1.25956434178259
4 1.25956261931933
```

To generate output from a loop in progress an explicit call to a function such as `print` or `cat` has to be used. For tracing output `cat` is usually convenient since it can combine several items.

Vectorized calculations and loops

Programmers coming to S from other languages are often slow to take advantage of the power of S to do vectorized calculations, that is calculations that operate on entire vectors rather than on individual components in sequence. This often leads to unnecessary loops. For example consider calculating the Pearson chi-squared statistic for testing independence in a two-way contingency table. This is defined as

$$X_P^2 = \sum_{i=1}^{r} \sum_{j=1}^{s} \frac{(f_{ij} - e_{ij})^2}{e_{ij}}$$

where $e_{ij} = f_{i.}f_{.j}/f_{..}$ are the expected frequencies. Two nested `for` loops may seem to be necessary, but in fact no explicit loops are needed. Assuming the frequencies f_{ij} are held as a matrix the most efficient calculation in S uses matrix operations:

```
fi. <- f %*% rep(1,ncol(f))
f.j <- rep(1, nrow(f)) %*% f
e <- (fi. %*% f.j)/sum(fi.)
X2p <- sum((f-e)^2/e)
```

Explicit loops in S should be regarded as potentially expensive in time and memory use and ways of avoiding them should be considered. (Note that this will be impossible with genuinely iterative calculations such as our Newton scheme.)

The functions `apply`, `tapply`, `sapply` and `lapply` offer a way around explicit loops. For example another way of calculating the expected frequencies e_{ij} would be:

```
e <- outer(apply(f, 1, sum), apply(f, 2, sum))/sum(f)
```

However, prior to S-PLUS 3.2 these functions were written in S and used `for` loops themselves internally. They may still be preferable to user-written loops on other grounds such as safety, brevity and transparency. From S-PLUS 3.2 they all use the internal function `lapply`.

4.2   Writing your own functions

Functions in S are created by assignment using the keyword `function`:

```
fname <- function(arg1, arg2, etc.) statement
```

where `arg1`, `arg2`, ...[1] are arguments to be satisfied on the call. The `statement` defining the body of the function is any S statement, but is usually a grouped statement and so enclosed within braces, `{ }`.

A function is called by giving its name with an argument sequence in parentheses:

```
fname(val1, val2, etc.)
```

Several protocols are available for specifying both the arguments in the function definition and their values on a call.

The value returned by the function is the value of the statement, which is usually an unassigned final expression within the grouped statement. A function may be terminated at any stage by executing the `return` function where the argument specifies the value returned by the function. For example if one of the arguments x is empty it might be appropriate to return the missing value marker as the function value immediately.

```
    ....
    if(length(x) == 0) return(NA)
    ....
```

If `return` is given several arguments the value returned is a list of these components, which seems to be an undocumented feature.

As a simple example consider a function to perform a two-sample t-test the way it is often taught in elementary courses.

---

[1] Note that we are using `etc.` to denote an indefinite number of similar items, as " ... " is an allowable literal argument with a special meaning. We always use *four* dots when indicating an omission.

```
ttest <- function(y1, y2, test = "two-sided", alpha = 0.05)
{
  n1 <- length(y1)
  n2 <- length(y2)
  ndf <- n1 + n2 - 2
  s2 <- ((n1 - 1) * var(y1) + (n2 - 1) * var(y2))/ndf
  tstat <- (mean(y1) - mean(y2))/sqrt(s2 * (1/n1 + 1/n2))
  tail.area <- switch(test,
    "two-sided" = 2 * (1 - pt(abs(tstat), ndf)),
    lower = pt(tstat, ndf),
    upper = 1 - pt(tstat, ndf),
    {
      warning("test must be 'two-sided', 'lower' or 'upper'")
      NULL
    }
  )
  list(tstat = tstat, df = ndf,
       reject = if(!is.null(tail.area)) tail.area < alpha,
       tail.area = tail.area)
}
```

This function requires two arguments to be specified, `y1` and `y2`, the two sample vectors. It also allows two more, `test` and `alpha`, but if no values for these are given on the call, the default values are used. Thus if `test` is not specified a two-sided test is assumed, and if `alpha` is not specified a significance level of 5% is used.

The result is a list of components giving information about the test result. This is conventional, although giving a result as a primary object with additional information carried as attributes is sometimes appropriate.

It is easy to test the function using simulated data:

```
> x1 <- round(rnorm(10),1); x1
>  [1]  0.5 -1.4  1.5  0.5 -0.9 -0.9 -0.1 -0.5 -1.2  0.1
> x2 <- round(rnorm(10)+1,1); x2
>  [1]  0.4 -0.3  0.9  2.2  1.8  0.7  2.7  1.0  1.1  2.9
> ttest(x1, x2)
$tstat:
[1] -3.6303
$df:
[1] 18
$reject:
[1] T
$tail.area:
[1] 0.0019138
```

A more compact way to print the result is to coerce it to a numeric vector using `unlist`:

```
> unlist(ttest(x1, x2))
   tstat df reject tail.area
 -3.6303 18      1 0.0019138
```

Users of S-PLUS have a similar function available, t.test, which we can use to check ours:

```
> t.test(x1, x2)

            Standard Two-Sample t-Test

data:   x1 and x2
t = -3.6303, df = 18, p-value = 0.0019
    ....
```

Lazy evaluation

When an S function is called the argument expressions are parsed but not evaluated. When they are required they are evaluated and the values are used by the function. In particular if they are not used by the function on a particular call, they are not evaluated, and the default expressions for arguments of functions are only evaluated when they are needed.

This protocol is called *lazy evaluation* and it has some important consequences for S programming. Default expressions for function arguments may involve not only other arguments to the function but also variables in the search path, including local variables in the function itself. The rule is that the expression must be capable of evaluation when it is needed; in particular any local variables involved must have values at that point.

It is possible to retrieve within a function the argument expressions used in its call and coerce them to character mode. The function substitute can be used on an argument to make the necessary substitutions from the call, leaving the result as a parsed but unevaluated expression. The function deparse takes an object of mode expression and coerces it to mode character. Hence if x is an argument deparse(substitute(x)) gives whatever expression was used for argument x (or the default if none was) and returns it as a character string. This is often used to generate labels on graphs. For example:

```
myplot <- function(x,y) {
  lab <- deparse(substitute(y))
    ....
  title(main=paste("A plot of", lab))
    ....
}
```

Editing S functions and objects

It is often necessary to correct S functions while they are being developed and corrected. This can be done in several ways.

1. Cut-and-paste can be used between the S session and an editable window. Sometimes it is helpful to save the function definition in a file (from the editor or with `dump("obj", "obj.q")`) and use `source("obj.q")` to read it into S. Note that `.q` is the preferred filename extension, as `.s` is used for assembler code in Unix.

2. The function `fix` takes as argument an object to edit:

```
> fix(obj)
```

This initiates an editing session with a text version of `obj` available for correction. On completion of the editing session the corrected version is assigned to the object and the S session is resumed. The editor used is a system default (usually `vi`), but a different editor can be specified as an `options` argument:

```
> options(editor="emacs")
```

The most recently edited object can be re-edited by invoking `fix()`.

3. The function `ed` can achieve the same outcome by assignment. It takes an object to edit as its principal argument and the name of an editor to use as another optional argument:

```
> obj <- ed(obj, editor="emacs")
```

If `vi` is the editor chosen we can use `obj <- vi(obj)`. Again, the function can be re-edited by invoking `ed` or `vi` without specifying an object.

The functions `warning`, `stop` and `missing`

The functions `warning` and `stop` are used inside functions to handle unexpected situations; `warning` arranges for a warning to be issued when control returns to the session level but the action of the function continues. For example, if we call our `ttest` function with an invalid character string for `test` a warning message is issued and the default test performed. We use `unlist` again for a compact display:

```
> unlist(ttest(x1, x2, test="left"))
   tstat df
-3.6303 18
Warning messages:
  test must be 'two-sided', 'lower' or 'upper' in: \
       switch(test, ....
```

The function `stop` terminates the action of the function, issues an error message and returns control to the session level immediately. It does not terminate the S session as `q()` does. As we shall see in Section 4.3 it can also be made to precipitate a dump of information on the state of the calculation that can often help in tracing errors.

The function `missing` allows a function to check whether a value was specified for an argument. For example we could modify our `ttest` function to include a check for `y1` and `y2`:

```
ttest <- function(y1, y2, test = "two-sided", alpha = 0.05)
{
    if(missing(y1) || missing(y2))
        stop("two samples needed")
    ....

> ttest(x1)
Error in ttest(x1): two samples needed
Dumped
```

It is often tempting to use `missing` to specify default values of arguments, but the lazy evaluation mechanism is often a better way to do so.

A common idiom is to use `stop("some message")` as a default value when some argument for a function must be supplied. For example the function `rpois` for generating artificial Poisson samples requires that the mean parameter be specified:

```
> args(rpois)
function(n, lambda = stop("no lambda arg"))
NULL
```

Notice that the default value for an argument is only evaluated if it is needed, that is if no value has been supplied for the argument on the call.

Argument matching and abbreviations

When arguments are given in the `name=value` form, the `name` may be abbreviated in a similar way to component names of lists. For example, in our `ttest` function the argument names are `y1`, `y2`, `test` and `alpha`. The last two arguments may be abbreviated to the single `t` and `a` respectively:

```
> unlist(ttest(y1=x2, y2=x1, a=0.99, t="upper"))
  tstat df reject  tail.area
 3.6303 18      1 0.00095688
```

With character string arguments such as `test` in this example it is possible to give a set of allowable values from which the user may select by partial matching. The function `match.arg(arg, choices)` uses the character string `arg` to select that value from the character string vector `choices` for which a unique partial match occurs. If there is no unique partial match the result is an error.

If the call to `match.arg` is made from within a function and `arg` is an argument then `choices` may be omitted. The default value for `arg` must then be a character string vector and this default value is used as the `choices` argument.

If the function is called and no value for `arg` is supplied, the first entry in the default character string vector is the value used. To see this in action, consider the changes we could make to `ttest`:

```
ttest <- function(y1, y2,
         test = c("two-sided", "lower", "upper"), alpha = 0.05)
{
    test <- match.arg(test)
    ....
    tail.area <- switch(test,
        "two-sided" = 2 * (1 - pt(abs(tstat), ndf)),
        lower = pt(tstat, ndf),  upper = 1 - pt(tstat, ndf))
       list(tstat=tstat, df=ndf, reject=tail.area < alpha,
                    tail.area = tail.area)
}
```

The call `ttest(x1, x2)` will perform a two-sided *t*-test. To specify a lower-tailed one-sided *t*-test we could use:

```
ttest(x1, x2, tes="low")
```

Now `test="lower"` is selected by partial matching. An invalid choice for the argument `test` now precipitates an error from `match.arg`:

```
> ttest(x1, x2, tes="left")
Error in call to "ttest1": Argument "test" should be one of:\
        "two-sided", "lower", "upper"
Dumped
```

The function `match.arg` should be distinguished from the related functions `match`, `pmatch`, `amatch` and `charmatch` which can be used for finding if and where the elements of its first vector argument occur in its second. Consult the online help for full details.

The special argument "…"

The "three dots" argument, "…", is special in that any number of named arguments may be specified for it on the call. Inside the function, normally the only thing done with it is to use it as an argument to some other function, in which case any such arguments are passed on as they were specified in the original call.

Continuing our *t*-test example, we allow the option of producing a boxplot. To customize the boxplot the user has to be able to pass on additional arguments to `plot`. We need to make two changes to our function. The first is to add two more arguments to the definition, `boxp` and `...`:

```
ttest <- function(y1, y2, test = "two-sided", alpha = 0.05,
                boxp = T, ...)
{
    ....
```

A better default value for boxp would be T if a graphics device other than the null device is currently open, otherwise F. This is easy to do within S-PLUS by setting the default boxp=dev.cur() > 1. A default that works for any release of S is more complicated:

```
boxp = exists(".Device", frame=0) && .Device != "null device"
```

This will be more understandable after we discuss frames in Section 4.5.

We also need to do the boxplot. A convenient place for this is just before the return expression:

```
    . . . .
    if(boxp) {
    sam <- paste("Sample",1:2)
    f <- factor(rep(sam, c(length(y1), length(y2))))
    plot(f, c(y1, y2), ...)
    }
    list(tstat = etc.
```

Any additional arguments given to ttest will now be passed on to the plot function. For example it might be a good idea to add a more informative label to the vertical axis.

```
    ttest(x1, x2, ylab="Water Quality")
```

Since ylab does not match any other argument name it is included in the " . . . " argument and passed on to plot.

The on.exit function for exit actions

A common use of the " . . . " argument is to allow the user to specify temporary changes to the graphical parameters for some plot done within a function. There is a conventional way to do this:

1. The " . . . " argument is included, but only arguments to par may be substituted for it on the call.

2. Before any plotting is done the following statements are executed:

```
    oldpar <- par(...)
    on.exit(par(oldpar))
```

The first statement records the state of the graphics parameters. The second statement using on.exit arranges for the previously current settings to be re-set when the action of the function is terminated (whether normally or by an error exit).

The on.exit function can be used in a similar way to allow temporary changes to be made to the options settings.

An argument add to the on.exit function allows additional actions to be added to those already specified. Calling on.exit() with no arguments cancels any actions previously requested.

Creating a help document

Functions or data frames that may be used by more than one user, or even by one user over a long period of time should be documented. This is easy to do in a machine readable form. The first step is to use the `prompt` function with the object to be documented as argument:

```
> prompt(area)
created file named  area.d  in the current directory
  edit the file and move it to the appropriate .Help directory,
  dropping the .d
```

This creates an outline help document file, here `area.d`, that should be completed using an editor. Although on Unix systems the document is written using a macro package under `nroff` it is easy to provide the required information using only the instructions in the prompt file itself. Further information can be obtained from the `prompt` help document.

By convention the file has the same name as the S object but with a file extension of ".d". To make it readable by the S help facility it should be moved or copied into the `.Help` subdirectory of the `.Data` directory and the file extension dropped.

```
> ?prompt
> ! vi area.d  # fill in the outline file using an editor
> ! cp area.d .Data/.Help/area  # place in .Help as 'area'
> ?area                 # check that it reads correctly.
```

(These instructions apply only to Unix systems; Windows users should consult the appropriate S-PLUS manual. On that system `prompt` creates a help file with the (mapped) name of the object in directory _DATA_HELP, and the files are plain text and not `nroff` files.)

4.3   Finding errors

When an abnormal exit occurs from a function it is often useful to find out the state of the variables when the error occurred. Two system functions, `dump.calls` and `dump.frames` can be used to help. These are only commonly used as the value for the `error` argument of `options`. The default setting is

```
options(error=dump.calls)
```

Any error during execution then causes details of the calls in progress when the error occurred to be recorded, or 'dumped'. These details can then be exhibited using the `traceback` function.

A simple if contrived example makes the process clear. The `sqrt` function is a simple function with the following definition:

```
sqrt <- function(x) x^0.5
```

A call to sqrt(-1) will generate an NA and issue a warning, but setting options(warn=2) will convert warnings into errors:

```
> options(warn=2)
> sqrt(-1)
Error in x^0.5: (warning converted to error) (-1)^(0.5) DOMAIN
          error
Dumped
> traceback()
Message: (warning converted to error) (-1)^(0.5) DOMAIN error

3: x^0.5
2: sqrt(-1)
1:
```

Setting the error option to dump.frames gives a more complete picture. (Frames are considered more formally in Section 4.5. For present purposes we can think of a frame as the environment within which a function is evaluated.) Consider the ttest function introduced in Section 4.2.

```
> options(error=dump.frames)
> ttest(x1, ylab="Yield")
Error in ttest(x1, ylab = "Yield"): two samples needed
Dumped
> traceback()
Message: two samples needed
3: stop("two samples needed")
2: ttest(x1, ylab = "Yield")
1:
```

The debugger function is a facility for inspecting the dumped information. In its simplest use it offers a menu selection for variables in any selected frame.

```
> debugger()
Message: two samples needed

1:
2: ttest(x1, ylab = "Yield")
3: stop("two samples needed")
Selection: 2
Frame of ttest(x1, ylab = "Yield")
d(2)> ?
1: y1
2: y2
3: test
4: alpha
5: boxplot
6: ...
d(2)> 2
.Argument(, y2 = )
d(2)> 6
```

```
.Argument((ylab = "Yield"), ... = )
d(2)> 0

1:
2: ttest(x1, ylab = "Yield")
3: stop("two samples needed")
Selection: 0
NULL
```

Notice that debugger operates at two levels, on a list of frames or on a selected frame within the list. At the top level the prompt is

```
Selection:
```

and at the inner level it changes to

```
d(2)>
```

where the 2 indicates that we are looking at frame 2. (At this level it is using browser; see below.)

A question mark, ?, provides a numbered list of variables whose values can be requested, and a zero, 0, passes control up one level or out of debugger entirely.

The function dump.frames should only be used as the error option when debugging is needed since it can cause very large objects to be deposited in the working directory.

Two more tools for detecting errors are trace and browser, which are often used together. Consider trying to find the source of an error in ttest that does not produce a stop message. This will happen, for example, if we specify an empty sample.

```
> x <- c(x1, x2)
> f <- factor(rep(0:1, rep(10,2)))
> ttest(x[f==1], x[f==2])
Error in dim<-: Invalid value for dimension 1: c(0, ..)
Dumped
```

The error message is not very enlightening. The traceback suggests the error occurred inside var while calculating the variance of the second sample:

```
> traceback()
Message: Invalid value for dimension 1
4: dim(xm) <- c(n, 1)
3: var(y2)
2: ttest(x[f == 1], x[f == 2])
1:
>
```

We could now insert a print or cat command in ttest just before the variances are calculated, or the more flexible tool browser to inspect the variables interactively. With this change ttest becomes:

```
    . . . .
    if(missing(y1) || missing(y2))
        stop("two samples needed")
    n1 <- length(y1)
    n2 <- length(y2)
    ndf <- n1 + n2 - 2
browser()                # debugging insert
    s2 <- ((n1 - 1) * var(y1) + (n2 - 1) * var(y2))/ ndf
    tstat <- (mean(y1) - mean(y2))/sqrt(s2 * (1/n1 + 1/n2))
    tail.area <- switch(test,
    . . . .
```

When we execute the modified function the action is stopped at that point, a special prompt, b(2)>, appears and we can work temporarily in the local frame. The obvious thing to do is to inspect the values of the variables, but any command may be given. Typing 0 at the menu interface exits the browser and returns control to the function.

```
> ttest(x[f==1], x[f==2])
Called from: ttest(x[f == 1], x[f == 2])
b(2)> ?
1: ndf
2: n2
3: y2
4: n1
5: y1
6: y1
7: y2
8: test
9: alpha
10: boxp
11: ...
b(2)> n2
[1] 0
b(2)> y2
numeric(0)
b(2)> y1
 [1]   0.4 -0.3  0.9  2.2  1.8  0.7  2.7  1.0  1.1  2.9
b(2)> 0
Error in dim<-: Invalid value for dimension 1: c(0, ..)
Dumped
```

We notice that n2 is 0, y2 is empty and y1 is actually the second sample. The problem was that factor f has levels 0 and 1, not 1 and 2. This is a common way for empty arguments to be generated.

Notice that browser can specify variables by name or by menu number. It can also be used to evaluate arbitrary expressions involving the variables. A question mark can be used to give a list of variables in the frame.

The function trace operates in a way that seems non-invasive to the user and allows function calls to be traced or other actions to be taken when errors occur.

The first and only required argument is the name of one or more functions whose calls are to be traced.

```
> trace(c("ttest", "var"))
> ttest(x[f==1], x[f==2])
On entry:               ttest(x[f == 1], x[f == 2])
On entry:                    var(y1)
On entry:                    var(y2)
Error in dim<-: Invalid value for dimension 1: c(0, ..)
Dumped
>
```

By default `trace` inserts a call to the function `std.trace` as an extra first step in any traced function. The effect is to print out details of the call. The function `std.trace` is then called the *tracer*. It is also possible to make the trace function to use `browser` as the tracer and to specify a particular place or places in the function for the call to `browser` to be located. To specify a place we need to know the number of each statement at the top level of our function. The function `tprint` prints a function with numbered statements, but before using it we should remove the trace.

```
> untrace(ttest)
> tprint(ttest)
{
1: if(missing(y1) || missing(y2)) stop("two samples needed")
2: n1 <- length(y1)
3: n2 <- length(y2)
4: s2 <- ((n1 - 1)*var(y1) + (n2 - 1)*var(y2))/(n1 + n2 - 2)
5: tstat <- (mean(y1) - mean(y2))/sqrt(s2 * (1/n1 + 1/n2))
6: tail.area <- switch(test,
     . . . .
}
```

The call to `browser` should go just before statement number 4; this is the `at` argument of `trace`:

```
> trace(ttest, tracer=browser, at=4)
> ttest(x[f==1], x[f==2])
At 4: Called from: ttest(x[f == 1], x[f == 2])
b(2)> n1
[1] 10
b(2)> n2
[1] 0
b(2)> 0
On entry:                    var(y1)
On entry:                    var(y2)
Error in dim<-: Invalid value for dimension 1: c(0, ..)
Dumped
> untrace()
> unlist(ttest(x[f==0], x[f==1]))  # using the right levels!
   tstat df reject tail.area
 -3.6303 18      1 0.0019138
```

Calling untrace() with no arguments removes all tracing.

Note that trace works by temporarily substituting a modified version of the function with tracing statements inserted, so it is in a sense invasive. It is necessary to turn tracing off before changing a function and to re-set the trace again afterwards if needed.

Since correcting errors in programs can be difficult it is important to become familiar with the tools that are available to help. We summarize those we have introduced here in Table 4.1.

Table 4.1: Tracing and debugging facilities in **S**.

print, cat	Printing key quantities from within a function may be all that is needed to locate an error.
traceback	Prints the calls in process of evaluation at the time of any error that causes a dump.
options(warn=2)	Changes a warning into an error, precipitating a dump.
options(error=FUN)	Specifies the dump action. The default FUN is dump.calls which dumps the details of calls, only, but dump.frames can be used to cause a dump of all evaluation frames in existence at the time of the error.
last.dump	The object in the .Data directory that contains a list of calls or frames after a dump.
debugger	Function to inspect last.dump after an error.
browser	Function that may be inserted to interrupt the action and allow variables in the frame to be inspected before the error occurs.
trace	Place tracing information at the head of, or inside, functions. May be used to insert calls to browser at specified positions.
tprint	Produces a numbered listing of the body of a function for use with the at argument of trace.
untrace	Turns some or all tracing off.
inspect	(S-PLUS 3.2 only.) An interactive debugger.

Using inspect

S-PLUS 3.2 introduced another debugging function, inspect, which works much more like interactive debuggers for C or FORTRAN. Consider again our t-test example:

```
> inspect(ttest(x[f == 1], x[f == 2]))
entering function ttest
stopped in ttest (frame 3), at:
        if(missing(y1) || missing(y2))
                stop("ttest requires two samples")
d>
```

Here d> is the inspect prompt, and typing help gives a help screen (Table 4.2). Inspection reveals the problem:

Table 4.2: Main options for the S-PLUS 3.2 function `inspect`.

Individual help entries are available for the following.

Advance evaluation:
    step – walk through expressions
    do – do expressions atomically
    complete – a loop or function
    resume – continue to next mark
    enter – descend into a function call
    quit – abandon evaluation

Display:
    where – current calls and expr'n
    objects – local frame objects
    show – installed tracks and marks
    find – location of an S-PLUS object
    return.value – function return value
    on.exit – scheduled on.exit expr'ns
    fundef – function definition

Halt evaluation; track functions:
    mark, unmark – arrange to stop in
      a function or at an expression
    track, untrack – install or change
      function call reporting

Examine other current frames:
    up, down

Miscellaneous:
    eval – S-PLUS expressions
    help – syntax and description
    debug.options – option settings

Informational help entries:
    names – when to quote
    keywords – list of reserved words

```
d> eval y2
numeric(0)
d> quit
```

without either inserting a call to `browser` or remembering a call to `options(error=dump.frames)`.

4.4 Calling the operating system

Sometimes there is an operating-system command that will do exactly what you want. The functions `unix`, `win3` and `dos` are provided to let you communicate with operating-system commands.

Unix

The syntax of the `unix` command is

```
unix(command, input=NULL, output.to.S=T)
```

The argument `command` is a character string which is passed to a Bourne shell to execute (but could invoke another shell itself). If `input` is specified, it is written to a file used for standard input to the command. By default the standard output is returned line-by-line as a character vector, but if the command needs to interact with the user, set output.to.S=F, when the exit status of the command is returned.

As an example, let us test if a file exists and is readable, before attempting to read data from it.

```
file.exists <- function(name)
    unix(paste("test -r", name), output.to.S=F) == 0
```

Useful functions in conjunction with unix are tempfile which creates a temporary file with a unique name, dput and dget to write and read objects and unlink to remove files. The functions tempfile and unlink are simple calls to unix:

```
tempfile <- function(pattern = "file")
    paste("/tmp/", pattern, unix("echo $$"), sep = "")
unlink <- function(x)
    invisible(unix(paste("rm", paste("'", x,, "'", sep = "",
    collapse = " ")), output = F))
```

We use this to write a function to compare two objects (a simpler version of the S-PLUS function objdiff):

```
objchk <- function(x, y)
{
    old <- tempfile("old"); dput(x, old)
    new <- tempfile("new"); dput(y, new)
    unix(paste("diff", old, new), output = F)
    unlink(c(old, new))
    invisible()
}
```

One good use of unix is to manipulate character strings, for example to convert to upper and to lower case we could use the functions

```
to.upper <- function(str) unix('tr "[a-z]" "[A-Z]"',str)
to.lower <- function(str) unix('tr "[A-Z]" "[a-z]"',str)
```

One very useful function based on operating system calls is unix.time, which times its argument and returns a 5-element vector given by the Unix command time , the times (in seconds) of the user, system and elapsed times, plus the user and system times taken by child processes (if any). Further, although this is a function, it is arranged so that assignments within it are done at the level of its call. Note that if the argument returns an invisible result, this results in the return value from unix.time being marked as invisible and not printed.

Windows and DOS

The **Windows** interface function is

```
win3(command, multi=F, trans=F)
```

and the arguments `multi` determine whether the command is multitasked, and `trans` whether file paths are converted between Unix–style `"/test/file"` and DOS–style `"\test\file"`.

The DOS interface function (also available under **Windows**) is

```
dos(command, input, output=T,  multi=F, trans=F, redirection=F)
```

By default the `input` is added to the command, but if `redirection=T`, redirection is used (as with the unix command). Using `multi=T` sets up a DOS box for the command which must be closed explicitly.

We can use `dos` to write a `file.exists` function:

```
file.exists <- function(name)
    length(dos(paste("attrib", name))) > 0
```

The function `dos.time` is the equivalent of `unix.time`, but measures elapsed time (in units of a second) only.

4.5   Some more advanced features. Recursion and frames

Functions in S are allowed to be recursive, that is they are allowed to call themselves. This idea is both interesting in itself and useful for an understanding of how S organizes its calculations when functions are called, but in practice it should be used cautiously as it can lead to slow and memory-intensive code.

An adaptive numerical integration example

Consider a simple function to integrate a function $f(x)$ of one real variable numerically over a finite interval $[a, b]$. Users of **S-PLUS** already have such a function, `integrate`, and a public domain package is available from the `statlib` archive. To show some of the features of S programming we will use an adaptive procedure based on a simple idea. Let $d = (a+b)/2$ be the mid-point of the interval. We first calculate two approximations to the integral a_1 and a_2 using the trapezoidal rule and Simpson's rule respectively:

$$a_1 = [f(a) + f(b)](b - a)/2, \qquad a_2 = [f(a) + 4f(d) + f(b)](b - a)/6$$

If $|a_1 - a_2| < \epsilon$ then we use a_2 as the value of the integral, otherwise apply the same procedure to the two sub-intervals $[a, d]$ and $[d, b]$ and add the results.

This idea will fail badly for integrands like $f(x) = \sin^2(x)$ over $[0, 2\pi]$, since the first step will terminate the procedure and report the value 0, but with some care the effect will be for the algorithm to adapt to the integrand and concentrate function evaluations in regions where they appear to be most necessary for the integral. Another potentially serious disadvantage is that the algorithm is very selective in where it evaluates the integrand and cannot take much advantage of vectorized functions, whereas conventional methods such as Gaussian quadrature are in a good position to do so.

When writing functions it is a good idea to start simply, get something working immediately and build capabilities gradually and interactively. A first version of the function might be:

```
> area1 <- function(f, a, b) {
    d <- (a + b)/2
    a1 <- (f(a) + f(b)) * (b-a)/2
    a2 <- (f(a) + 4*f(d) + f(b)) * (b-a)/6
    if(abs(a1-a2) < .0001)
        return(a2)
    area1(f,a,d) + area1(f,d,b)
}
> area1(sin, 0, pi)
[1] 2
```

Although it seems to work the obvious flaws are that it evaluates the function more often than necessary and the tolerance is fixed in the code. A second version fixes these problems. We also use Recall, which calls the current function, whatever its name. This ensures the function body will work regardless of the name to which it is assigned.

```
> area2 <- function(f,a,b, fa=f(a), fb=f(b), eps=1.0e-4) {
    d <- (a + b)/2
    fd <- f(d)
    a1 <- (fa + fb)*(b-a)/2
    a2 <- (fa + 4*fd + fb)*(b-a)/6
    if(abs(a1-a2) < eps)
        return(a2)
    Recall(f,a,d, fa, fd, eps) + Recall(f,d,b, fd, fb, eps)
}
> area2(sin, 0, pi)
[1] 2
```

Notice that the default values for arguments make use of other arguments.

Two further ways in which we can improve the function are:

- As a safety measure we can impose a limit on the depth to which the recursion is allowed to go. This is easily done by an extra argument limit which decreases by one each time a recursive call is made.

- We can allow the integrand function f to have additional arguments that are not involved in the integration.

Our final version incorporates these changes.

```
area <- function(f, a, b, ..., fa = f(a, ...), fb = f(b, ...),
                   limit = 10, eps = 100*.Machine$single.eps)
{
  h <- b - a
  d <- (a + b)/2
  fd <- f(d, ...)
  a1 <- ((fa + fb) * h)/2
  a2 <- ((fa + 4 * fd + fb) * h)/6
  if(abs(a1 - a2) < eps)
    return(a2)
  if(limit == 0) {
    warning(paste("recursion limit reached near x = ", d))
    return(a2)
  }
  Recall(f, a, d, ..., fa=fa, fb=fd, limit=limit-1, eps=eps) +
    Recall(f, d, b, ..., fa=fd, fb=fb, limit=limit-1, eps=eps)
}
```

The system dataset `.Machine` is a list of machine constants.

We can now test our function by calculating a beta function by integration,

$$
B(\alpha, \beta) = \int_0^1 x^{\alpha-1}(1-x)^{\beta-1}\, dx = \frac{\Gamma(\alpha)\Gamma(\beta)}{\Gamma(\alpha+\beta)}
$$

```
> fbeta <- function(x, alpha, beta) {
    x^(alpha-1) * (1-x)^(beta-1)
  }
> b0 <- area(fbeta, 0, 1, alpha=3.5, beta=1.5)
> b1 <- exp(lgamma(3.5) + lgamma(1.5) - lgamma(5))  # a check
> c(b0, b1, b0-b1)
[1]  1.2272e-01  1.2272e-01 -1.4474e-06
```

It is interesting to see where `area` has chosen to evaluate the integrand. The easiest way is to include printing statements within `fbeta` to trace the arguments used. However, in this case we would like to record the values for future use (a plot), so we will arrange for `fbeta` to record the values given to it. We return to this problem on page 109 after discussing *frames*.

Frames

An *evaluation frame* is a list that associates names with values. Their purpose is much the same as dictionaries on the search list. Evaluation frames and object dictionaries are collectively called *databases*.

When an S expression is evaluated in a frame any values needed are obtained by reference to that frame *first*. This first point of reference is called the *local frame* of the expression. To show this in action we can use the function `eval` which takes an S object of mode `expression` and (optionally) a local frame:

```
> a <- 1
> b <- 2
> d <- 3
> eval(expression(a + b + d), local=list(a=10, b=20))
[1] 33
> a + b + d
[1] 6
```

Any names (like d in this example) not satisfied by the local frame are next referred to the frame of the top-level expression, frame 1, then the frame of the session, frame 0, and then on to the search list. The sequence of databases consisting of the local frame, frame 1, frame 0 and the search list may be called the *search path* for an object. Notice that the search path for an object at, say, frame 6, does *not* include frames 2 to 5; this can cause unexpected problems if not borne in mind. In particular, objects created in parent functions will *not* be found in this search.

Frame 0 is born with the session and dies with the session. New entries may be added to this frame, but unlike ordinary assignments made during the session they are not permanent. To be permanent an object must reside in an object dictionary on the search list. It is used by the system for storing options set in .First in the example below.

When an expression is evaluated at the interactive level it initiates a new frame called frame 1, usually containing only the unevaluated expression and an auto-print flag specifying whether the result is to be printed or not.

```
> objects(frame=0)
[1] ".Device"              ".Devices"              ".Options"
[4] ".PostScript.Options"
> objects(frame=1)
[1] ".Auto.print" ".Last.expr"
```

If the expression that generated frame 1 contains calls to functions they generate frames at higher levels as the calls are processed. These frames contain the names and values of arguments and local variables within the function. Normal assignments within functions only change the local frame, and so do not affect values for variables outside that frame. This is how recursion is accommodated. Each recursive call establishes its own frame without disturbing values for the same variables at different recursion levels.

The contents of any database can be inspected and manipulated using the functions in Table 4.3, each of which has an argument frame to specify a particular frame number and where to specify (alternatively) a position number in the search list.

The functions get and exists by default consult the entire search path. The others consult the local frame (or the working directory at the interactive level) unless another focus is specified.

There is a special assignment operator, <<-, which can be used within a function to make an assignment to the working directory.

As an example, the get function can be made to "look at itself":

Table 4.3: Functions to access arbitrary databases

assign	Creates a new name=value pair in a specified database.
exists	Tests whether a given object exists in the search path.
get	Returns a copy of the object, if it exists; otherwise returns an error.
objects	Returns a character vector of the names of objects in the specified database.
remove	Deletes specified objects, if they exist, from the specified database.

```
> get(".Last.expr", frame=1)
expression(get(".Last.expr", frame = 1))
> get(".Auto.print",frame=1)
[1] T
```

In frame 1 automatic printing is normally turned on. If an expression or the return value of a function is given as invisible(value) it is not automatically printed. The invisible function itself is quite short:

```
> invisible
function(x = NULL)
{
    assign(".Auto.print", F, frame = sys.parent(2))
    x
}
```

The function sys.parent takes an integer argument n and returns the index of the frame n generations behind the present. The assignment inside invisible simply turns off auto-printing in the frame of the parent of the function or expression that called invisible. Other functions whose names begin with "sys." are available to determine other aspects of the current memory frames. See the on-line help for sys.parent for more details.

Sometimes it is necessary for variables in a function to be made available to functions it calls. The easiest way to do this is to assign the variables in frame 1, since this is on the search path throughout the evaluation of the current top-level expression.

Communicating through frame 0 or 1

We now return to our numerical integration example and record the values at which area evaluates the integrand. An easy way to do this is to use a variable, say val, in frame 0. Before the call we set it to NULL, and we modify fbeta so that it adds a new value to val every time it is called:

```
> assign("val", NULL, frame=0)
NULL
> fbeta.tmp <- function(x, alpha, beta) {
    assign("val", c(val,x), frame=0)
```

```
      x^(alpha - 1) * (1 - x)^(beta - 1)
   }
> b0 <- area(fbeta.tmp, 0, 1, 3.5, 1.5); b0
> [1] 0.12272
> plot(val, fbeta(val, 3.5, 1.5))
```

The result is shown in Figure 4.1.

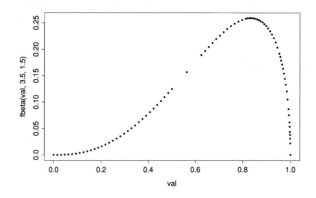

Figure 4.1: The choice of points in an adaptive numerical integration.

If variables are only needed during the duration of a function call, it is better to assign them to frame 1 (which is on the search path) than to frame 0, as frame 1 exists only during the execution of the current top-level expression.

4.6   Generic functions and object-oriented programming

Generic functions are used by S to provide families of related functions. For example, the function `plot` is the primary facility for producing graphical output. The graphical output produced depends on the kind of object given as the first argument to `plot`. The first step taken by `plot(x, ...)` is to check the `class` vector of the principal argument, `x`. This is a character string vector of *class names*, possibly null. Every object implicitly has class `"default"`, which can be regarded as appended to every class vector. The evaluator checks each class name in turn to see if there is a function, `plot.name`. When such a function is found it is used to produce the graphical output. If the class vector is NULL, `plot.default` is used.

The function `plot` is called a *generic function* since it consults the class of its first argument and allows the object itself to determine the action taken. The actual functions used, `plot.default`, `plot.tree`, and so on are called *method functions* for the generic function `plot`. In the case of `plot` the list of method functions includes those shown in Table 4.4.

These are not all intended to be used as method functions. For example, both `plot.data.frame` and `plot.dataframe` can be used to plot data frames. The

Table 4.4: Some `plot` method functions.

`plot.data.frame`	`plot.formula`	`plot.preplot.gam`	`plot.tree`
`plot.dataframe`	`plot.gam`	`plot.preplot.loess`	`plot.tree.sequence`
`plot.default`	`plot.glm`	`plot.profile`	`plot.varcomp`
`plot.design`	`plot.lm`	`plot.stl`	`plot.xy`
`plot.factor`	`plot.loess`	`plot.surv.fit`	

first produces a sequence of plots for each column with a pause between plots; the second produces the same graphical output but compactly, with all plots in one figure region. Since the class name for a data frame is `data.frame` the first will be used in response to a call to `plot`. For example if `mydata` is a data frame

```
plot(mydata)
```

will use `plot.data.frame` to produce the graphical output. To use the function `plot.dataframe` we can either call the function directly or else add the name `dataframe` to the class vector for `mydata` and call `plot`:

```
class(mydata) <- c("dataframe", class(mydata))
plot(mydata)
```

In this second case the object is said to be of class `dataframe` and to *inherit* from class `data.frame`. It will still respond to most generic functions like an ordinary data frame, a process known as *inheritance*. More generally, an object is said to have the class of the first element of its class vector and to inherit from the classes specified by the remaining elements (if any) and from the class `"default"`.

Most generic functions are short: `plot` is

```
plot <- function(x, ...) UseMethod("plot")
```

A generic function normally uses the `UseMethod` function, which chooses the method function to be used. When the function `UseMethod` is executed, the frame of the generic function is used to provide the arguments to the method function selected. The method matches its arguments in that local frame and hence the arguments do not need to be included explicitly on the call. This process is sometimes called the *method dispatch mechanism*.

The function `NextMethod` is similar to `UseMethod` but is normally called from within some method function. It determines the next inherited method for this function and evaluates its method function, returning the value to the calling method. Calling `NextMethod` is the usual way to define a new method that extends the actions of an inherited method for the same function.

Consider a simple example. We saw in our *t*-test function how it was convenient to have the result printed in compact form using the `unlist` function on the result. In some situations it is useful to have a result in list form but displayed in compact form when printed automatically. To do this we can give the object a class, say `abbrev`, and define a print method function to accommodate it:

```
print.abbrev <- function(obj) {
        obj <- unlist(obj)
        NextMethod("print")
    }
```

Of course the next method to be selected will usually be the default print method, but in principle an object could be given membership of the abbrev class to have some other print method invoked after unlisting.

We can now arrange for automatic printing to be done by this function:

```
> tt <- ttest(x1, x2)
> class(tt) <- "abbrev"
> tt
   tstat df reject tail.area
 -3.6303 18      1 0.0019138
```

(Of course the class assignment would normally be done in the ttest function itself.) To remove the class vector and use the default printing method we can use the function unclass:

```
> unclass(tt)
$tstat:
[1] -3.6303
$df:
[1] 18
$reject:
[1] T
$tail.area:
[1] 0.0019138
```

It is dangerous to assign a new class to an object, and unless you know that it is of the default class, it is better to add the class and rely on inheritance.

Examples

Our libraries contain many examples of the use of the object-oriented programming paradigm. Most often we just add method functions for generic functions such as print, summary and predict, as in classes "lda", "qda" and "nnet". Similarly objects of our class "negbin" inherit from class "glm" but need their own method functions for anova, summary and family to cater for the few special needs of negative binomial models.

The robust linear regression functions discussed in Chapter 8 define a new class "rlm", and also inherit from the linear regression class "lm". Both print and summary functions are provided. However, summary.rlm produces a result of class "summary.lm", so printing the summary invokes print.summary.lm. We could have used inheritance for the print method too, with a function something like

```
print.rlm <- function(x, ...)
{
   NextMethod("print")
   cat("Scale estimate:", format(signif(x$s, 3)), "\n")
   if(x$converged)
      cat("Converged in", length(x$conv), "iterations\n")
   else cat("Ran", length(x$conv),
                      "iterations without convergence\n")
   invisible(x)
}
```

In general inheritance works well only when the classes are designed in parallel with inheritance in mind. For example, it might appear that we could use inheritance from summary.lm to build summary.rlm, but scale estimation by the root mean square of the residuals is so embedded in summary.lm that this proves to be impossible. (Similarly, it is embedded into summary.glm, which gives nonsensical results for its own robust option.)

4.7   Using C and FORTRAN routines

The S program does much of its computations through routines written in either FORTRAN or C. User-written routines can be called from the S language using the .Fortran or .C functions with a protocol that follows. (The function polynom on page 114 provides an example.)

- The first argument to .C or .Fortran is a character string giving the name of the routine.

- Each further argument must match the argument of the routine. In particular the data passed through to the routine must have the correct storage.mode and must match the argument in length. Unlike S neither FORTRAN nor C can deduce the length or mode of arguments.

- The arguments may be given name fields. These do not match anything in the routine itself, but will be retained as name fields in the result.

- The value returned by the call to .C or .Fortran is a list containing all the arguments passed to the routine. The components of the list will reflect any changes made by the routine. Any attributes of the arguments will be retained so that arrays will return as arrays, for example.

The storage modes for arguments and their C and FORTRAN counterparts are given in Table 4.5 taken from Table 7.1 of Becker, Chambers & Wilks (1988, p. 197).

This is the only place in S where a distinction must be made between storage modes double, single and integer all of which correspond to S mode numeric and the coercion functions double, single and integer are all specializations of numeric used only in this context, to ensure that arguments

Table 4.5: Argument storage modes in S and corresponding data types for C and FORTRAN routines where applicable.

S storage mode	C	FORTRAN
"logical"	long *	LOGICAL
"integer"	long *	INTEGER
"single"	float *	REAL
"double"	double *	DOUBLE PRECISION
"character"	char **	CHARACTER (*)
"complex"	struct { double re, im;} *	DOUBLE COMPLEX
"list"	void **	

have the correct storage mode. Functions such as `as.integer` create new objects of the correct mode but strip all attributes and so reduce arrays to vectors, for example. If preserving attributes is important it is necessary to use the assignment form:

```
storage.mode(x) <- "single"
```

Notice from Table 4.5 that the allowable argument types in C routines are all *pointers*. This is because the quantities manipulated are S vectors and so must be accessed by C indirectly.

To illustrate the process consider a simple example. Figure 4.2 shows a pair of C routines that together evaluate a polynomial using a Horner scheme. Notice that the C function `horner` cannot be called directly from S since its arguments are not all pointers, but `poly` does conform. Notice also that `poly` evaluates the polynomial for a vector of x values and returns via p a vector of results. This is in line with the preferred style of S functions.

Supposing these functions have been loaded with our copy of S (see below) we can write a function to use them. The argument b is the vector of polynomial coefficients (including the constant).

```
polynom <- function(x, b) {
  m <- as.integer(length(x))
  n <- as.integer(length(b)-1)
  storage.mode(x) <- "double"
  p <- x
  storage.mode(b) <- "double"
  .C("poly", m, val = p, x, n, b)$val
}
```

Since only the result component is needed, we only return that component. Note that by returning the result 'in place' we retain the attributes of x, such as its dimensions.

To evaluate the polynomial $x^2 - 1$ for each element of a 3×3 matrix we can call polynom:

```
double horner(x, b, n)
  double x, *b;
  int n;
{
    int i;
    double p = b[n];
    for(i = n-1; i >= 0; i--)
        p = b[i] + x * p;
    return p;
}

poly(m, p, x, n, b)
  double *p, *x, *b;
  long *m, *n;
{
    long i;
    for (i = 0; i < *m; i++)
        p[i] = horner(x[i], b, (int)*n);
}
```

Figure 4.2: File horner.c : C functions to evaluate a polynomial using Horner's scheme.

```
> mat <- matrix(1:9,3,3)
> polynom(mat, -1:1)
     [,1] [,2] [,3]
[1,]    0   15   48
[2,]    3   24   63
[3,]    8   35   80
```

It is important to note that matrices and arrays in S will be passed to the C function as vectors, not as multiply subscripted arrays, although as we have noted the dim attribute will be preserved in the result.

Our libraries provide several useful examples of calls to C and FORTRAN routines, for example in the functions sammon and nnet.

Static and dynamic loading

S already has many FORTRAN and C functions loaded for use by functions such as svd, qr and eigen. There are three main ways of including other functions, such as those from our horner.c. (S-PLUS for Windows 3.2 introduced a fourth, dll.load, which we do not describe.)

1. *Static loading.* This option creates a private copy of S for the user which includes all the S routines together with any others that may be needed. Since S is large (around 3–4Mb) this should only be done in special circumstances, such as when a group of users will use the same copy, or when the alternative facilities listed below are not available.

There is a LOAD facility for this option. For our function it amounts to running S-PLUS with an initial command line argument

```
$ Splus LOAD horner.c
```

This does several things. First it compiles the external routines using the compiler appropriate to the file extension. Then it creates an executable file called local.Sqpe (nsplus.exe in Windows, for which the command is just LOAD) in the local directory containing all of S together with the extra routines. Once this executable is made, when S (or S-PLUS) is invoked from that directory it uses this file instead of the system executable file.

This method is the least desirable, but sometimes is the only feasible route (especially under Windows).

2. *Incremental loading* with dyn.load. This is possibly the most convenient but not available on some machines because of software limitations. The dyn.load function dynamically loads an object file and makes the functions available to the version of S presently running. Each file should be compiled in the normal way (but see below). If more than one file is to be loaded they should be accumulated into a single relocatable object file using

```
$ ld -r -d -o objects.o horner.o bessel.o utilities.o
```

after which the file may be dynamically loaded within S by using

```
dyn.load("objects.o")
```

The result is an invisible character string vector giving the names of the routines loaded. (Note: multiple files can be specified directly to dyn.load, but they can not then reference later files in the same load.)

Libraries can also be specified on the ld command, but on dynamically-loading systems these should be the *static* versions (such as /lib/libm.a).

This method is (at the time of writing) available on all Unix S-PLUS systems except for the HP 700/800 series, and for Windows.

3. *Augmented loading* with dyn.load2. This is similar to dyn.load except that it works by loading the extra routines into central memory, temporarily augmenting the symbol table, recording the locations of the new routines and finally discarding the augmented symbol table. The result is that each time new routines are loaded they cannot connect with previously loaded routines. All routines needed for a task must be loaded in one operation; this is not usually a serious restriction. The advantages of dyn.load2 are that it is more widely available than dyn.load, and it is easier to use libraries. Most Unix S-PLUS systems have dyn.load2, and a public domain version of dyn.load2 is available from the statlib archive. The disadvantages of dyn.load2 are that it can be very much slower, take up considerably more memory and it is much less forgiving.

The S-PLUS system has a COMPILE facility that handles some of the compilation details, so our horner.c file was compiled and loaded as follows:

```
> !Splus COMPILE horner.c
targets= horner.o
make -f /usr/local/newsplus/newfun/lib/S_makefile  horner.o
cc -c -I${SHOME-'Splus SHOME'}/include -02    horner.c
> print(dyn.load("horner.o"))
[1] "_poly"   "_horner"
```

Note that this will consult the Makefile (or makefile) in the current directory, which can be used to set or change the compiler flags. (Beware: despite its name, this is a make facility. Thus it will not recompile if name.o is newer than its source file, and if both name.c and name.f are present, Splus COMPILE name.c will compile name.f !)

Ideally, routines should be dynamically loaded only once, although the latest version loaded will be used, which can be useful when developing code. The function is.loaded may be used to determine if some particular function has been loaded. The argument to is.loaded is the name of a function as a character string, but the name as it is known to the symbol table rather than as it may be known to the programmer. From the above display it can be seen that the C function poly has been given the name "_poly" on the particular system used. The function symbol.C may be used to find the symbol table name corresponding to any C function name. Similarly symbol.For can be used for FORTRAN subroutines. For example consider extending our polynom function so that if the C function poly is not available the file horner.o is dynamically loaded first:

```
polynom <- function(x, b) {
  if(!is.loaded(symbol.C("poly"))) dyn.load("horner.o")
    ....
}
```

Another possibility is to include the dynamic loading in the session startup function .First, and for libraries to use .First.lib (see Appendix D).

Note that if the object file is not in the working directory its full path name must be given on the call to dyn.load or dyn.load2.

If the C routine dynamically allocates storage it should use S_alloc which is similar to malloc except that S automatically frees the space afterwards. Note that the call is

```
S_alloc(n, size)
```

like calloc rather than malloc. There are also functions Calloc, Realloc and Free to replace the usual un-capitalized versions. *Do not use* free *with these functions.* Declarations for these and other functions available to the C programmer are given in the include file S.h which resides in the include subdirectory of the SHOME directory. Using Splus COMPILE will ensure that the correct include files are found; other users can use the form of cc function in the example of its use above.

Since this area is system-dependent users may need to consult their implementation literature for further details. It is possible to call routines within S-PLUS to
report warnings and errors; some simple examples are given in our library nnet.
The S random-number generator can be called from C, as in the point process
functions in our library spatial.

Calling S from C

As well as calling C routines from S it is also possible to do the inverse operation
of calling an S function from a C routine with the C routine call_S provided by
S. This is necessary, for example, in a routine for numerical integration where the
integrand may be an S function. Details and an example are given in the online
help and a fuller discussion in Becker, Chambers & Wilks (1988, §7.2.4). Note
that it is only possible to call S functions from within C functions called from S,
and that the process may be extremely slow; it is often convenient to generate a
look-up table from the S function and pass that to the C or FORTRAN function.

FORTRAN input/output

S does not use Fortran input/output routines, and these may not work correctly if
included in your FORTRAN code. Instead, the output functions DBLEPR, INTPR
and REALPR are provided. These all have the form

```
SUBROUTINE name(LABEL,NCHAR,DATA,NDATA)
```

where LABEL is a character string used for the printout, NCHAR is the length of
the label (or zero for no label) and there are NDATA items to be printed from
array DATA. The names correspond to the FORTRAN data types of DOUBLE
PRECISION, INTEGER and REAL. These functions often suffice, especially
for debugging. If elaborate layouts are needed, information can be written to a
character string, and any of these routines used to print the string as a label.

A further problem arises with libraries which may call FORTRAN output for
error messages. Even if it is possible to ensure that the messages are never
produced (as it often is by setting a flag), the routines will still be referenced in
the object module. With dyn.load this is no problem, as setting its argument
undefined to one (for a warning) or greater (for no warning) allows missing
external code to be ignored. However, dyn.load2 is not as forgiving! It will
normally search the FORTRAN libraries, and include the FORTRAN output routines.
Unfortunately, these will often include names which conflict with the C routines
included in S, and dyn.load2 will then refuse to proceed. *In extremis* dummy
routines may have to be written in C to replace the unused FORTRAN routines.

Consider the following example, under S-PLUS 3.1. (There are no problems
with version 3.2 on our systems.) First the file test.f:

```
          subroutine test(x, ifail)
          real x
          x = 1
          if(ifail .ge. 0) write(*, 1000) x
          return
1000      format(" X is ",f6.3)
          end
```

and the S-PLUS session:

```
> testf <- function() .Fortran("test", x=single(1), ifail=-1)$x
> dyn.load("test.o",1)
Warning messages:
1: No definition for symbol "_s_wsFe" in: dyn.load("test.o",1)
2: No definition for symbol "_do_f_out" in: dyn.load("test.o",1)
3: No definition for symbol "_e_wsfe" in: dyn.load("test.o",1)
> testf()
[1] 1
> dyn.load2("test.o")
ld: .... /libm.a(econvert.o): _econvert: multiply defined
ld: .... /libm.a(econvert.o): _fconvert: multiply defined
     ....
Error in dyn.load2("test.o"): 'ld -A ....' failed
```

Writing dummy.c:

```
_s_wsFe() {}
_do_f_out() {}
_e_wsfe() {}
```

and compiling it saves the day:

```
> dyn.load2(c("test.o", "dummy.o"))
> testf()
[1] 1
```

It may be better to put warning messages in the dummy routines, in case they *are* called, and indeed this is a good idea for production code using dyn.load or dyn.load2.

4.8 Exercises

4.1. Write a function to determine which integers (less than 10^{10}) in a given vector are prime.

4.2. Given two vectors a and b, compute s, where s[i] is the number of a[j] <= b[i], for a much longer than b. Then consider a[j] >= b[i]. (Based on a question of Frank Harrell.)

4.3. Implement print, summary and plot method functions for the class "lda" of Chapter 12. (Perhaps the plot method should give a barchart of the singular values or plot the data on the first few linear discriminants, identifying the groups and marking their means.)

4.4.   The Wichmann-Hill (1982) pseudo-random number generator can be defined in Fortran as

```
real function wh(x,y,z)
integer x,y,z
x = mod(171*x, 30269)
y = mod(172*y, 30307)
z = mod(170*z, 30323)
wh = mod(x/30269.0 + y/30307.0 + z/30323.0, 1.0)
end
```

(a) Implement in S a version of `runif` using the Wichmann-Hill generator. Consider where the seed vector should be stored, and how it should be initialized.

(b) Re-implement the function calling C or FORTRAN code.

4.5.   Write an S function to increment its argument object. [Not as easy as it looks: based on an example of David Lubinsky, *Statistical Science* **6**, p. 356.]

4.6.   Write an S function to produce a matrix whose rows contain the subsets of size r of the first n elements of the vector v. Ignore the possibility of repeated values in v and give this argument the default value 1:n.

4.7.   For designed experiments it is often useful to predict the response at each combination of levels of the treatment factors. Write a function `expand.factors` which takes any number of factor arguments and returns a data frame with the factors as columns and each combination of levels occurring as exactly one row.

Chapter 5

Distributions and Data Summaries

In this chapter we cover a number of topics from classical univariate statistics. Many of the functions used are S-PLUS extensions to S.

5.1 Probability distributions

In this section we confine attention to *univariate* distributions, that is, the distributions of random variables X taking values in the real line $\mathbb{R}$.

The standard distributions used in statistics are defined by probability density functions (for continuous distributions) or probability functions $P(X = n)$ (for discrete distributions). We will refer to both as *densities* since the probability functions can be viewed mathematically as densities (with respect to counting measure). The *cumulative distribution function* or CDF is $F(x) = P(X \leqslant x)$ which is expressed in terms of the density by a sum for discrete distributions and as an integral for continuous distributions. The *quantile function* $Q(u) = F^{-1}(u)$ is the inverse of the CDF where this exists. Thus the quantile function gives the percentage points of the distribution. (For discrete distributions the quantile is the smallest integer m such that $F(m) \geqslant u$.)

S has built-in functions to compute the density, cumulative distribution function and quantile function for many standard distributions. In many cases the functions can not be written in terms of standard mathematical functions, and the built-in functions are very accurate approximations. (But this is also true for the numerical calculation of cosines, for example.)

The first letter of the name of the S function indicates the function, so for example dnorm, pnorm, qnorm are respectively, the density, CDF and quantile functions for the normal distribution. The rest of the function name is an abbreviation of the distribution name. The distributions available are listed in Table 5.1.

The first argument of the function is always the observation value q (for quantile) for the densities and CDF functions, and the probability p for quantile functions. Additional arguments specify the parameters, with defaults for 'standard' versions of the distributions where appropriate. Precise descriptions of the parameters are given in the on-line help pages.

Table 5.1: S function names and parameters for standard probability distributions. Those marked by † are only in S-PLUS.

Distribution	S name	parameters
beta	beta	shape1, shape2
binomial	binom	size, prob
Cauchy	cauchy	location, scale
chisquared	chisq	df
exponential	exp	rate
F	f	df1,df2
gamma	gamma	shape
geometric	geom	prob
hypergeometric	hyper	m, n, k
log-normal	lnorm	meanlog, sdlog
logistic	logis	location, scale
negative binomial	nbinom	size, prob
normal	norm	mean, sd
normal range †	nrange	size
Poisson	pois	lambda
stable	stab	index, skewness
T	t	df
uniform	unif	min, max
Weibull †	weibull	shape
Wilcoxon †	wilcox	m, n

These functions can be used to replace statistical tables. For example, the 5% critical value for a (two-sided) t test on 11 degrees of freedom is given by qt(0.975, 11), and the p-value associated with a Poisson(25)-distributed count of 32 is given by 1 - ppois(32, 25). The functions can be given a vector first argument to calculate several p-values or quantiles.

Q-Q Plots

One of the best ways to compare the distribution of a sample x with a distribution is to use a Q-Q plot. The normal probability plot is the best-known example. For a sample x the quantile function is the inverse of the empirical CDF, that is

$$\text{quantile}\,(p) = \min\,\{z \mid \text{proportion } p \text{ of the data } \leqslant z\,\}$$

The function qqplot(x, y, ...) plots the quantile functions of two samples x and y against each other, and so compares their distributions. The function qqnorm(x) replaces one of the samples by the quantiles of a standard normal distribution. This idea can be applied quite generally. For example, to test a sample against a t_9 distribution we might use

```
plot( qt(ppoints(x),9), sort(x) )
```

where the function `ppoints` computes the appropriate set of probabilities for the plot (the values of $(i - 1/2)/n$ sorted in the same order as `x`, but for $n \leqslant 10$, $(i - 3/8)/(n + 1/4)$).

The function `qqline` helps assess how straight a `qqnorm` plot is by plotting a straight line through the upper and lower quartiles. To illustrate this, we generate 250 points from a t distribution and compare them to a normal distribution (Figure 5.1):

```
x <- rt(250, 9)
qqnorm(x); qqline(x)
```

The greater spread of the extreme quantiles for the data is indicative of a long-tailed distribution.

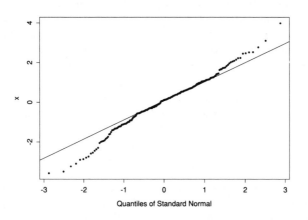

Figure 5.1: Normal probability plot of 250 simulated points from the t_9 distribution.

This method is most often applied to residuals from a fitted model. Some people prefer the roles of the axes to be reversed, so that the data goes on the x–axis and the theoretical values on the y–axis; this is achieved by the argument `datax=T`.

5.2   Generating random data

There are S functions to generate independent random samples from all the probability distributions listed in Table 5.1. These have prefix `r` and first argument `n`, the size of the sample required. For most of the functions the parameters can be specified as vectors, allowing the samples to be non-identically distributed. For example, we can generate 100 samples from the contaminated normal distribution in which a sample is from $N(0, 1)$ with probability 0.95 and otherwise from $N(0, 9)$, by

```
contam <- rnorm( 100, 0, (1 + 2*rbinom(100, 1, 0.05)) )
```

The function `sample` re-samples from a data vector, with or without replacement. It has a number of quite different forms. Here `n` is an integer, `x` is a data vector and `p` is a probability distribution on `1, ..., length(x)` :

`sample(n)`	select a random permutation from $1, \ldots, n$.
`sample(x)`	randomly permute `x`.
`sample(x, replace=T)`	a bootstrap sample.
`sample(x, n)`	sample n items from `x` without replacement.
`sample(x, n, replace=T)`	sample n items from `x` with replacement.
`sample(x, n, replace=T, prob=p)`	probability sample of n items from `x`

The last of these provides a way to sample from an arbitrary (finite) discrete distribution; set `x` to the vector of values and `prob` to the corresponding probabilities.

The numbers produced by these functions are of course *pseudo*-random rather than genuinely random. The state of the generator is controlled by a set of integers stored in the S object `.Random.seed`. Whenever `sample` or an 'r' function is called, `.Random.seed` is read in, used to initialize the generator, and then its current value is written back out at the end of the function call. If there is no `.Random.seed` in the current working directory, one is created with default values. To re-run a simulation the initial value of `.Random.seed` needs to be recorded. For some purposes it may be more convenient to use the function `set.seed` with argument between 1 and 1000, which selects one of 1000 preselected seeds.

Details of the pseudo-random number generator

Some users will want to understand the internals of the pseudo-random number generator; others should skip the rest of this section. Background knowledge on pseudo-random number generators is given by Ripley (1987, chapter 2). We describe the generator which was current at the time of writing and which had been in use for several years; it is of course not guaranteed to remain unchanged.

The current generator is based on George Marsaglia's "Super-Duper" package from about 1973. The generator produces a 32-bit integer whose top 31 bits are divided by 2^{31} to produce a real number in $[0, 1)$. The 32-bit integer is produced by a bitwise exclusive-or of two 32-bit integers produced by separate generators. The C code for the S random number generator is

```
unsigned long congrval, tausval;
static double xuni()
{
        unsigned long n, lambda = 69069;
        congrval = congrval * lambda;
        tausval ^= tausval >> 15;
        tausval ^= tausval << 17;
        n = tausval ^ congrval;
        return ( (((n>>1) & 017777777777)) / 2147483648.);
}
```

The integer `congrval` follows the congruential generator

$$X_{i+1} = 69069 X_i \bmod 2^{32}$$

as unsigned integer overflows are (silently) reduced modulo 2^{32}, that is, the overflowing bits are discarded. As this is a multiplicative generator (no additive constant), its period is 2^{30} and the bottom bit must always be odd (including for the seed).

The integer `tausval` follows a 32-bit Tausworthe generator (of period 4 292 868 097 $< 2^{32} - 1$):

$$b_i = b_{i-p} \operatorname{xor} b_{i-(p-q)}, \qquad p = 32, \quad q = 15$$

This follows from a slight modification of Algorithm 2.1 of Ripley (1987, p. 28). (In fact, the period quoted is true only for the vast majority of starting values; for the remaining 2 099 198 non-zero initial values there are shorter cycles.)

For most starting seeds the period of the generator is $2^{30} \times$ 4 292 868 097 $\approx 6.6 \times 10^{14}$, which is quite sufficient for calculations which can be done in a reasonable time in S. The current values of `congrval` and `tausval` are encoded in the vector `.Random.seed`, a vector of 12 integers in the range $0, \ldots, 63$. If x represents `.Random.seed`, we have

$$\texttt{congrval} = \sum_{i=1}^{6} x_i \, 2^{6(i-1)} \quad \text{and} \quad \texttt{tausval} = \sum_{i=1}^{6} x_{i+6} \, 2^{6(i-1)}$$

5.3 Data summaries

Standard univariate summaries such as `mean`, `median` and `var` are available. The `summary` function returns the mean, quartiles and the number of missing values, if non-zero.

The `var` function will take a data <u>matrix</u> and give the variance-covariance matrix, and `cor` computes the correlations, either from two vectors or a data matrix. There are also standard functions `max`, `min` and `range`.

The function `quantile` computes quantiles at specified probabilities, by default $(0, 0.25, 0.5, 0.75, 1)$ giving a "five number summary" of the data vector. This function linearly interpolates, so if $x_{(1)}, \ldots, x_{(n)}$ is the ordered sample,

$$\texttt{quantile(x, p)} = [1 - (p(n-1) - \lfloor p(n-1) \rfloor)] \, x_{1+\lfloor p(n-1) \rfloor}$$
$$+ [p(n-1) - \lfloor p(n-1) \rfloor] \, x_{2+\lfloor p(n-1) \rfloor}$$

where $\lfloor \; \rfloor$ denotes the 'floor' or integer part of. (This differs from the definitions of a quantile given earlier.)

The functions `mean` and `cor` will compute trimmed summaries using the argument `trim`. More sophisticated robust summaries are discussed in Chapter 8.

Histograms and stem-and-leaf plots

The standard histogram function is `hist(x, ...)` which plots a conventional histogram. More control is available via the extra arguments; `probability=T` gives a plot of unit total area rather than of cell counts, and `style="old"` gives the more conventional representation by outlines of bars rather than grey bars.

The argument `nclass` sets the number of bins, and `breaks` specifies the breakpoints between bins. The default for `nclass` is $\log_2 n + 1$. This is known as Sturges' formula, and was based on a histogram of a normal distribution (Scott, 1992, p. 48). Non-normal distributions need more bins; for example Doane's rule is to use $\log_2 n + 1 + \log_2(1 + \hat{\gamma}\sqrt{n/6})$ where $\hat{\gamma}$ is an estimate of the kurtosis.

Sturges' formula corresponds to a bin width of $\text{range}(x)/(\log_2 n + 1)$, and this is often too wide. Note that outliers may inflate the range dramatically and so increase the bin width in the centre of the distribution. Two rules based on compromises between the bias and variance of the histogram for a reference normal distribution are to choose bin width as

$$h = 3.5\hat{\sigma}n^{-1/3} \qquad (5.1)$$

$$h = 2Rn^{-1/3} \qquad (5.2)$$

due to Scott (1979) and Freedman & Diaconis (1981) respectively. Here $\hat{\sigma}$ is the estimated standard deviation and R the inter-quartile range. The Freedman–Diaconis formula is immune to outliers, and chooses rather smaller bins than the Scott formula. We program these to give the number of classes and leave `hist` to choose 'pretty' breakpoints.

```
nclass.scott <- function(x)
{
    h <- 3.5 * sqrt(var(x)) * length(x)^{-1/3}
    ceiling(diff(range(x))/h)
}
nclass.FD <- function(x)
{
    r <- quantile(x, c(0.25, 0.75))
    names(r) <- NULL # to avoid the label 75%
    h <- 2 * (r[2] - r[1]) * length(x)^{-1/3}
    ceiling(diff(range(x))/h)
}
hist.scott <- function(x, ...)
    invisible(hist(x, nclass.scott(x),
    xlab = deparse(substitute(x)), ...))
hist.FD <- function(x, ...)
    invisible(hist(x, nclass.FD(x),
    xlab = deparse(substitute(x)), ...))
```

These are not always satisfactory, as Figure 5.2 shows. Column `eruptions` of our data frame `faithful` gives the duration (in minutes) of 272 eruptions of the Old Faithful geyser in the Yellowstone National Park (from Härdle, 1991); chem is discussed later in this section and tperm in Section 5.6. (S-PLUS has a different version of this dataset as geyser, from Azzalini & Bowman, 1990.)

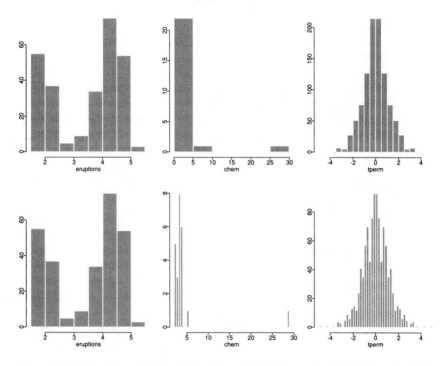

Figure 5.2: Histograms with bin widths chosen by the Scott rule (5.1) (top row) and the Freedman-Diaconis rule (5.2) (bottom row) for datasets faithful$eruptions, chem and tperm.

Scott (1992, §3.3.1) suggests a lower bound $\sqrt[3]{2n}$ for the number of bins and the upper bound $2.61Rn^{-1/3}$ for the bin width. More sophisticated rules for choosing the bin widths are available which make use of the 'shape' of the histogram; see Scott (1992, §3.3.2). However, we regard it as more important to use a better estimator of the density function and so defer further discussion to Section 5.5.

A *stem-and-leaf* plot is an enhanced histogram. The data are divided into bins, but the 'height' is replaced by the next digits in order. We apply this to swiss.fertility, the standardized fertility measure for each of 47 French-speaking provinces of Switzerland at about 1888.

```
> stem(swiss.fertility)

N = 47    Median = 70.4
Quartiles = 64.4, 79.3

Decimal point is 1 place to the right of the colon
```

```
3 : 5
4 : 35
5 : 46778
6 : 024455555678899
7 : 00222345677899
8 : 0233467
9 : 222
```

Apart from giving a visual picture of the data, this gives more detail. The actual data, in sorted order, are 35, 43, 45, 54, ... and this can be read off the plot. Sometimes the pattern of numbers (all odd? many 0s and 5s?) gives clues. Quantiles can be computed (roughly) from the plot. If there are outliers, they are marked separately:

```
> stem(chem)

N = 24    Median = 3.385
Quartiles = 2.75, 3.7

Decimal point is at the colon

    2 : 22445789
    3 : 00144445677778
    4 :
    5 : 3

High: 28.95
```

(This dataset on measurements of copper in flour, Analytical Methods Committee (1989a), is discussed further in Chapter 8.) Sometimes the result is less successful, and manual override is needed:

```
> stem(abbey)

N = 31    Median = 11
Quartiles = 8, 16

Decimal point is at the colon

     5 : 2
     6 : 59
     7 : 0004
     8 : 00005
     ....
    26 :
    27 :
    28 : 0

High:    34   125
```

```
> stem(abbey, scale=-1)

N = 31   Median = 11
Quartiles = 8, 16

Decimal point is 1 place to the right of the colon

   0 : 56777778888899
   1 : 011224444
   1 : 6778
   2 : 4
   2 : 8

High:   34  125
```

Here the `scale` argument sets the backbone to be tens rather than units. The `nl` argument controls the number of rows per backbone unit as 2, 5 or 10. The details of the design of stem-and-leaf plots are discussed by Mosteller & Tukey (1977), Velleman & Hoaglin (1981) and Hoaglin, Mosteller & Tukey (1983).

Boxplots

A *boxplot* is a way to look at the overall shape of a set of data. The central box shows the data between the 'hinges' (roughly quartiles), with the median represented by a line. 'Whiskers' go out to the extremes of the data, and very extreme points are shown by themselves.

```
par(mfrow=c(1,4))
boxplot(chem, sub="chem", range=0.5)
boxplot(abbey, sub="abbey")
boxplot(chem, sub="chem: ATT",  style.bxp="att")
boxplot(abbey, sub="abbey: ATT", style.bxp="att")
```

There is a bewildering variety of optional parameters to `boxplot` documented in the on-line help page. Two basic styles from S-PLUS and 'pure' S (but available in S-PLUS) are shown in Figure 5.3. Note how these plots are dominated by the outliers. It is also possible to plot boxplot for groups side-by-side; see Figure 5.4.

5.4   Classical univariate statistics

S-PLUS has a section on classical statistics. The same S-PLUS functions are used to perform tests and to calculate confidence intervals.

Table 5.2 shows the amounts of shoe wear in an experiment reported by Box, Hunter & Hunter (1978). There were two materials (A and B) that were randomly assigned to the left and right shoes of 10 boys. We will use these data to illustrate one-sample and paired and unpaired two-sample tests. (The rather voluminous output has been edited.)

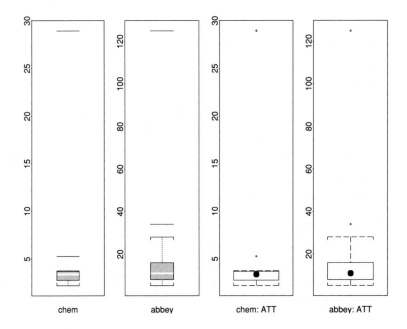

Figure 5.3: Boxplots for the chem and abbey data. The left pair are S-PLUS style, the right pair AT&T style.

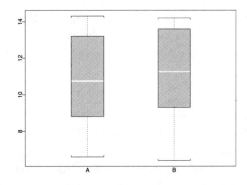

Figure 5.4: Boxplots of the shoe-wear dataset.

```
> shoes <- scan(, list(A=0, B=0))
1: 13.2 14.0
    ....
21:
> attach(shoes)
> boxplot(shoes)
```

This reads in the data and attaches the dataset so we can refer to A and B directly. First we test for a mean of 10 and give a confidence interval for the mean, both for material A.

Table 5.2: Data on shoe wear from Box, Hunter & Hunter (1978).

boy	A		B	
1	13.2	(L)	14.0	(R)
2	8.2	(L)	8.8	(R)
3	10.9	(R)	11.2	(L)
4	14.3	(L)	14.2	(R)
5	10.7	(R)	11.8	(L)
6	6.6	(L)	6.4	(R)
7	9.5	(L)	9.8	(R)
8	10.8	(L)	11.3	(R)
9	8.8	(R)	9.3	(L)
10	13.3	(L)	13.6	(R)

```
> t.test(A, mu=10)

           One-sample t-Test

data:   A
t = 0.8127, df = 9, p-value = 0.4373
alternative hypothesis: true mean is not equal to 10
95 percent confidence interval:
  8.8764 12.3836
sample estimates:
 mean of x
      10.63

> t.test(A)$conf.int
[1]   8.8764 12.3836
attr(, "conf.level"):
[1] 0.95

> wilcox.test(A, mu=10)

          Exact Wilcoxon signed-rank test

data:   A
signed-rank statistic V = 34, n = 10, p-value = 0.5566
alternative hypothesis: true mu is not equal to 10
```

Next we consider two-sample paired and unpaired tests, the latter assuming
equal variances (the default) or not. Note that we are using this example for
illustrative purposes; only the paired analyses are really appropriate.

```
> var.test(A, B)

          F test for variance equality
```

```
data:   A and B
F = 0.9474, num df = 9, denom df = 9, p-value = 0.9372
95 percent confidence interval:
 0.23532 3.81420
sample estimates:
 variance of x variance of y
         6.009          6.3427


> t.test(A, B)

        Standard Two-Sample t-Test

data:   A and B
t = -0.3689, df = 18, p-value = 0.7165
95 percent confidence interval:
 -2.7449  1.9249
sample estimates:
 mean of x mean of y
     10.63      11.04


> t.test(A, B, var.equal=F)

        Welch Modified Two-Sample t-Test

data:   A and B
t = -0.3689, df = 17.987, p-value = 0.7165
95 percent confidence interval:
 -2.745  1.925
     ....


> wilcox.test(A, B)

        Wilcoxon rank-sum test

data:   A and B
rank-sum normal statistic with correction Z = -0.5293,
   p-value = 0.5966


> t.test(A, B, paired=T)

        Paired t-Test

data:   A and B
t = -3.3489, df = 9, p-value = 0.0085
95 percent confidence interval:
 -0.68695 -0.13305
sample estimates:
 mean of x - y
         -0.41
```

```
> wilcox.test(A, B, paired=T)

        Wilcoxon signed-rank test

data:  A and B
signed-rank normal statistic with correction Z = -2.4495,
    p-value = 0.0143
```

The sample size is rather small, and one might wonder about the validity of
the t–distribution. An alternative for a randomized experiment such as this is to
base inference on the permutation distribution of d = B-A. Figure 5.5 shows that
the agreement is very good. Its computation is discussed in Section 5.6.

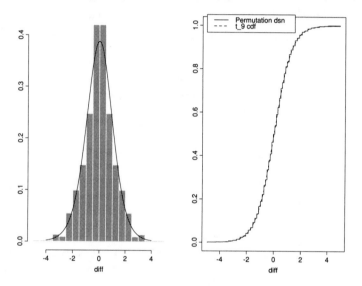

Figure 5.5: Histogram and empirical CDF of the permutation distribution of the paired
t–test in the shoes example. The density and CDF of t_9 are shown overlaid.

The full list of classical tests is:

binom.test	chisq.test	cor.test	fisher.test
friedman.test	kruskal.test	mantelhaen.test	mcnemar.test
prop.test	t.test	var.test	wilcox.test

Many of these have alternative methods—for cor.test there are methods
"pearson", "kendall" and "spearman". We have already seen one- and
two-sample versions of t.test and wilcox.test, and var.test which compares the variances of two samples. The function cor.test tests for non-zero
correlation between two samples, either classically or via ranks.

The Kruskal–Wallis test implemented by kruskal.test is a non-parametric
version of a one-way analysis of variance. The Friedman test is a non-parametric

version of the analysis of an unreplicated randomized block designed experiment. The remaining tests are for count data.

5.5   Density estimation

The non-parametric estimation of probability density functions is a large topic; several books have been devoted to it, notably Silverman (1986), Härdle (1991) and Scott (1992). The methods used are closely related to some of the smoothing methods discussed in Sections 10.1 and 14.1.

The histogram with `probability=T` is of course an estimator of the density function. We would do better to plot it as a *frequency polygon*, joining the mid-points of the bars by straight lines:

```
frequency.polygon <- function(x, nclass = nclass.freq(x),
    xlab="", ylab="", ...)
{
    hst <- hist(x, nclass, probability=T, plot=F, ...)
    midpoints <- 0.5 * (hst$breaks[-length(hst$breaks)]
                   + hst$breaks[-1])
    plot(midpoints, hst$counts, type="l", xlab=xlab, ylab=ylab)
}
nclass.freq <- function(x)
{
    h <- 2.15 * sqrt(var(x)) * length(x)^{-1/5}
    ceiling(diff(range(x))/h)
}
```

As the frequency polygon is smoother than the histogram on which it is based, we can afford to use larger bins. The equivalent of Scott's rule becomes (Scott, 1992, p. 99)

$$h = 2.15\hat{\sigma}n^{-1/5} \tag{5.3}$$

Because of the different power, this formula only gives larger h for $n \geqslant 39$, and the difference is small until n reaches 1000.

The histogram and consequently the frequency polygon depend on the starting point of the grid of bins. The effect can be surprisingly large; see Figure 5.6. The figure also shows that by averaging the histograms and then drawing a frequency polygon we can obtain a much clearer view of the distribution.

```
attach(faithful)
par(mfrow=c(2,3))
hist(eruptions, , seq(1, 6, 0.5), prob=T,
    xlim=c(1, 6), ylim=c(0, 0.6))
hist(eruptions, , 0.1+seq(1, 6, 0.5), prob=T,
    xlim=c(1, 6), ylim=c(0, 0.6))
hist(eruptions, , 0.2+seq(1, 6, 0.5), prob=T,
    xlim=c(1, 6), ylim=c(0, 0.6))
hist(eruptions, , 0.3+seq(1, 6, 0.5), prob=T,
```

```
    xlim=c(1, 6), ylim=c(0, 0.6))
hist(eruptions, , 0.4+seq(1, 6, 0.5), prob=T,
    xlim=c(1, 6), ylim=c(0, 0.6))

breaks <- seq(1, 6.4, 0.1)
counts <- numeric(length(breaks))
for(i in (0:4)) counts[i+(1:50)] <- counts[i+(1:50)] +
    rep(hist(eruptions, , 0.1*i+ seq(1, 6, 0.5),
    prob=T, plot=F)$counts, rep(5,10))
plot(breaks+0.05, counts/5, type="l", xlab="eruptions",
    ylab="averaged", bty="n", xlim=c(1, 6), ylim=c(0, 0.6))
```

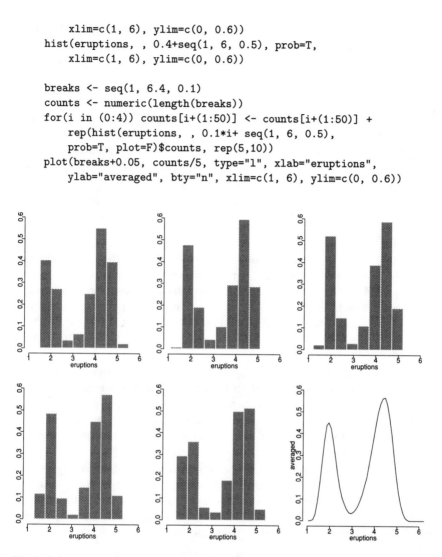

Figure 5.6: Five shifted histograms with bin width 0.5 and the frequency polygon of their average, for the Old Faithful geyser `eruptions` data.

This idea of an *average shifted histogram* or *ASH* density estimate is a useful one and is discussed in detail in Scott (1992). Suppose we average over m shifts by $\delta = h/m$. Another way to look at the result is to take a histogram with bin width δ, and to compute the histograms of bin width h by aggregating counts in m adjacent bins. Specifically, if the counts are ν_k, the ASH for a point x in small bin k is

$$\hat{f}(x) = \frac{1}{nh} \sum_{i=1-m}^{m-1} \left[1 - \frac{|i|}{m}\right] \nu_{k+i} = \frac{1}{nh} \sum_{i} \left[1 - \frac{|i\delta|}{h}\right]_{+} \nu_{k+i}$$

so is a weighted average of the bin counts with a triangular weighting function. Now suppose we let $m \to \infty$. The counts in bins reduce to zero or one, so the limit is

$$\hat{f}(x) = \frac{1}{nh} \sum_{j=1}^{n} \left[1 - \frac{|x - x_j|}{h} \right]_+$$

This is a *kernel density* estimate with a triangular kernel. As well as motivating kernel estimators, ASH and its extension *WARP* discussed in Härdle (1991) provide computationally attractive approximate algorithms for density estimation.

The only built-in S function specifically for density estimation is `density`. This implements a kernel-density smoother of the form

$$\hat{f}(x) = \frac{1}{b} \sum_{j=1}^{n} K \left(\frac{x - x_j}{b} \right) \tag{5.4}$$

for a sample $x_1, \ldots, x_n$, a fixed kernel $K()$ and a bandwidth b. The kernel is normally chosen to be a probability density function; the default is the normal (argument `window="g"` for Gaussian), with alternatives "rectangular", "triangular" and "cosine" (the latter being $(1 + \cos \pi x)/2$ over $[-1, 1]$).[1] The bandwidth is the length of the non-zero section for the alternatives, and four times the standard deviation for the normal. (*Note* that these definitions are twice and four times those most commonly used.)

The choice of bandwidth is a compromise between smoothing enough to remove insignificant bumps and not smoothing too much to smear out real peaks. Mathematically there is a compromise between the bias of $\hat{f}(x)$, which increases as b is increased, and the variance which decreases. Theory suggests that the bandwidth should be proportional to $n^{-1/5}$, but the constant of proportionality depends on the unknown density.

We can illustrate this effect with the permutation distribution of Figure 5.5, which is stored in vector `tperm`. Under-smoothing (`width=0.2`) shows spurious bumps (Figure 5.7); over-smoothing (`width=1.5`) has no bumps in the tails but reduces the height of the main peak. By default the density is computed at 50 equally-spaced points, which may be rather too few.

```
tperm <- test.t.2(B-A)   # see Section 5.6
par(mfrow=c(2,2))
x <- seq(-5, 5, 0.1)
plot(c(-5, 5), c(0, 0.45), type="n", bty="l",
    sub="default", xlab="", ylab="")
lines(x, dt(x,9), lty=2)
lines(density(tperm))
plot(c(-5, 5), c(0, 0.45), type="n", bty="l",
    sub="width=0.2", xlab="", ylab="")
lines(x, dt(x,9), lty=2)
lines(density(tperm, width=0.2, from=-5, to=5))
    . . . .
```

---

[1] established by plotting; Härdle (1991) and Scott (1992) have a different cosine kernel $(\pi/4) \cos(\pi x/2) I(|x| \leqslant 1)$.

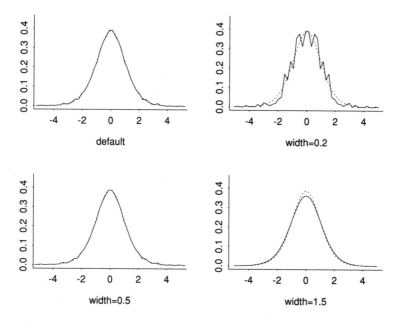

Figure 5.7: Density plots for the permutation distribution for the t statistic for the difference in shoe materials. In each case the reference t_9 density is shown dotted.

The default width b is given by a rule of thumb

$$\frac{\text{range}\,(x)}{2(1 + \log_2 n)}$$

which happens to work quite well in this example (with $b \approx 0.45$). (This seems to be based on Sturges' rule and a rough equivalence with a histogram density estimate of bin width equal to the bandwidth.) There are better-supported rules of thumb such as

$$\hat{b} = 1.06 \min(\hat{\sigma}, R/1.34)n^{-1/5} \qquad (5.5)$$

for the IQR R and the Gaussian kernel of bandwidth the standard deviation, to be quadrupled for use in `density` (Silverman, 1986, pp. 45–47).

We can relate the bandwidths of different kernels by the constants given in Table 5.3; kernels with bandwidth $c \times h$ will give estimates of comparable smoothness. This enables rules such as (5.5) to be used for other kernels. Since the rectangular kernel is non-differentiable, its estimates will be subjectively rougher. (Table 5.3 is calculated from results in Härdle (1991) and Scott (1992); it includes other commonly-used kernels implemented in S by Härdle (1991).)

We return to the geyser eruptions data. This is an example in which the default bandwidth, 0.19, is too small, about 12% of the value suggested by (5.5), which is too large (Figure 5.8). Scott (1992, §6.5.1.2) suggests that the value

$$b_{OS} = 1.144\hat{\sigma}n^{-1/5}$$

Table 5.3: Bandwidth constants for various kernels scaled as used in `density`; kernels with bandwidth $c \times h$ will give comparable smoothness.

Rectangular	1	Triweight	1.711
Triangular	1.393	Gaussian	1.15
Epanechnikov	1.272	Cosine	1.577
Quartic	1.507		

provides an upper bound on the (Gaussian) bandwidths one would want to consider, again to be quadrupled for use in `density`. The normal reference (5.5) is only slightly below this upper bound, and will be too large for densities with multiple modes.

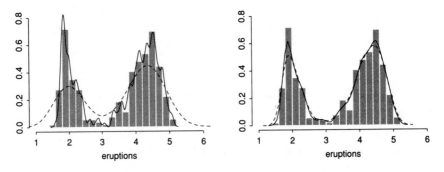

Figure 5.8: Density plots for the Old Faithful eruptions data. Superimposed on a histogram are the default width (left, solid), normal reference bandwidth (left, dashed) and the UCV (right, solid) and BCV (right, dashed) kernel estimates with a Gaussian kernel.

```
attach(faithful)
hist(eruptions, 15, prob=T, xlim=c(1,6), ylim=c(0, 0.8))
lines(density(eruptions, n=200))
bandwidth.nrd <- function(x)
{
    r <- quantile(x, c(0.25, 0.75))
    h <- (r[2] - r[1])/1.34
    4 * 1.06 * min(sqrt(var(x)), h) * length(x) ^ {-1/5}
}
bandwidth.nrd(eruptions)
[1] 1.5772
lines(density(eruptions, width=1.6), lty=3)
bandwidth.nrd(tperm)
[1] 1.1536
```

Note that the data in this example are (negatively) serially correlated, so the theory for independent data must be viewed as only a guide.

Ways to find automatically compromise value(s) of b are discussed by Härdle (1991) and Scott (1992). These are based on estimating the *mean integrated square error*

$$MISE = E \int |\hat{f}(x;b) - f(x)|^2 \, dx = \int E|\hat{f}(x;b) - f(x)|^2 \, dx$$

and choosing the smallest value as a function of b. We can expand $MISE$ as

$$MISE = \int \hat{f}(x;b)^2 \, dx - 2E\hat{f}(X;b) + \int f(x)^2 \, dx$$

where the third term is constant and so can be dropped.

The estimates are based on *cross-validation*, that is using the remaining observations to fit a density at each observation in turn. (See. for example, Stone, 1974.) Let the fitted value be $\hat{f}(x_i; b)_{-i}$. Then the *unbiased cross-validation* estimates $E\hat{f}(X;b)$ by the mean of $\hat{f}(x_i; b)_{-i}$ to give

$$UCV(b) = \int \hat{f}(x;b)^2 \, dx - \frac{2}{n} \sum_{j=1}^{n} \hat{f}(x_i; b)_{-i}$$

The minimizer $\hat{b}$ is found to be accurate but rather variable. A more involved approximation gives the *biased cross-validation* criterion $BCV(b)$ which is minimized over $b \leqslant b_{OS}$, and is usually less variable.

Scott (1992, §6.5.13) provides formulae for the Gaussian kernel which we can evaluate via in the functions ucv and bcv of our library. (These contain typographic errors which we have corrected.) As these involve calculations over all pairs of points, they should not be used for very large datasets (say thousands of observations). ASH and WARP approximations could be used, but these functions take another approach, and estimate the bandwidth from a sub-sample, then scale the bandwidth proportionally to $n^{-1/5}$. (For the results here we turned off subsampling to get repeatable results. This is not recommended unless the C version of these routines is installed.)

```
> c( ucv(eruptions,0),  bcv(eruptions,0) )
[1] 0.41062 0.63201
> hist(eruptions, 15, prob=T, xlim=c(1, 6), ylim=c(0, 0.8))
> lines(density(eruptions, width=0.41))
> lines(density(eruptions, width=0.63), lty=3)
```

which shows both criteria in this example provide visually sensible values, Figure 5.8.

Another approach is to make an asymptotic expansion of $MISE$ and estimate its terms, which involve the integral $\int (f'')^2$. This in turn is estimated using the second derivative of a kernel estimator with a different bandwidth, chosen by repeating the process, this time using a reference bandwidth. Details are given by Sheather & Jones (1991) and implemented for the Gaussian kernel in our function width.SJ.

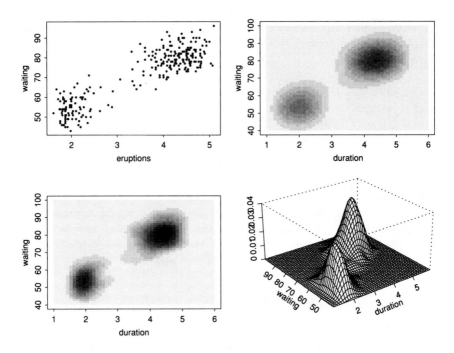

Figure 5.9: Scatter plot and two-dimensional density plots of the bivariate Old Faithful geyser data.

```
> width.SJ(eruptions, 0)
[1] 0.61003
> c( ucv(tperm, 0), bcv(tperm, 0), width.SJ(tperm, 0) )
[1] 1.1706 1.0809 1.1286
```

End effects

Most density estimators will not work well when the density is non-zero at an end of its support, such as the exponential and half-normal densities. (They are designed for continuous densities and this is discontinuity.) One trick is to reflect the density and sample about the endpoint, say a. Thus we compute the density for the sample c(x, 2a-x), and take double its density on $[a, \infty)$ (or $(-\infty, a]$ for an upper end). This will impose a zero derivative on the estimated density at a, but the end effect will be much less severe. For details and further tricks see Silverman (1986, §3.10).

Two-dimensional data

It is often useful to look at densities in two dimensions. Visualizing in more dimensions is difficult, although Scott (1992) provides some examples of visualization of three-dimensional densities.

   The dataset on the Old Faithful geyser has two components, the `eruptions` duration which we have studied, and `waiting`, the waiting time in minutes until the next eruption. There is also evidence of non-independence of the durations. S provides a function `hist2d` for two-dimensional histograms, but its output is too rough to be useful. We can apply the *ASH* device here too, using a two-dimensional grid of bins; we use code from David Scott's library `ash`[2] available from `statlib`. The results are shown in Figures 5.9 and 5.10.

```
attach(faithful)
library(ash)
par(mfrow=c(2,2))
plot(eruptions, waiting)
faithful.bin <- bin2(cbind(x=eruptions, y=waiting),
    matrix( c(1,6,40,100), 2, 2, byrow=T), c(50,50))
faithful.1 <- ash2(faithful.bin, m = c(10,10))
image(faithful.1, zlim=c(0, 0.04), xlab="duration",
    ylab="waiting")
faithful.2 <- ash2(faithful.bin, m = c(5,5))
image(faithful.2, zlim=c(0, 0.04), xlab="duration",
    ylab="waiting")
persp(faithful.2, xlab="duration", ylab="waiting", zlab="")

plot(eruptions[-272], eruptions[-1],
    xlab="previous duration", ylab="duration")
faithful.b2 <- bin2(cbind(x=eruptions[-272],
    y=eruptions[-1]),  matrix( c(1,6,1,6), 2, 2, byrow=T),
    c(90,90))
contour(ash2(faithful.b2, m=c(15,15)),xlab="previous duration",
    ylab="duration", levels =c(0.05, 0.1, 0.2, 0.4) )
contour(ash2(faithful.b2, m=c(7,7)), xlab="previous duration",
    ylab="duration", levels =c(0.05, 0.1, 0.2, 0.4) )
contour(ash2(faithful.b2, m=c(5,5)), xlab="previous duration",
    ylab="duration", levels =c(0.05, 0.1, 0.2, 0.4) )
```

An alternative approach is sketched in Exercise 5.2.

5.6   Bootstrap and permutation methods

Several modern methods of what is often called *computer-intensive statistics* make use of extensive repeated calculations to explore the sampling distribution of a parameter estimator $\hat{\theta}$. Suppose we have a random sample $x_1, \ldots, x_n$ drawn independently from one member of a parametric family $\{F_\theta \mid \theta \in \Theta\}$ of distributions. Suppose further that $\hat{\theta} = T(\mathbf{x})$ is a *symmetric* function of the sample, that is does not depend on the sample order.

   The *bootstrap* procedure (Efron, 1982; Efron & Tibshirani, 1993) is to take m samples from $\mathbf{x}$ *with replacement* and to calculate $\hat{\theta}_j$ for these samples. Note that

---

[2] which uses FORTRAN code.

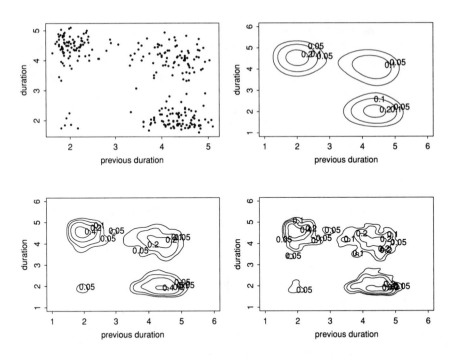

Figure 5.10: Scatter plot and two-dimensional density plots of previous and current duration for the Old Faithful geyser data. The contour plots show the effect of varying the amount of smoothing.

the new samples consist of an integer number of copies of each of the original data points, and so will normally have ties. Efron's idea was to assess the variability of $\hat{\theta}$ about the unknown true θ by the variability of $\hat{\theta}_i$ about $\hat{\theta}$. In particular, the bias of $\hat{\theta}$ may be estimated by the mean of $\hat{\theta}_i - \hat{\theta}$.

As an example, suppose that we needed to know the median m of the eruptions data. The obvious estimator is the sample median, which is 4. How accurate is this estimator? The large-sample theory says that the median is asymptotically normal with mean m and variance $1/4n\,f(m)^2$. But this depends on the unknown density at the median. We can use our best density estimators to estimate $f(m)$, but as we have seen we can find considerable bias and variability if we are unlucky enough to encounter a peak (as in a unimodal symmetric distribution). Let us try the bootstrap:

```
density(eruptions, n=1, from=4, to=4.01, width=0.41)$y
[1] 0.42081
density(eruptions, n=1, from=4, to=4.01, width=0.63)$y
[1] 0.41028
1/(2*sqrt(length(eruptions))*0.415)
[1] 0.073053
set.seed(101); m <- 1000
res <- numeric(m)
```

```
for (i in 1:m) res[i] <- median(sample(eruptions, replace=T))
mean(res - median(eruptions))
[1] -0.014813
sqrt(var(res))
[1] 0.079443
```

which takes about a minute and confirms the adequacy of the large-sample mean
and variance for our example (assuming independence).

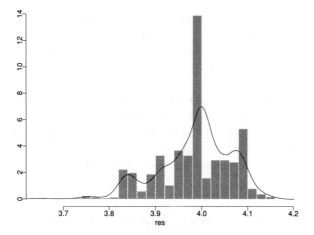

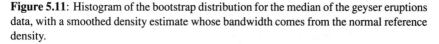

Figure 5.11: Histogram of the bootstrap distribution for the median of the geyser eruptions
data, with a smoothed density estimate whose bandwidth comes from the normal reference
density.

The bootstrap distribution of $\hat{\theta}_i$ about $\hat{\theta}$ is far from normal (see Figure 5.11).
It is a discrete distribution, but its 'shape' is quite different from that of a normal:

```
> hist.FD(res, prob=T)
> lines(density(res, n=200, width=bandwidth.nrd(res)))
> c(ucv(res, 0), bcv(res, 0), width.SJ(res, 0))
[1] 0.0095 0.0908 0.0056
> quantile(res, c(0.025, 0.975))
  2.5%  97.5%
 3.833 4.1085
```

which will take several minutes (but sampling could be used), with a modified
function width.SJ. Note the great discrepancy in the suggested bandwidths: the
normal reference gives 0.0727 and b_{OS} is 0.0913. The difficulty arises from the
discreteness of the bootstrap distribution, and also affects data with tied values
(for example, after rounding).

In fact, in this example the bootstrap resampling can be avoided, for the
bootstrap distribution of the median can be found analytically (Efron 1982, Chapter
10; Staudte & Sheather, 1990, p. 84), at least for odd n.

One approach to a confidence interval for the median is to use the quantiles
given here; this is termed the *percentile confidence interval* and sometimes a *Monte*

Carlo confidence interval (see Ripley, 1987, p. 176). More complex intervals are discussed in the references, and S code from Efron & Tibshirani (1993) is available from `statlib`[3] to compute these.

Permutation tests

Inference for designed experiments is often based on the distribution over the random choices made during the experimental design, on the belief that this randomness alone will give distributions which can be approximated well by those derived from normal-theory methods. There is considerable empirical evidence that this is so, but with modern computing power we can check it for our own experiment, by selecting a large number of re-labellings of our data and computing the test statistics for the re-labelled experiments.

Consider again the shoe-wear data of Section 5.4. The most obvious way to explore the permutation distribution of the t-test of d = B-A is to select random permutations. The supplied function `t.test` computes much more than we need, and so is rather slow (about 0.2 secs on a Sun SparcStation IPC). In this case it is simple to write a replacement function to do exactly what we need:

```
attach(shoes)
d <- B-A
ttest <- function(x) mean(x)/sqrt(var(x)/length(x))
n <- 1000
res <- numeric(n)
for(i in 1:n) {
        x <- d*sign(runif(10)-0.5)
        res[i] <- ttest(x)
}
```

which takes around 2 minutes on a Sun SparcStation IPC.

As the permutation distribution has only $2^{10} = 1024$ points we can explore it directly for Figure 5.5:

```
test.t.1 <- function(d) {
    binary <- function(x, digits = if(x > 0) 1 +
                floor(log(x, base = 2))      else 1)
    {
            ans <- 0:(digits - 1)
            (x %/% 2^ans) %% 2
    }
        digits <- length(d)
        n <- 2^digits
        perm.res <- numeric(n)
        for(i in 1:n) {
                x <- d * 2 * (binary(i, digits = digits) - 0.5)
                perm.res[i] <- ttest(x)
        }
```

---

[3] See Appendix D.

```
                perm.res
}
tperm <- test.t.1(B-A)
par(mfrow=c(1,2))
hist.scott(tperm, probability=T, xlab="diff")
# 25 bins corresponds to Scott's formula
x <- seq(-4,4, 0.1)
lines(x, dt(x,9))
sres <- c(sort(tperm), 4)
yres <- (0:1024)/1024
plot(sres, yres, type="S", xlab="diff", ylab="")
lines(x, pt(x,9), lty=3)
legend(-5, 1.05, c("Permutation dsn","t_9 cdf"), lty=c(1,3))
```

This is slower than one would like (3 minutes) despite having been streamlined considerably.[4] However, as the t-test is the ratio of linear operations, we can do many at once via matrix operations, taking only 1.2 sec! We need a vector form of binary:

```
binary.v <- function(x, digits) {
    if(missing(digits)) {
    mx <- max(x)
        digits <- if(mx > 0) 1 + floor(log(mx, base = 2))
        else 1
    }
    ans <- 0:(digits - 1)
    lx <- length(x)
    x <- rep(x, rep(digits, lx))
    x <- (x %/% 2^ans) %% 2
    dim(x) <- c(digits, lx)
    x
}
test.t.2 <- function(d) {
    # ttest is function(x) mean(x)/sqrt(var(x)/length(x))
    digits <- length(d)
    n <- 2^digits
    x <- d * 2 * (binary.v(1:n, digits) - 0.5)
    s <- matrix(1/digits, 1, digits)
    mx <- s %*% x
    s <- matrix(1/(digits - 1), 1, digits)
    vx <- s %*% (x - matrix(mx, digits, n, byrow=T))^2
    as.vector(mx/sqrt(vx/digits))
}
```

The speed-up in this example is unusually large, but it is always worth looking for simplifying ideas which allow a large number of calculations to be done by one function.

---

[4] We are grateful to Bill Dunlap of StatSci for his suggestions for this problem.

5.7   Exercises

5.1.   Experiment with our dataset `galaxies`. How many modes do you think there are in the underlying density?

5.2.   We can apply two-dimensional kernel analysis directly, as well as using *ASH*. This is most straightforward for the normal kernel aligned with axes, that is with variance $\mathrm{diag}(h_x^2, h_y^2)$. Then the kernel estimate is

$$f(x, y) = \frac{\sum_s \phi\big((x - x_s)/h_x\big)\phi\big((y - y_s)/h_y\big)}{n\, h_x h_y}$$

which can be evaluated on a grid as XY^T where $X_{is} = \phi\big((gx_i - x_s)/h_x\big)$ and (gx_i) are the grid points, and similarly for Y. Write a function `kde2d` to implement this, and try it on the `faithful` examples. How would you choose the bandwidths?

Chapter 6

Linear Statistical Models

Linear models form the core of classical statistics, and S provides extensive facilities to fit and manipulate them. These work with a version of the Wilkinson-Rogers syntax (Wilkinson & Rogers, 1973) for specifying models which we discuss in the Section 6.2, and which is also used for generalized linear models, models for survival analysis and tree-based models in later chapters. The main function for fitting linear models is `lm`, which provides our first example of a style of S functions we shall see repeatedly in later chapters, producing a fitted model object which is then analysed by generic functions.

The analysis of designed experiments by least-squares can be seen as a special case of linear regression, but there are different conventions for displaying the results. The S function `aov` follows that tradition.

We will consider a simple regression example before exploring the full complexity of formulae for linear models.

6.1 A linear regression example

Fisher (1947) published a dataset of H. G. O. Holck giving the heart weights, body weights and sex of 47 female and 97 male adult cats. These had been used in bioassay experiments with the muscle-relaxing drug digitalis. The dataset is given in data frame `cats` in our library with three columns, `Sex`, `Bwt` and `Hwt`.

It appears that cats with a body weight of at least 2 kilograms were selected. Fisher's interest in the data was to study the variation of heart weight with body weight and to discover if this differs by sex. Since these questions pertain to the conditional distribution of heart weight given body weight, the lower limit on body weight does not cause an inferential problem. It might be considered a design flaw, particularly for the female cats which are on average smaller than the males.

The ratio of average heart weight to average body weight and the average ratio are both remarkably close for the two sexes:

```
> attach(cats)
> tapply(Hwt, Sex, mean)/tapply(Bwt, Sex, mean)
     F      M
3.8999 3.9044
```

```
> tapply(Hwt/Bwt, Sex, mean)
      F       M
  3.9151  3.8945
```

Fisher pointed out that for separate linear regressions of heart weight on body weight the intercept term is non-significant for males but (borderline) significant for females, so can heart weight be considered proportional to body weight, at least for females over 2 kg? We investigate this and related questions as an introduction to the linear model function `lm`.

The main arguments for the function `lm` are

```
lm(formula, data, weights, subset, na.action)
```

where

`formula` is the model formula (the only required argument),

`data` is an optional data frame,

`weights` is a vector of positive weights, if non-uniform weights are needed,

`subset` is a vector specifying a subset of the data is to be used (by default all rows are used),

`na.action` is a function specifying how missing values are to be handled (by default, missing values are not allowed).

If argument `data` is specified it gives a data frame from which variables are selected ahead of the search list. Working with data frames and using this argument is strongly recommended. For regressions, the formula is of the form

dependent variable ~ regressor(s) separated by +.

Consider fitting the separate linear regression models for male and female cats:

```
catsF <- lm(Hwt ~ Bwt, data=cats, subset = Sex=="F")
catsM <- update(catsF, subset = Sex=="M")
```

The first line fits a linear regression for the females. The males could be fitted in the same way, but we illustrate the function `update`. This has first argument a *fitted model object*, the result of a call to one of the fitting functions. The remaining arguments specify changes to the call which generated the first argument.

Fitted model objects are lists of the appropriate class, in this case `lm`. Generic functions are available to perform further calculations using the information in the object. These include

`print` for a simple display,

`summary` for a conventional regression analysis output,

`coef` (or `coefficients`) for regression coefficients,

`resid` (or `residuals`) for residuals,

`fitted` (or `fitted.values`) for fitted values,

`deviance` for the residual sum of squares,

anova for a sequential analysis of variance table,

predict for predicting means for new data, optionally with standard errors, and

plot for diagnostic plots.

The only component of the fitted object likely to be used at all often for which there is no method function is df.residual, the residual degrees of freedom.

The output from summary is self-explanatory. Edited results for the cats data are:

```
> summary(catsF)
Residuals:
   Min     1Q  Median    3Q  Max
 -3.01 -0.686 -0.0451 0.796 2.22

Coefficients:
             Value Std. Error t value Pr(>|t|)
(Intercept) 2.981     1.485     2.007    0.051
        Bwt 2.636     0.625     4.215    0.000

Residual standard error: 1.16 on 45 degrees of freedom
> summary(catsM)
Residuals:
   Min    1Q Median    3Q  Max
 -3.77 -1.05 -0.298 0.984 4.86

Coefficients:
             Value Std. Error t value Pr(>|t|)
(Intercept) -1.184    0.998    -1.186    0.239
        Bwt  4.313    0.340    12.688    0.000

Residual standard error: 1.56 on 95 degrees of freedom
```

The five-number summaries of the residuals mildly suggest some skewness. It is not stated why Fisher used separate regressions rather than pooling the two samples, but this may be because of the difference in variance. The textbook test for equality of variances does show significance:

```
> vF <- deviance(catsF)/(nF <- catsF$df.resid)
> vM <- deviance(catsM)/(nM <- catsM$df.resid)
> c(Male=vM, Female=vF, F=(f <- vF/vM), Pr=pf(f, nF, nM))
   Male Female      F         Pr
 2.4238 1.3508 0.55733 0.015521
```

but it is now known this test is highly non-robust to non-normality (see for example Hampel *et al.* (1986, pp. 55, 188)).

A plot of the two samples together with the regression lines is given in Figure 6.1. The regression lines are plotted for the weight range for each sex.

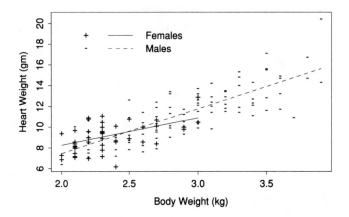

Figure 6.1: Plots of Holck's cat weight data with separate regression lines for each sex.

```
plot(Bwt, Hwt, xlab="Body Weight (kg)",
               ylab="Heart Weight (gm)", type="n")
text(Bwt, Hwt, c("+","-")[Sex])
legend(2.0, 20, c("Females","","Males"), pch="+ -",
               lty=c(2,-1,3), bty="n")
lines(Bwt[Sex=="F"], fitted(catsF), lty=1)
lines(Bwt[Sex=="M"], fitted(catsM), lty=2)
```

The models for the two sexes may be fitted together using the `Sex` factor. Using the `/` formula operator fits separate regression models parametrized as two separate slopes, and removing the implicit intercept term (by `- 1`) overrides the constraints on main effects and parametrizes the model with separate parameters for each intercept. The summary output confirms the slope and intercept estimates, but the standard errors differ as they are based on a pooled estimate of the error variance. Neither intercept is significantly different from zero in this pooled model.[1]

```
> catsMF <- lm(Hwt ~ Sex/Bwt - 1, data=cats)
> summary(catsMF)
    ....
Coefficients:
          Value Std. Error t value Pr(>|t|)
    SexF  2.981    1.843      1.618   0.108
    SexM -1.184    0.925     -1.281   0.202
 SexFBwt  2.636    0.776      3.398   0.001
 SexMBwt  4.313    0.315     13.700   0.000

Residual standard error: 1.44 on 140 degrees of freedom
    ....
```

---

[1] S-PLUS 3.2 rel 1 for Sun Sparc and for Windows incorrectly label the coefficients in this and related models; a correction is in the `fixes` directory on the disk.

We can test for curvature in the regression of heart weight on body weight, as there are many repeated values. We do so by including a term for every weight that occurs, most conveniently by including body weight as a factor in the model:

```
> catsMF1 <- lm(Hwt ~ Sex/factor(Bwt), data=cats,
      singular.ok=T)
> anova(catsMF, catsMF1)
Analysis of Variance Table
    ....
   Resid. Df     RSS     Test Df Sum of Sq F Value   Pr(F)
1           140 291.05
2           114 228.78 1 vs. 2 26   62.262  1.1932 0.25871
```

As this shows anova applied to two or more nested models gives an analysis of variance table for those models. This shows insignificant evidence of lack of fit of the linear regression. Note that singular.ok=T is needed as some weights do not occur for female cats and so the corresponding parameter cannot be fitted.

In a reappraisal of Fisher's analysis, Aitchison (1986, pp. 227ff) suggested a log transform of both heart and body weights, in part to remove (or at least diminish) the impact of skewness and variance heterogeneity. On the log scale a natural hypothesis to investigate is the simple model implied by Fisher's original calculation, namely $H \propto B$. One way to do this is to fit the model

$$\log (H/B) = \beta_0 + \beta_1 \log B + \epsilon$$

for each sex and test the hypothesis $\beta_1 = 0$:

```
> logcats.lm2 <- lm(log(Hwt/Bwt) ~ Sex/log(Bwt), data=cats)
> summary(logcats.lm2)
    ....
Coefficients:
              Value Std. Error t value Pr(>|t|)
(Intercept)   1.429    0.088     16.248   0.000
        Sex  -0.183    0.088     -2.083   0.039
SexFlog(Bwt) -0.301    0.177     -1.702   0.091
SexMlog(Bwt)  0.099    0.084      1.186   0.237

Residual standard error: 0.134 on 140 degrees of freedom
Multiple R-Squared: 0.0303
F-statistic: 1.46 on 3 and 140 degrees of freedom,
                              the p-value is 0.229
    ....
```

The final F-statistic shown suggests the total contribution to the model of all terms additional to the constant term is negligible, suggesting that in both sexes heart weight is proportional to body weight, *and* that the constant of proportionality

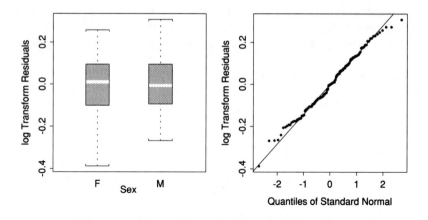

Figure 6.2: Diagnostic plots for the log transformed cats data.

is the same for both. This can be checked formally with a two-step sequential
analysis of variance:

```
> logcats.lm1 <- lm(log(Hwt/Bwt) ~ Sex, data=cats)
> logcats.lm0 <- update(logcats.lm1, . ~ . - Sex)
> anova(logcats.lm0, logcats.lm1, logcats.lm2)
    ....

          Terms Resid. Df    RSS                 Test
1             1         143 2.5818
2           Sex         142 2.5807
3 Sex/log(Bwt)         140 2.5037 +log(Bwt) %in% Sex

    Df Sum of Sq F Value    Pr(F)
1
2    1  0.001119  0.0626  0.80284
3    2  0.077003  2.1529  0.11998

> plot(Sex, resid(logcats.lm0), ylab="log Transform Residuals")
> qqnorm(resid(logcats.lm0), ylab="log Transform Residuals")
> qqline(resid(logcats.lm0))
```

When the update function is used to change the formula, the single dot name,
' . ', may be used to mean "whatever was in the previous model formula at that
point".

Diagnostic plots for the transformed data show little evidence of departures
from the assumptions. Some plots using the final fitted model are shown in
Figure 6.2. Similar plots for the untransformed data show some evidence of both
skewness and variance heterogeneity.

6.2   Model formulae

Consider a multiple regression model with three regressors. In the usual algebraic representation this would be

$$y = \beta_0 + \beta_1 x_1 + \beta_2 x_2 + \beta_3 x_3 + \epsilon$$

where the error term ϵ is modelled as a random variable with zero mean and constant variance, independent for each observed case. To specify such a model it is necessary to give the *response*, y and the *regressors* 1, x_1, x_2 and x_3. The regressor 1 with regression coefficient β_0 is called an *intercept*. Such a linear model is specified in S by a *model formula*, in this case

```
y ~ x1 + x2 + x3
```

to be read as "`y` is modelled as `x1 + x2 + x3`". The intercept (as usual in regression) is implicit.

The left-hand side of a model formula may be a vector or matrix (or an expression that evaluates to one of these). If it is a matrix, multiple linear models are fitted, one to each column, but we do not discuss such models.

The right-hand side terms are used to construct the *model matrix*, X ; they may be vectors or matrices (or factors, discussed below), separated by `+`. Matrices contribute one regression variable per column. The normally implicit intercept term may be explicitly included as a term, `1`:

```
y ~ 1 + x1 + x2 + x3
```

Terms are removed from a model by using a `-` instead of the `+` operator. Thus to remove the intercept we use:

```
y ~ -1 + x1 + x2 + x3
```

Factors (usually) generate columns in X to allow a separate parameter for each level. If `a` is a factor and `x` a numerical vector then `y ~ a` specifies a model for the one-way layout determined by `a`, and `y ~ a + x` is a model for parallel regressions of `y` on `x` within the levels of `a`, that is a model of the form

$$y = \alpha_a + \beta x + \epsilon$$

where the intercept parameters depend on the level of `a`. There are many ways to parametrize a model including factors which give the same fitted values but different coefficients. The details are deferred to a later subsection, but note that the default choices in S are not conventional ones.

Operators used inside arguments to functions within a formula have their usual arithmetic meaning. There is an identity function, `I`, which evaluates its argument and returns the result; this is used within model formulae to allow operators to be used with their arithmetic meaning. For example

```
profit ~ I(dollar.inc + 1.55*pound.inc)
```

regresses profit on total income (at a fixed exchange rate), and avoids the special interpretation of + and, as we shall see, *.

When the calling function has a data argument, ' . ', (a single dot) may be used in the formula to stand for "all other variables in the data frame, joined by plus signs and enclosed in parentheses". This is especially useful with regression models, to indicate the regression of one variable on all the others.

Crossed and nested classifications

For factors a and b the term a:b defines their interaction. Thus a model for a two-way layout may be specified as

    y ˜ a + b + a:b

The term a*b is shorthand for a + b + a:b, that is a full two-factor model with the additive model as an explicit submodel. In algebraic notation this is

$$y = \mu + \alpha_a + \beta_b + \gamma_{ab} + \epsilon$$

The constraints imposed on the coefficients are determined by the coding used (see below).

Colon products may also be used between a factor a and a quantitative variable x, implying a separate parameter for x within each level of a. Thus in algebraic notation y ˜ a + a:x is

$$y = \mu + \alpha_a + \beta_a x + \epsilon$$

where the parameters are indexed by the levels of factor a. The model y ˜ a*x corresponds to

$$y = \mu + \alpha_a + \beta x + \gamma_a x + \epsilon$$

again with constraints on the parameters α_a and γ_a. It is often more natural to use an equivalent nested model (below).

Colon products of more than two terms may be used to define higher-order interaction terms in the model. Star products provide a shorthand for full multi-way models. A term such as a*b*c can be thought of as

    (1 + a):(1 + b):(1 + c) - 1

where the terms are expanded like an algebraic product and the terms arranged by degree in the result giving a + b + c + a:b + a:c + b:c + a:b:c. Further, terms such as (a + b + c)^2 will be expanded as star products, in this example to give all main effects and second-order interactions. (With just three factors this could also be specified as a*b*c - a:b:c, but with more factors the power form is simpler to use and to interpret.) In such expansions repeated symbols such as a:a are regarded as equivalent to a. This is the desired behaviour for factors, but not for numerical vectors, so (x + y)^3 does *not* expand to the general polynomial in x and y of degree 3. The latter may be obtained using orthogonal polynomials as poly(x,y,3).

Nested models are defined by the slash operator, `/`. If `a` is a factor and `t` is any formula expression then `a/t` defines separate submodels of type `1 + t` within the levels of `a`. Thus `a/x` can be used to define separate regressions on `x` within the classes defined by `a`; in algebraic notation

$$y = \alpha_{\mathrm{a}} + \beta_{\mathrm{a}}\, x + \epsilon$$

Formally `s/t` expands to `s + t %in% s`, where the second term is equivalent to `t + s:t` but uses the individual slopes as the parameters rather than a single slope plus adjustments.

Coding of factors

This subsection is rather technical, and can be skimmed on first reading. The coding of factors determines how linear models are parametrized through the model matrix, but not the space spanned by the columns of the model matrix, hence the fitted values, residuals, analysis of variance tables and most other quantities of statistical interest which are independent of the coding.

When a factor with p levels is added to a model, normally $p - 1$ columns are added to the model matrix. The protocol that determines the columns generated by a factor is called the *coding* of that factor. Conventionally, codings are specified by giving constraints on the parameters of an algebraic formulation of the model. The default coding in S is the so-called Helmert coding for factors and orthogonal polynomials on $\{1, 2, \ldots\}$ for ordered factors. The Helmert coding does not correspond to any commonly-used parameter constraint.

The default codings can be changed by resetting the `contrasts` option. This is a character vector of length two giving the names of the functions that generate the codings for factors and ordered factors respectively. For example

```
options(contrasts=c("contr.treatment", "contr.poly"))
```

sets the default coding function for ordinary factors to `contr.treatment` and for ordered factors to `contr.poly` (which is the original default). Four suitable contrast functions are

`contr.helmert` for the Helmert coding.

`contr.treatment` for a coding in which each coefficient represents a comparison of that level with level 1 (omitting level 1 itself). This corresponds to the constraint $\alpha_1 = 0$.

`contr.sum` where the coefficients are constrained to add to zero (and the last is omitted).

`contr.poly` for orthogonal polynomial coding.

and others can be written using these as templates. Most published analyses correspond to either sum or treatment contrasts. We recommend the use of the treatment coding for unbalanced layouts, including generalized linear models and survival models.

The numbers placed in each column of the model matrix for each level of the factor form the *contrast matrix* for the factor. If the factor has p levels this will normally be a $p \times (p - 1)$ matrix. The pattern for the Helmert coding is clear from the following example.

```
> N <- factor(c(0,1,2,4))
> contrasts(N)
   [,1] [,2] [,3]
0   -1   -1   -1
1    1   -1   -1
2    0    2   -1
4    0    0    3
```

For an ordered factor the contrast matrix is

```
> contrasts(ordered(N))
         .L    .Q       .C
0 -0.67082  0.5 -0.22361
1 -0.22361 -0.5  0.67082
2  0.22361 -0.5 -0.67082
4  0.67082  0.5  0.22361
```

Note that these contrasts may not be appropriate if (as here) the numerical levels of N are unequally spaced.

The contrast matrix may also be set as an attribute of the factor itself. This can be done either by assigning to contrasts(N) or by using the function C which takes three arguments, the factor, the matrix from which contrasts are to be taken (or the abbreviated name of a function which will generate such a matrix) and the number of contrasts. (A contrast matrix for a p-level factor may have fewer than $p - 1$ columns and thus as a term in a model use fewer than $p - 1$ degrees of freedom.) For example, suppose we wish to create a factor N2 which would generate orthogonal linear and quadratic polynomial terms, only:

```
> N2 <- C(N, poly(as.numeric(levels(N))), 2)
> contrasts(N2)
            1
0 -0.591608 -0.47557
1 -0.253546  0.12320
2  0.084515  0.76635
4  0.760639 -0.41397
```

Translating between parameters for different codings is straightforward, if sometimes tedious. Consider the cats example again. First we fit separate regressions using the Helmert coding:

```
> options(contrasts=c("contr.helmert","contr.poly"))
> catsH <- lm(Hwt ~ Sex*Bwt, data=cats)
> coef(catsH)
 (Intercept)     Sex     Bwt Sex:Bwt
     0.89861 -2.0827 3.4745 0.83813
```

To reconstruct the separate intercepts and slopes we multiply the first and last two components by the contrast matrix with an initial column of 1s:

```
> cbind(1, contrasts(cats$Sex)) %*% matrix(coef(catsH),2)
       [,1]   [,2]
F   2.9813 2.6364
M  -1.1841 4.3127
```

which agrees with the model fitted earlier. There we removed the intercept term, which changes the constraint on the first main effect so that a separate parameter (or intercept) for each class is fitted, whatever contrast options are in force. However, if the factor has an explicit contrast matrix attribute assigned to it this will be used (and the intercept term is retained).

6.3 Regression diagnostics

The cats example is relatively 'clean' and we did not have to work hard to find an adequate linear model. There is an extensive literature (e.g. Atkinson, 1985) on examining the fit of linear models to consider whether one or more points are not fitted as well as they should be or have undue influence on the fitting of the model. This can be contrasted with the robust regression methods we discuss in Chapter 8, which automatically take account of anomalous points.

The basic tool for examining the fit is the residuals, and we have already looked for patterns in residuals and assessed the normality of their distribution. The residuals are not independent (they sum to zero if an intercept is present) and they do not have the same variance. Indeed, their variance-covariance matrix is

$$\text{var}(e) = \sigma^2[I - H] \tag{6.1}$$

where $H = X(X^TX)^{-1}X^T$ is the *hat* matrix. This is said to measure the *leverage* of an observation, for if h_{ii} is large, changing Y_i will move the fitted surface appreciably towards the altered value. The trace of H is p, the number of regressors, so 'large' is taken as greater than two or three times p/n.

Having large leverage has two consequences for the corresponding residual. First, its variance will be lower than average from (6.1). We can compensate for this by re-scaling the residuals to have unit variance. The *standardized residuals* are

$$e_i' = \frac{e_i}{s\sqrt{1 - h_{ii}}}$$

where as usual we have estimated σ^2 by s^2, the residual mean square. Second, if one error is very large, the variance estimate s^2 will be too large, and this deflates all the standardized residuals. Let us consider fitting the model without observation i. We get a prediction $\hat{y}_{(i)}$ of y_i. The *studentized residuals* are

$$e_i^* = \frac{y_i - \hat{y}_{(i)}}{\sqrt{\text{var}(y_i - \hat{y}_{(i)})}}$$

but with σ replaced by $s_{(i)}$. Fortunately, it is not necessary to re-fit the model each time an observation is deleted, for

$$e_i^* = e_i' \Big/ \left[\frac{n - p - e_i'^2}{n - p - 1} \right]^{\frac{1}{2}}$$

Notice that this implies that the standardized residuals must be bounded by $\pm\sqrt{n-p}$.

The terminology used here is not universally adopted; in particular studentized residuals are sometimes called *jackknifed* residuals.

It is usually better to compare studentized residuals rather than residuals; in particular we recommend that they are used for normal probability plots.

There are no system functions to compute studentized or standardized residuals, but we have provided functions `studres` and `stdres`. There is a function `hat`, but this expects the model matrix as its argument. The diagonal of the hat matrix can be obtained by `lm.influence(lmobject)$hat`.

Scottish hill races

As an example of regression diagnostics, let us return to the data on 35 Scottish hill races in our data frame `hills` considered in Chapter 1. The data come from Atkinson (1986) and are discussed further in Atkinson (1988) and Staudte & Sheather (1990). The columns are the overall race distance, the total height climbed and the record time. In Chapter 1 we considered a regression of `time` on `dist`. We can now include `climb`:

```
> hills.lm <- lm(time ~ dist + climb, hills)
> hills.lm
    ....
Coefficients:
 (Intercept)  dist    climb
      -8.992 6.218 0.011048

Degrees of freedom: 35 total; 32 residual
Residual standard error: 14.676
> plot(fitted(hills.lm), studres(hills.lm))
> identify(fitted(hills.lm), studres(hills.lm),
      row.names(hills))
> qqnorm(studres(hills.lm))
> qqline(studres(hills.lm))
> hills.hat <- lm.influence(hills.lm)$hat
> cbind(hills, lev=hills.hat)[hills.hat > 3/35, ]
               dist climb    time     lev
 Bens of Jura    16  7500 204.617 0.42043
 Lairig Ghru     28  2100 192.667 0.68982
   Ben Nevis     10  4400  85.583 0.12158
Two Breweries    18  5200 170.250 0.17158
 Moffat Chase    20  5000 159.833 0.19099
```

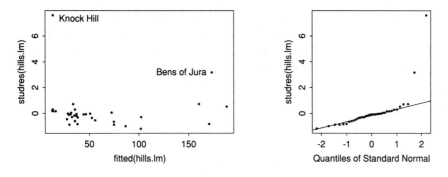

Figure 6.3: Diagnostic plots for Scottish hills data, unweighted model.

so two points have very high leverage, two points have large residuals, and Bens of Jura is in both sets. (See Figure 6.3.)

If we look at Knock Hill we see that the prediction is over an hour less than the reported record:

```
> cbind(hills, predict(hills.lm))["Knock Hill", ]
            dist climb  time      X
Knock Hill     3   350 78.65 13.529
```

and Atkinson (1988) suggests that the record is one hour out. We drop this observation to be safe:

```
> hills1.lm <- lm(time ~ dist + climb, hills[-18, ])
> hills1.lm
Call:
lm(formula = time ~ dist + climb)

Coefficients:
 (Intercept)    dist   climb
      -13.53  6.3646 0.011855

Degrees of freedom: 34 total; 31 residual
Residual standard error: 8.8035
```

Since Knock Hill did not have a high leverage, as expected deleting it did not change the fitted model greatly. On the other hand, Bens of Jura had both a high leverage and a large residual and so does affect the fit:

```
> lm(time ~ dist + climb, hills[-c(7,18), ])
    ....
Coefficients:
 (Intercept)    dist    climb
     -10.362  6.6921 0.0080468

Degrees of freedom: 33 total; 30 residual
Residual standard error: 6.0538
```

If we consider this example carefully we find a number of worrying features. First, the prediction is negative for short races. Extrapolation is often unsafe, but on physical grounds we would expect the model to be a good fit with a zero intercept; indeed hill-walkers use a prediction of this sort (3 miles/hour plus 20 mins per 1000 feet). We can see from the summary that the intercept is significantly negative:

```
> summary(hills1.lm)
    . . . .
Coefficients:
              Value Std. Error t value Pr(>|t|)
(Intercept) -13.530    2.649    -5.108   0.000
       dist    6.365    0.361    17.624   0.000
      climb    0.012    0.001     9.600   0.000
    . . . .
```

Further, we would not expect the predictions of times which range from 15 minutes to over 3 hours to be equally accurate, but rather that the accuracy be roughly proportional to the time. This suggests a log transform, but that would be hard to interpret. Rather we weight the fit using distance as a surrogate for time. We want weights inversely proportional to the variance:

```
> summary(lm(time ~ dist + climb, hills[-18, ],
      weight=1/dist^2))
    . . . .
Coefficients:
              Value Std. Error t value Pr(>|t|)
(Intercept)  -5.809    2.034    -2.855   0.008
       dist    5.821    0.536    10.858   0.000
      climb    0.009    0.002     5.873   0.000

Residual standard error: 1.16 on 31 degrees of freedom
```

The intercept is still significantly non-zero. If we are prepared to set it to zero on physical grounds, we can achieve the same effect by dividing the prediction equation by distance, and regressing inverse speed (time/distance) on gradient (climb/distance):

```
> lm(time ~ -1 + dist + climb, hills[-18, ], weight=1/dist^2)
Coefficients:
 (Intercept)        grad
         4.9 0.0084718

Degrees of freedom: 34 total; 32 residual
Residual standard error (on weighted scale): 1.2786
> hills$ispeed <- hills$time/hills$dist
> hills$grad <- hills$climb/hills$dist
> hills2.lm <- lm(ispeed ~ grad, hills[-18, ])
> hills2.lm
Coefficients:
```

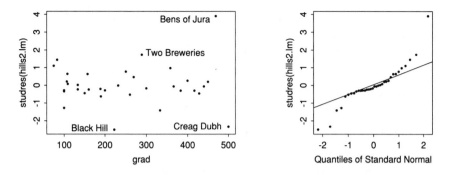

Figure 6.4: Diagnostic plots for Scottish hills data, weighted model.

```
(Intercept)        grad
        4.9 0.0084718

Degrees of freedom: 34 total; 32 residual
Residual standard error: 1.2786
> plot(hills$grad[-18], studres(hills2.lm))
> identify(hills$grad[-18], studres(hills2.lm),
    row.names(hills)[-18])
> qqnorm(studres(hills2.lm))
> qqline(studres(hills2.lm))
> hills2.hat <- lm.influence(hills2.lm)$hat
> cbind(hills[-18,], lev=hills2.hat)[hills2.hat > 1.8*2/34, ]
              dist climb    time  ispeed    grad      lev
Bens of Jura    16  7500 204.617 12.7886  468.75  0.11354
   Creag Dubh     4  2000  26.217  6.5542  500.00  0.13915
```

The two highest-leverage cases are now the steepest two races, and are outliers pulling in opposite directions. We could consider elaborating the model, but this would be to fit only one or two exceptional points; for most of the data we have the formula of 5 minutes/mile plus 8 minutes per 1000 feet. We return to this example in Section 8.4, where robust fits do support a zero intercept.

6.4   Safe prediction

A warning is needed on the use of the `predict` method function when polynomials are used (and also splines, see Chapter 10). We will illustrate this by the dataset `wtloss`, for which a more appropriate analysis is given in Chapter 9. This has a weight loss `Weight` against `Days`. Consider a quadratic polynomial regression model of `Weight` on `Days`. This may be fitted by either of

```
quad1 <- lm(Weight ~ Days + I(Days^2), wtloss)
quad2 <- lm(Weight ~ poly(Days, 2), wtloss)
```

The second form uses orthogonal polynomials and is usually preferred on numerical grounds.

Suppose we wished to predict future weight loss. The first step is to create a new data frame with a variable x containing the new values, for example

```
new.x <- data.frame(Days = seq(250, 300, 10),
                     row.names=seq(250, 300, 10))
```

The predict method may now be used:

```
> predict(quad1, newdata=new.x)
    250    260    270    280    290    300
 112.51 111.47 110.58 109.83 109.21 108.74
> predict(quad2, newdata=new.x)
    250    260    270    280    290    300
 244.56 192.78 149.14 113.64 86.29 67.081
```

The first form gives correct answers but the second does not; the predict method reconstructs the model matrix using the formula and the new data, thereby producing a basis of orthogonal polynomials for the new data, rather than evaluating the original orthogonal polynomials.

The remedy is to use the method function predict.gam:

```
>  predict.gam(quad2, newdata=new.x)
    250    260    270    280    290    300
 112.51 111.47 110.58 109.83 109.21 108.74
```

This constructs a new model matrix by putting old and new data together, re-estimates the regression using the old data only and predicts using these estimates of regression coefficients. This can involve appreciable extra computation, but the results will be correct for polynomials (but not exactly so for splines).

6.5 Factorial designs and designed experiments

Factorial designs are powerful tools in the design of experiments. Experimenters often cannot afford to perform all the runs needed for a complete factorial experiment, or they may not all be fitted into one experimental block. To see what can be achieved, consider the following N, P, K (*nitrogen, phosphate, potassium*) factorial experiment on the growth of peas conducted on 6 blocks. The response is yield (in lbs/(1/70)acre-plot).

pk	np	—	nk	n	npk	k	p
49.5	62.8	46.8	57.0	62.0	48.8	45.5	44.2
n	npk	k	p	np	—	nk	pk
59.8	58.5	55.5	56.0	52.0	51.5	49.8	48.8
p	npk	n	k	nk	np	pk	—
62.8	55.8	69.5	55.0	57.2	59.0	53.2	56.0

Half of the design (technically a fractional factorial design) is performed in each of six blocks, so each half occurs three times. (If we consider the variables to take values ± 1, the halves are defined by even or odd parity, equivalently product equal to $+1$ or -1.) Note that the NPK interaction cannot be estimated as it is confounded with block differences, specifically with $(b_2 + b_3 + b_4 - b_1 - b_5 - b_6)$. Suppose the data are entered on file npk.dat. Then the ANOVA table is computed by:

```
> npk1 <- read.table("npk.dat", header=T)
> npk <- data.frame(block=factor(rep(1:6, rep(4,6))),
      N=factor(npk1$N), P=factor(npk1$P), K=factor(npk1$K),
      yield=npk1$yield)
> npk.aov <- aov(yield ~ block + N*P*K, npk)
> npk.aov
    ....
Terms:
                   block      N       P      K     N:P     N:K
 Sum of Squares   343.30 189.28    8.40  95.20   21.28   33.13
 Deg. of Freedom       5      1       1      1       1       1

                  P:K Residuals
 Sum of Squares   0.48    185.29
 Deg. of Freedom     1        12

Residual standard error: 3.9294
1 out of 13 effects not estimable
Estimated effects are balanced
> summary(npk.aov)
            Df Sum of Sq Mean Sq F Value   Pr(F)
block        5    343.30   68.66   4.447 0.01594
N            1    189.28  189.28  12.259 0.00437
P            1      8.40    8.40   0.544 0.47490
K            1     95.20   95.20   6.166 0.02880
N:P          1     21.28   21.28   1.378 0.26317
N:K          1     33.13   33.13   2.146 0.16865
P:K          1      0.48    0.48   0.031 0.86275
Residuals   12    185.29   15.44
> alias(npk.aov)
    ....
Complete
        (Intercept) block1 block2 block3 block4 block5 N P K
N:P:K              1   0.33   0.17   -0.3   -0.2
        N:P N:K P:K
N:P:K
> coef(npk.aov)
 (Intercept) block1 block2  block3  block4 block5
      54.875 1.7125 1.6792 -1.8229 -1.0138  0.295
        N        P       K     N:P    N:K     P:K
   2.8083 -0.59167 -1.9917 -0.94167 -1.175 0.14167
```

Note how the `N:P:K` interaction is silently omitted in the summary, although its absence *is* mentioned in printing `npk.aov`. The `alias` command shows which effect is missing (the particular combinations corresponding to the use of Helmert coding of the factor `block`).

Only the N and K main effects are significant (we ignore blocks whose terms are there precisely because we expect them to be important). For two-level factors the Helmert coding is the same as the sum coding (up to sign) giving -1 to the first level and $+1$ to the second level. Thus the effects of adding nitrogen and potassium are 5.62 and -3.98 respectively. This interpretation is easier to see with treatment contrasts:

```
> options(contrasts=c("contr.treatment", "contr.poly"))
> npk.aov1 <- aov(yield ~ block + N + K, npk)
> summary.lm(npk.aov1)
   ....
Coefficients:
            Value Std. Error t value Pr(>|t|)
   ....
        N   5.617    1.609     3.490   0.003
        K  -3.983    1.609    -2.475   0.025

Residual standard error: 3.94 on 16 degrees of freedom
```

Note the use of `summary.lm` to give the standard errors. Standard errors of contrasts can also be found from the function `se.contrast`. The full form is quite complex, but a simple use is:

```
> se.contrast(npk.aov1, list(N==0, N==1), data=npk)
Refitting model to allow projection
[1] 1.6092
```

The function `aov.genyates` is very similar to `aov` but uses a generalized Yates' algorithm. It can only be used with balanced experimental designs, that is those in which each combination of factor levels occurs equally often, but it can be much faster than `aov` for very large designs. (Balance can be checked using the function `replications`.)

Generating designs

The three functions `expand.grid`, `fac.design` and `oa.design` can each be used to construct designs such as our example.

Of these, `expand.grid` is the simplest. It is used in a similar way to `data.frame`: the arguments may be named and the result is a data frame with those names. The columns contain all combinations of values for each argument. If the argument values are numeric the column is numeric; if they are anything else, for example character, the column is a factor. Consider an example:

```
> mp <- c("-","+")
> NPK <- expand.grid(N=mp, P=mp, K=mp)
> NPK
  N P K
1 - - -
2 + - -
3 - + -
4 + + -
5 - - +
6 + - +
7 - + +
8 + + +
```

Note that the first column changes fastest and the last slowest. This is a single complete replicate.

Our example used three replicates, each split into two blocks so that the block comparison is confounded with the highest order interaction. We can construct such a design in stages. First we find a half replicate to be repeated three times and form the contents of three of the blocks. The simplest way to do this is to use `fac.design`:

```
blocks13 <- fac.design(levels=c(2,2,2),
    factor=list(N=mp, P=mp, K=mp), rep=3, fraction=1/2)
```

The first two arguments give the numbers of levels and the factor names and level labels. The third argument gives the number of replications (default 1). The `fraction` argument may only be used for 2^p factorials. It may either be given as a small negative power of 2, as here, or as a *defining contrast formula*. When `fraction` is numerical the function chooses a defining contrast which becomes the `fraction` attribute of the result. For half replicates the highest order interaction is chosen to be aliased with the mean. To find the complementary fraction for the remaining three blocks we need to use the defining contrast formula form for `fraction`:

```
blocks46 <- fac.design(levels=c(2,2,2),
    factor=list(N=mp, P=mp, K=mp), rep=3, fraction=~ -N:P:K)
```

(This will be explained below.) To complete our design we put the blocks together, add in the `block` factor and randomize:

```
NPK <- design(block = factor(rep(1:6, rep(4,6))),
    rbind(blocks13, blocks46))
i <- order(runif(6)[NPK$block], runif(24))
NPK <- NPK[i,]  # Randomized
```

Using `design` instead of `data.frame` creates an object of class `design` that inherits from `data.frame`. For most purposes designs and data frames are equivalent, but some generic functions such as `plot`, `formula` and `alias` have useful `design` methods.

Defining contrast formulae resemble model formulae in syntax only; the meaning is quite distinct. There is no left hand side. The right hand side consists of colon products of factors only, separated by + or − signs. A plus (or leading blank) specifies that the treatments with *positive* signs for that contrast are to be selected and a minus those with *negative* signs. A formula such as ~A:B:C−A:D:E specifies a quarter replicate consisting of the treatments which have a positive sign in the ABC interaction and a negative sign in ADE.

Box, Hunter & Hunter (1978, §12.5) consider a 2^{7-4} design used for an experiment in riding up a hill on a bicycle. The seven factors are Seat (up or down), Dynamo (off or on), Handlebars (up or down), Gears (low or medium), Raincoat (on or off), Breakfast (yes or no) and Tyres (hard or soft). A resolution III design was used, so the main effects are not aliased with each other. Such a design cannot be constructed using a numerical fraction in fac.design so the defining contrasts have to be known. Box, Hunter & Hunter use the design relations:

$$D = AB, \quad E = AC, \quad F = BC, \quad G = ABC$$

which mean that ABD, ACE, BCF and $ABCG$ are all aliased with the mean, and form the defining contrasts of the fraction. Whether we choose the positive or negative halves is immaterial here.

```
> lev <- rep(2,7)
> factors <- list(S=mp, D=mp, H=mp, G=mp, R=mp, B=mp, Ty=mp)
> Bike <- fac.design(lev, factors, fraction =
    ~ S:D:G + S:H:R + D:H:B + S:D:H:Ty)
> Bike
  S D H G R B Ty
1 - - - - - -  -
2 - + + + + -  -
3 + - + + - +  -
4 + + - - + +  -
5 + + + - - -  +
6 + - - + + -  +
7 - + - + - +  +
8 - - + - + +  +

Fraction:   ~ S:D:G + S:H:R + D:H:B + S:D:H:Ty
```

Note that factor names have to be usable as variable names so the single letter T (and F) is invalid.

We may check the symmetry of the design using replications:

```
> replications(~.^2, data=Bike)
  S D H G R B Ty S:D S:H S:G S:R S:B S:Ty D:H D:G D:R D:B D:Ty
  4 4 4 4 4 4  4   2   2   2   2   2    2   2   2   2   2    2
  H:G H:R H:B H:Ty G:R G:B G:Ty R:B R:Ty B:Ty
    2   2   2    2   2   2    2   2    2    2
```

Fractions may either be specified in a call to `fac.design` or subsequently using the `fractionate` function.

The third function, `oa.design`, provides some resolution III designs (also known as *main effect plans* or *orthogonal arrays*) for factors at 2 or 3 levels. Only low order cases are provided, but these are the most useful in practice.

Random effects and variance components

The function `raov` may be used for balanced designs with only random effects, and gives a conventional analysis including the estimation of variance components. The function `varcomp` is more general, and may be used to estimate variance components for balanced or unbalanced mixed models.

To illustrate these functions we take a portion of the data from the data frame `coop` of our library. This is a co-operative trial in analytical chemistry taken from Analytical Methods Comittee (1987). Seven specimens were sent to 6 laboratories, each 3 times a month apart for duplicate analysis. The response is the concentration of (unnamed) analyte in g/kg. We use the data from specimen 1 shown in Table 6.1.

Table 6.1: Analyte concentrations (g/kg) from 6 laboratories and 3 batches.

Batch	Laboratory 1	2	3	4	5	6
1	0.29	0.40	0.40	0.9	0.44	0.38
	0.33	0.40	0.35	1.3	0.44	0.39
2	0.33	0.43	0.38	0.9	0.45	0.40
	0.32	0.36	0.32	1.1	0.45	0.46
3	0.34	0.42	0.38	0.9	0.42	0.72
	0.31	0.40	0.33	0.9	0.46	0.79

The purpose of the study was to assess components of variation in co-operative trials. For this purpose laboratories and batches are regarded as random, so a model for the response for laboratory i, batch j and duplicate k is

$$y_{ijk} = \mu + \xi_i + \beta_{ij} + \epsilon_{ijk}$$

where ξ, β and ϵ are independent random variables with zero means and variances σ_L^2, σ_B^2 and σ_e^2 respectively. For l laboratories, b batches and $r = 2$ duplicates a nested analysis of variance gives:

Source of variation	Degrees of freedom	Sum of squares	Mean square	E(MS)
Between laboratories	$l-1$	$br\sum_i(\bar{y}_i - \bar{y})^2$	$\mathrm{MS_L}$	$br\sigma_L^2 + r\sigma_B^2 + \sigma_e^2$
Batches within laboratories	$l(b-1)$	$r\sum_{ij}(\bar{y}_{ij} - \bar{y}_i)^2$	$\mathrm{MS_B}$	$r\sigma_B^2 + \sigma_e^2$
Replicates within batches	$lb(r-1)$	$\sum_{ijk}(y_{ij} - \bar{y}_{ij})^2$	MS_e	σ_e^2

So the unbiased estimators of the variance components are

$$\hat{\sigma}_L^2 = (br)^{-1}(\mathrm{MS_L} - \mathrm{MS_B}), \quad \hat{\sigma}_B^2 = r^{-1}(\mathrm{MS_B} - \mathrm{MS}_e), \quad \hat{\sigma}_e^2 = \mathrm{MS}_e$$

The model is fitted in the same way as an analysis of variance model, but with `raov` replacing `aov`:

```
> summary(raov(Conc ~ Lab/Bat, data=coop, subset=Spc=="S1"))
                Df Sum of Sq Mean Sq Est. Var.
Lab              5    1.8902 0.37804  0.060168
Bat %in% Lab 12      0.2044 0.01703  0.005368
Residuals    18      0.1134 0.00630  0.006297
```

The same variance component estimates can be found using `varcomp`, but as mixed models are allowed we need first to declare which factors are random effects using the `is.random` function. All factors in a data frame are to be declared random effects by

```
is.random(coop) <- T
```

If we use `Spc` as a factor it may need to be treated as fixed. Individual factors can also have their status declared in the same way. When used as an unassigned expression `is.random` reports the status of (all the factors in) its argument:

```
> is.random(coop$Spc) <- F
> is.random(coop)
 Lab Spc Bat
   T   F   T
```

We can now estimate the variance components:

```
> varcomp(Conc ~ Lab/Bat, data=coop, subset= Spc=="S1")
 Variances:
       Lab Bat %in% Lab Residuals
  0.060168    0.0053681 0.0062972
```

The fitted values for such a model are technically the grand mean, but the `fitted` method function calculates them as if it were a fixed effects model, so giving best linear unbiased predictors (BLUPs). Residuals are also calculated as differences of observations from the BLUPs. An examination of the residuals and fitted values points up laboratory 4 batches 1 & 2 as possibly suspect, and these will inflate the estimate of σ_e^2. (However laboratory 4 overall and laboratory 6 batch 3 are outliers for the other variance components, but a simple residual analysis will not expose this. See Analytical Methods Committee, 1987.)

Variance component estimates are known to be very sensitive to aberrant observations; we can get some check on this by repeating the analysis with a robust estimating method. The result is very different:

```
> varcomp(Conc ~ Lab/Bat, data=coop, subset=Spc=="S1",
    method = c("winsor", "minque0"))
Variances:
        Lab Bat %in% Lab Residuals
  0.0040043   -0.00065681 0.0062979
```

(A related robust analysis is given in Analytical Methods Committee (1989b).)

The possible estimation methods are `"minque0"` for minimum norm quadratic estimators (the default), `"ml"` for maximum likelihood and `"reml"` for residual (or reduced or restricted) maximum likelihood. (See, for example, Rao, 1971a, b; Rao & Kleffe, 1988). Method `"winsor"` specifies that the data are 'cleaned' before further analysis. As in our example, a second method may be prescribed as the one to use for estimation of the variance components on the 'cleaned' data. The method used, Winsorizing, is discussed in Chapter 8; unfortunately the tuning constant is not adjustable, and is set for rather mild data cleaning.

6.6   An unbalanced four-way layout

Aitkin (1978) discussed an observational study of S. Quine. The response is the number of days absent from school in a year by children from a large town in rural New South Wales, Australia. The children were classified by four factors, namely

Eth   Ethnic group: 2 levels, namely aboriginal or non-aboriginal.
Sex   Sex: 2 levels, male and female.
Age   Age group: 4 levels, primary, first form, second form or third form.
Lrn   Learner class: 2 levels, slow learner or average learner.

The dataset is included in the paper of Aitkin (1978) and is available as data frame quine in our library.

There were 146 children in the study. The frequencies of the combinations of factors are

```
> attach(quine)
> table(Lrn, Age, Sex, Eth)
, , F, A               , , F, N
   F0 F1 F2 F3             F0 F1 F2 F3
AL  4  5  1  9         AL   4  6  1 10
SL  1 10  8  0         SL   1 11  9  0

, , M, A               , , M, N
   F0 F1 F2 F3             F0 F1 F2 F3
AL  5  2  7  7         AL   6  2  7  7
SL  3  3  4  0         SL   3  7  3  0
```

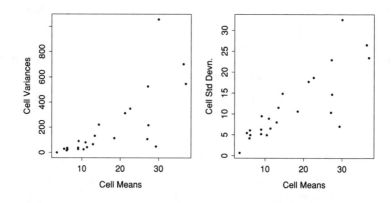

Figure 6.5: Two diagnostic plots for the Quine data.

(The output has been slightly rearranged to save space.) The classification is unavoidably very unbalanced. There are no slow learners in the fourth form, but all other cells apart from these four are non-empty. In his paper Aitkin considers a normal analysis on the untransformed response, but in the reply to the discussion he considers using a transformed response $\log(D+1)$.

An obvious first step is to plot the cell variances and standard deviations against the cell means.

```
Means <- tapply(Days, list(Eth, Sex, Age, Lrn), mean)
Vars  <- tapply(Days, list(Eth, Sex, Age, Lrn), var)
SD <- sqrt(Vars)
par(mfrow=c(1,2))
plot(Means, Vars, xlab="Cell Means", ylab="Cell Variances")
plot(Means, SD, xlab="Cell Means", ylab="Cell Std Devn.")
```

Notice how missing values are silently omitted from the plot. Interpretation of the result in Figure 6.5 requires some caution because of the small and widely different degrees of freedom on which each variance is based. The approximate linearity of the standard deviations against the cell means suggests something like a logarithmic transformation is appropriate. (See, for example, Rao, 1973, §6g).

Some further insight on the transformation needed is provided by considering a model for the transformed observations

$$y^{(\lambda)} = \begin{cases} (y^\lambda - 1)/\lambda & \lambda \neq 0, \\ \log y & \lambda = 0. \end{cases}$$

where here $y = D + \alpha$. (The positive constant α is added to avoid problems with zero entries.) Rather than include α as a second parameter we first consider Aitkin's choice of $\alpha = 1$. Box & Cox (1964) show that the profile likelihood function for λ is

$$\widehat{L}(\lambda) = \text{const} - \tfrac{n}{2} \log \text{RSS}(z^{(\lambda)})$$

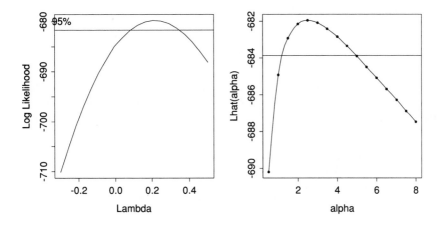

Figure 6.6: Profile likelihood for a Box–Cox transformation model with displacement $\alpha = 1$, left, and a displaced log transformation model, right.

where $z^{(\lambda)} = y^{(\lambda)}/\dot{y}^{\lambda-1}$, $\dot{y}$ is the geometric mean of the observations and $\mathrm{RSS}(z^{(\lambda)})$ is the residual sum of squares for the regression of $z^{(\lambda)}$.

Box & Cox suggest using the profile likelihood function for the largest linear model to be considered as a guide in choosing a value for λ, which will then remain fixed for any remaining analyses. Ideally other considerations from the context will provide further guidance in the choice of λ, and in any case it is desirable to choose easily interpretable values such as square-root, log or inverse.

Our library function `boxcox` calculates and (optionally) displays the Box–Cox profile likelihood function, together with a horizontal line showing what would be an approximate 95% likelihood ratio confidence interval for λ. The arguments are (1) a fitted model object to define the response variable and regression model and (2) optionally, a sequence of values over which to evaluate the profile likelihood.

Since the dataset has four empty cells the full model `Eth*Sex*Age*Lrn` will generate a rank-deficient model matrix. Rank-deficient models can be fitted using `lm` provided the argument `singular.ok=T` is used, and with `aov` which has a fitting algorithm that automatically detects and accommodates rank deficiency.

```
quine.aov <- aov(Days+1 ~ Eth*Sex*Age*Lrn, data=quine)
boxcox(quine.aov, lambda=seq(-0.3,0.5,len=20), singular.ok=T)
```

Note that the additional `singular.ok=T` argument is passed on to `lm` inside `boxcox`. The result is shown on the left hand side of Figure 6.6. Clearly a log transformation is not optimal when $\alpha = 1$ is chosen. An alternative one-parameter family of transformations that could be considered in this case is

$$t(y, \alpha) = \log(y + \alpha)$$

Using the same analysis as presented in Box & Cox (1964) the profile log likelihood for α is easily seen to be

$$\widehat{L}(\alpha) = \text{const} - \tfrac{n}{2} \log \text{RSS}\{\log(y + \alpha)\} - \sum \log(y + \alpha)$$

It is interesting to see how this may be calculated directly using low-level tools, in particular the functions `qr` for the QR-decomposition and `qr.resid` for orthogonal projection onto the residual space. (Our function `boxcox` works in the same way.)

```
n <- nrow(quine)
Xqr <- qr(model.matrix(quine.aov))
alpha <- seq(0.5,8,by=0.5)
prof.lik <- NULL
attach(quine)
for(a in alpha) {
    yt <- log(Days+a)
    rs <- qr.resid(Xqr, yt)
    prof.lik <- c(prof.lik, -n/2*log(sum(rs^2)) - sum(yt))
}
plot(alpha, prof.lik, xlab="alpha", ylab="Lhat(alpha)")
lines(spline(alpha, prof.lik))
abline(h=max(prof.lik) - qchisq(0.95,1)/2)
```

The result is shown in the right-hand panel of Figure 6.6. If a displaced log transformation is chosen a value $\alpha = 2.5$ is suggested, and we adopt this in our analysis. Note that $\alpha = 1$ is outside the notional 95% confidence interval.

Model selection

We now consider choosing a parsimonious model for the Quine data. The `anova` function applied to a single fitted model object shows the contribution of each term additional to those terms that precede it in the table. For an orthogonal design these contributions would be independent of order (except where terms are aliased) but for unbalanced designs, as here, the contributions are order dependent, so we only consider the high-order interactions at first.

```
> quine.aov0 <- aov(log(Days+2.5) ~ Eth*Sex*Age*Lrn,
        data=quine)
> anova(quine.aov0)
    ....
Terms added sequentially (first to last)
                Df Sum of Sq Mean Sq F Value   Pr(F)
    ....
Eth:Sex:Age      3     1.939   0.646   1.205 0.31116
Eth:Sex:Lrn      1     3.741   3.741   6.972 0.00940
Eth:Age:Lrn      2     2.146   1.073   2.000 0.13988
Sex:Age:Lrn      2     1.466   0.733   1.366 0.25903
Eth:Sex:Age:Lrn  2     0.789   0.394   0.735 0.48171
Residuals      118    63.310   0.537
```

Now consider removing terms. A term such as A:B is said to be *marginal* to any higher order term that includes it, such as A:B:C. If both occur in the model then discarding marginal terms does not affect the statistical meaning of the model, only the way it is parametrized. So when removing terms from a factorial model we can confine our attention to the currently non-marginal terms.

The only non-marginal term in the full model is Eth:Sex:Age:Lrn which may be eliminated since it is clearly non-significant.

```
quine.aov1 <- update(quine.aov0, . ~ . - Eth:Sex:Age:Lrn)
```

In discarding terms it is often useful to use another model-fitting function, drop1. This function takes a fitted model and removes each term separately, showing the contribution of each term additional to *all* others. For objects of class lm it gives Mallows' C_p and for class aov it provides F-tests:[2]

```
> drop1(quine.aov1)
Single term deletions
    ....
            Df Sum of Sq    RSS F Value   Pr(F)
   <none>                 64.099
Eth:Sex:Age  3    0.9739 65.073  0.6077 0.61125
Eth:Sex:Lrn  1    1.5788 65.678  2.9557 0.08816
Eth:Age:Lrn  2    2.1284 66.227  1.9923 0.14087
Sex:Age:Lrn  2    1.4662 65.565  1.3725 0.25743
```

Note that only the non-marginal terms are listed, as is appropriate.

The first argument to drop1 is a fitted model object and second (optional) argument can be used to specify the scope of the deletions to perform. If it is omitted, each non-marginal term is omitted in turn, as in the example above. The scope may be specified either as a character string vector of term names, or as a formula. As a formula only the right hand side needs to be specified, and each term in the expanded form is omitted in turn. Thus to drop all terms individually we could use:

```
drop1(quine.aov1, ~ .)
```

Note that when the scope is specified special care is needed if marginal terms are dropped, since the results will depend on the coding used. In our opinion these marginal terms are therefore of little statistical interest and in some cases misleading, although some authors disagree with this view. What are sometimes termed "Type 3 sums of squares" can be got this way provided the contrast matrices have column sums of zero; this is the case for helmert, poly and sum contrast matrices, which all give the same answers, but not for treatment which may give very different answers. (Note that drop1 and its companion add1 may use the single dot abbreviation in the same way as update.)

Of the four non-marginal terms the next term to consider removing from the model is Eth:Sex:Age. Proceeding in this way suggests a sequence of deletions:

---

[2] S-PLUS 3.2 rel 1 for Windows has an error in drop1.lm; a corrected function is in the fixes directory on the disk.

```
quine.aov2 <- update(quine.aov1, . ~ . - Eth:Sex:Age)
quine.aov3 <- update(quine.aov2, . ~ . - Sex:Age:Lrn)
quine.aov4 <- update(quine.aov3, . ~ . - Eth:Age:Lrn)
quine.aov5 <- update(quine.aov4, . ~ . - Eth:Age)
quine.aov6 <- update(quine.aov5, . ~ . - Age:Lrn)
```

and from this final model no non-marginal term is non-significant:

```
> drop1(quine.aov6)
Single term deletions

Model:
log(Days + 2.5) ~ Eth + Sex + Age + Lrn + Eth:Sex + Sex:Age +
        Eth:Lrn + Sex:Lrn + Eth:Sex:Lrn
            Df Sum of Sq    RSS F Value      Pr(F)
    <none>                74.639
   Sex:Age  3    9.9002 84.539   5.836 0.0008944
Eth:Sex:Lrn  1    6.2988 80.937  11.140 0.0010982
```

Note that `Sex/(Age + Eth*Lrn)` is a different parametrization for the same model, which indicates that the factors `Age` and `Eth*Lrn` behave additively within the levels of `Sex`. After fitting the model in this form the sequential analysis of variance is:

```
> quine.aov6a <- aov(log(Days+2.5) ~ Sex/(Age + Eth*Lrn),
    data=quine)
> anova(quine.aov6a)
    ....
Terms added sequentially (first to last)
                  Df Sum of Sq Mean Sq F Value  Pr(F)
Sex                1     0.597  0.5969  1.0556 0.30611
Age %in% Sex       6    12.836  2.1394  3.7835 0.00165
Eth %in% Sex       2     9.156  4.5778  8.0960 0.00048
Lrn %in% Sex       2     2.530  1.2650  2.2371 0.11080
Eth:Lrn %in% Sex   2     7.030  3.5150  6.2163 0.00263
Residuals        132    74.639  0.5654
```

Plotting the fitted model object itself, `plot(quine.aov6)`, produces two diagnostic plots which the reader can explore, but which in this case do not seem to us to be particularly informative. However two standard diagnostic plots do not suggest any problem with our modelling assumptions:

```
plot(fitted(quine.aov6), studres(quine.aov6),
    xlab="Fitted values", ylab="Studentized residuals")
qqnorm(studres(quine.aov6),ylab="Sorted Studentized residuals")
qqline(studres(quine.aov6))
```

The results are shown in Figure 6.7.

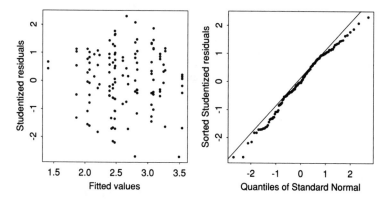

Figure 6.7: Two standard diagnostic plots for the Quine data.

Automated model selection

Constructing models by backward elimination, as we have done here, has some dangers and should be done with care and with as much regard for the context as possible. Aitkin (1978) in his discussion of this dataset offers some techniques for making the process more guarded, to which the reader is referred. Apart from the displacement constant α, Aitkin's recommended model is the same as ours. (However when we consider a negative binomial model for the same data in Section 7.4 a more extensive model seems to be needed.)

A less cautious approach is to use the `step` function which automates the model selection procedure using what its authors call AIC (or equivalently here, Mallows' C_p) to decide upon acceptance or rejection of terms. This penalizes the residual sum of squares by twice the number of parameters times the residual mean square for the initial model, and chooses the model in a sequence of steps which minimizes this criterion. Its effect is similar to conventional stepwise selection alternately adding or deleting terms, with an F test value of 2. As this is smaller than conventional F values, `step` tends to be cautious in pruning models and generous in adding variables.

For interest we present the results. Since the default output is large we turn trace off. The `anova` component prints details of the path of models selected.

```
> quine.step <- step(quine.aov0, trace=F)
> quine.step$anova
    ....
Initial Model:
log(Days + 2.5) ~ Eth * Sex * Age * Lrn

Final Model:
log(Days + 2.5) ~ Eth + Sex + Age + Lrn + Eth:Sex + Sex:Age +
        Eth:Lrn + Sex:Lrn + Age:Lrn + Eth:Sex:Lrn
```

```
                 Step Df Deviance Resid. Df Resid. Dev    AIC
1                                    118       63.310 93.356
2 - Eth:Sex:Age:Lrn  2   0.7887      120       64.099 91.998
3      - Eth:Sex:Age  3   0.9739      123       65.073 89.753
4      - Sex:Age:Lrn  2   1.5268      125       66.600 89.134
5      - Eth:Age:Lrn  2   2.0960      127       68.696 89.084
6          - Eth:Age  3   3.0312      130       71.727 88.896
> drop1(quine.step)
Single term deletions

    . . . .
             Df Sum of Sq     RSS F Value    Pr(F)
    <none>                 71.727
   Sex:Age  3    11.566 83.292   6.987 0.000215
   Age:Lrn  2     2.912 74.639   2.639 0.075279
Eth:Sex:Lrn  1     6.818 78.545  12.357 0.000605
```

The final model here has one more term, Age:Lrn, than the final model above, but this reflects the tendency of step to err on the side of large models.

Akaike Information Criterion (AIC)

It is helpful to relate this definition of AIC to that most commonly used and given by Akaike (1974), namely

$$\text{AIC} = -2 \text{ maximized log likelihood} + 2 \,\#\, \text{parameters}$$

Since the log-likelihood is defined only up to a constant depending on the data, this is also true of AIC.

For a regression model with n observations, p parameters and normally-distributed errors the log-likelihood is

$$L(\beta, \sigma^2; Y) = \text{const} - \frac{n}{2} \log \sigma^2 - \frac{1}{2\sigma^2} \|Y - X\beta\|^2$$

and on maximizing over β we have

$$L(\widehat{\beta}, \sigma^2; Y) = \text{const} - \frac{n}{2} \log \sigma^2 - \frac{1}{2\sigma^2} \text{RSS}$$

Thus if σ^2 is *known*, we can take

$$\text{AIC} = \frac{\text{RSS}}{\sigma^2} + 2p$$

but if σ^2 is *unknown*,

$$L(\widehat{\beta}, \widehat{\sigma}^2; Y) = \text{const} - \tfrac{n}{2} \log \widehat{\sigma}^2 - n/2, \qquad \widehat{\sigma}^2 = \text{RSS}/n$$

and so

$$\text{AIC} = n \log(\text{RSS}/n) + 2p.$$

If we now expand this about the residual sum of squares, s^2, for an initial model, we find

$$\text{AIC} = \text{AIC}_0 + n \log \text{RSS}/ns^2 + 2(p - p_0) \approx \text{AIC}_0 + \left[\frac{\text{RSS}}{s^2} - n\right] + 2(p - p_0)$$

so we can regard $\text{RSS}/s^2 + 2p$ as an approximation to the variable part of AIC. The value used by `step` is s^2 times this.

6.7   Multistratum models

Multistratum models occur where there is more than one source of random variation in an experiment, as in a split-plot experiment, and have been most used in agricultural field trials. The sample information and the model for the means of the observations may be partitioned into so-called *strata*. In orthogonal experiments such as split-plot designs each parameter may be estimated in one and only one stratum, but in non-orthogonal experiments such as a balanced incomplete block design with "random blocks", treatment contrasts are estimable both from the between-block and the within-block stratum. Combining the estimates from two strata is known as "the recovery of interblock information", for which the interested reader should consult a comprehensive reference such as Scheffé (1959, pp. 170ff). In this section we consider the separate stratum analyses only.

The details of the computational scheme employed for multistratum experiments are given by Heiberger (1989).

A split-plot experiment

Our example was first used by Yates (1935) and is discussed further in Yates (1937). It has also been used by many authors since; see, for example, John (1971, §5.7). The experiment involved varieties of oats and manure (nitrogen), conducted in 6 blocks of 3 whole plots. Each whole plot was divided into 4 subplots. Three varieties of oats were used in the experiment with one variety being sown in each whole plot (this being a limitation of the seed drill used), while four levels of manure (0, 0.2, 0.4 and 0.6 cwt per acre) were used, one level in each of the four subplots of each whole plot.

The data are shown in Table 6.2, where the blocks are labelled I-VI, V_i denotes the ith variety and N_j denotes the jth level of nitrogen. The dataset is available in our library as the data frame `oats`, with variables `B`, `V`, `N` and `Y`. Since the levels of `N` are ordered it is appropriate to create an ordered factor:

```
oats$Nf <- ordered(oats$N, levels=sort(levels(oats$N)))
```

The strata for the model we will use here are:

1. A 1-dimensional stratum corresponding to the total of all observations,
2. A 5-dimensional stratum corresponding to contrasts between block totals,

Table 6.2: A split-plot field trial of oat varieties.

		N_1	N_2	N_3	N_4			N_1	N_2	N_3	N_4
	V_1	111	130	157	174		V_1	74	89	81	122
I	V_2	117	114	161	141	IV	V_2	64	103	132	133
	V_3	105	140	118	156		V_3	70	89	104	117
	V_1	61	91	97	100		V_1	62	90	100	116
II	V_2	70	108	126	149	V	V_2	80	82	94	126
	V_3	96	124	121	144		V_3	63	70	109	99
	V_1	68	64	112	86		V_1	53	74	118	113
III	V_2	60	102	89	96	VI	V_2	89	82	86	104
	V_3	89	129	132	124		V_3	97	99	119	121

3. A 12-dimensional stratum corresponding to contrasts between variety (or equivalently whole plot) totals within the same block, and

4. A 54-dimensional stratum corresponding to contrasts within whole plots.

(We use the term 'contrast' here to mean a linear function with coefficients adding to zero, and thus representing a comparison.)

Only the overall mean is estimable within stratum 1 and since no degrees of freedom are left for error, this stratum is suppressed in the printed summaries (although a component corresponding to it is present in the fitted model object). Stratum 2 has no information on any treatment effect. Information on the V main effect is only available from stratum 3, and the N main effect and N × V interaction are only estimable within stratum 4.

Multistratum models may be fitted using aov (only), and are specified by a model formula of the form

response ~ mean.formula + Error(*strata.formula*)

In our example the *strata.formula* is B/V, specifying strata 2 and 3; the fourth stratum is included automatically as the "within" stratum, the residual stratum from the strata formula.

The fitted model object is of class aovlist, which is a list of fitted model objects corresponding to the individual strata. Each component is an object of class aov (although in some respects incomplete). There is no anova method function for class aovlist. The appropriate display function is summary which displays separately the anova tables for each stratum (except the first).

```
> oats.aov <- aov(Y ~ Nf*V + Error(B/V), data=oats)
> summary(oats.aov)
Error: B
```

```
                   Df Sum of Sq Mean Sq F Value Pr(F)
    Residuals  5      15875   3175.1

    Error: V %in% B
                   Df Sum of Sq Mean Sq F Value   Pr(F)
    V               2    1786.4  893.18  1.4853 0.27239
    Residuals 10      6013.3  601.33

    Error: Within
                   Df Sum of Sq Mean Sq F Value Pr(F)
    Nf              3    20020  6673.5  37.686 0.0000
    Nf:V            6      322    53.6   0.303 0.9322
    Residuals 45      7969   177.1
```

Polynomial components

There is a clear nitrogen effect, but no evidence of variety differences nor interaction. Since the levels of Nf are quantitative, it is natural to consider partitioning the sums of squares for nitrogen and its interactions into polynomial components. Here the details on factor coding become important.

Since the levels of nitrogen are equally spaced, and we chose an ordered factor, the default contrast matrix is appropriate. We can obtain an analysis of variance table with the degrees of freedom for nitrogen partitioned into components by giving an additional argument to the summary function:

```
> summary(oats.aov, split=list(Nf=list(L=1, Dev=2:3)))
    ....
Error: Within
               Df Sum of Sq Mean Sq F Value   Pr(F)
Nf              3    20020    6673   37.69 0.00000
    Nf: L       1    19536   19536  110.32 0.00000
    Nf: Dev     2      484     242    1.37 0.26528
Nf:V            6      322      54    0.30 0.93220
    Nf:V: L     2      168      84    0.48 0.62476
    Nf:V: Dev   4      153      38    0.22 0.92786
Residuals      45     7969     177
```

The split argument is a list with the name of each component that of some factor appearing in the model. Each component is also a list with components of the form name=int.seq where name is the name for the table entry and int.seq is an integer sequence giving the contrasts to be grouped under that name in the ANOVA table.

Residuals in multistratum analyses: Projections

Residuals and fitted values from the individual strata are available in the usual way by accessing each component as a fitted model object. Thus fitted(oats.aov[[4]]) and resid(oats.aov[[4]]) are vectors of length

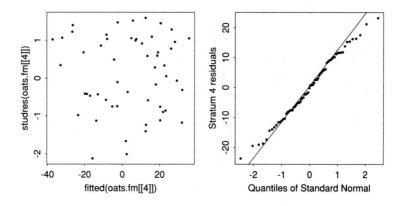

Figure 6.8: Two diagnostic plots for the oats data.

54 representing fitted values and residuals from the last stratum, based on 54 orthonormal linear functions of the original data vector. It is not possible to associate them uniquely with the plots of the original experiment, which for field trials is sometimes very useful.

The function `proj` takes a fitted model object and finds the projections of the original data vector onto the subspaces defined by each line in the analysis of variance tables (including, for multistratum objects, the suppressed table with the grand mean only). The result is a list of matrices, one for each stratum, where the column names for each are the component names from the analysis of variance tables. If the argument `qr=T` has been set when the model was initially fitted this calculation is considerably faster. Two diagnostic plots are shown in Figure 6.8, computed by

```
oats.fm <- update(oats.aov, qr=T)   # not strictly necessary
plot(fitted(oats.fm[[4]]), studres(oats.fm[[4]]))
oats.pr <- proj(oats.fm)
qqnorm(oats.pr[[4]][,"Residuals"], ylab="Stratum 4 residuals")
qqline(oats.pr[[4]][,"Residuals"])
```

Tables of means and components of variance

A better appreciation of the results of the experiment comes from looking at the estimated marginal means. The S-PLUS function `model.tables` calculates tables of the effects, means or residuals and optionally their standard errors. Tables of means are only available for balanced designs, and the standard errors calculated are for differences of means. We now refit the model omitting the quadratic and cubic components for `N` and the `V:N` interaction, but retaining the `V` main effect (since that might be considered a blocking factor). Since the `model.tables` calculation also requires the projections we should fit the model with either `qr=T` or `proj=T` set to avoid later refitting.

```
> oats$Nf1 <- C(oats$Nf, poly, 1)
> oats.aov1 <- aov(Y ~ Nf1 + V + Error(B/V), data=oats, qr=T)
> model.tables(oats.aov1, type="means", se=T)
Tables of means
    ....
 Nf1
 0.0cwt 0.2cwt 0.4cwt 0.6cwt
  81.87  96.61  111.3   126.1
 V
 Golden.rain Marvellous Victory
        104.5      109.8   97.63

Standard errors for differences of means
            Nf1       V
          4.289   7.079
replic. 18.000  24.000
```

When interactions are present the table is a little more complicated.

Finally we use the S-PLUS function `varcomp` to estimate the variance components associated with the lowest three strata. To do this we need to declare B to be a random factor:

```
> is.random(oats$B) <- T
> varcomp(Y ~ Nf1 + V + B/V, data=oats)
Variances:
      B V %in% B Residuals
 214.48    108.94    165.56
    ....
```

In this simple balanced design the estimates are the same as those obtained by equating the residual mean squares to their expectations. The result suggests that the main blocking scheme, at least, has been reasonably effective.

Chapter 7

Generalized Linear Models

Generalized linear models (GLMs) extend linear models to accommodate both non-normal response distributions and transformations to linearity. (We will assume that Chapter 6 has been read before this chapter.) The essay by Firth (1991) gives a good introduction to GLMs; the comprehensive reference is McCullagh & Nelder (1989).

A generalized linear model may be described by the following assumptions:

- There is a response, y, observed independently at fixed values of stimulus variables $x_1, \ldots, x_p$.
- The stimulus variables may only influence the distribution of y through a single linear function called the *linear predictor* $\eta = \beta_1 x_1 + \cdots + \beta_p x_p$.
- The distribution of y has density of the form

$$f(y_i; \theta_i, \varphi) = \exp\left[A_i \left\{ y_i \theta_i - \gamma(\theta_i) \right\} / \varphi + \tau(y_i, \varphi/A_i) \right] \qquad (7.1)$$

  where φ is a *scale parameter* (possibly known), A_i is a known prior weight and parameter θ_i controls the distribution of y_i.
- The mean, μ, is a smooth invertible function of the linear predictor:

$$\mu = m(\eta), \qquad \eta = m^{-1}(\mu) = \ell(\mu)$$

  The inverse function, $\ell(.)$, is called the *link function*.

Note that θ is also an invertible function of μ, in fact $\theta = (\gamma')^{-1}(\mu)$ as we show below. If φ were known the distribution of y would be a one-parameter canonical exponential family. An unknown φ is handled as a nuisance parameter by moment methods.

GLMs allow a unified treatment of statistical methodology for several important classes of models. We will consider a few examples.

Gaussian For a normal distribution $\varphi = \sigma^2$ and we can write

$$\log f(y) = \frac{1}{\varphi} \left\{ y\mu - \tfrac{1}{2}\mu^2 - \tfrac{1}{2}y^2 \right\} - \tfrac{1}{2} \log[2\pi\varphi]$$

so $\theta = \mu$ and $\gamma(\theta) = \theta^2/2$.

Poisson For a Poisson distribution with mean μ we have

$$\log f(y) = y \log \mu - \mu - \log(y!)$$

so $\theta = \log \mu$, $\varphi = 1$ and $\gamma(\theta) = \mu = e^\theta$.

Binomial For a binomial distribution with fixed number of trials a and parameter p we take the response to be $y = s/a$ where s is the number of "successes". The density is

$$\log f(y) = a \left[y \log \frac{p}{1-p} + \log(1-p) \right] + \log \binom{a}{ay}$$

so we take $A_i = a_i$, $\varphi = 1$, θ to be the logit transform of p and $\gamma(\theta) = -\log(1-p) = \log(1+e^\theta)$.

The generalized linear model families handled by the functions supplied in S include `gaussian`, `binomial`, `poisson`, `inverse.gaussian` and `Gamma` response distributions.

Each response distribution allows a variety of link functions to connect the mean with the linear predictor. Those automatically available are given in Table 7.1. The combination of response distribution and link function is called the *family* of the generalized linear model.

Table 7.1: Families and link functions. The default link is denoted by D.

			Family Name		
Link	binomial	Gamma	gaussian	inverse.- gaussian	poisson
logit	D				
probit	•				
cloglog	•				
identity			• D		•
inverse		D			
log		•			D
1/mu^2				D	
sqrt					•

For n observations from a GLM the log-likelihood is

$$l(\theta, \varphi; Y) = \sum_i [A_i \{y_i \theta_i - \gamma(\theta_i)\} / \varphi + \tau(y_i, \varphi/A_i)] \tag{7.2}$$

and this has score function for θ of

$$U(\theta) = A_i \{y_i - \gamma'(\theta_i)\} / \varphi \tag{7.3}$$

From this it is easy to show that

$$\mathrm{E}(y_i) = \mu_i = \gamma'(\theta_i) \qquad \mathrm{var}(y_i) = \frac{\varphi}{A_i} \gamma''(\theta_i)$$

(See, for example, McCullagh & Nelder, 1989 §2.2.) It follows immediately that

$$E\left(\frac{\partial^2 l(\theta, \varphi; y)}{\partial\theta\,\partial\varphi}\right) = 0$$

Hence θ and φ, or more generally β and φ, are *orthogonal parameters*.

The function defined by $V(\mu) = \gamma''(\theta(\mu))$ is called the *variance function*.

For each response distribution the link function $\ell = (\gamma')^{-1}$ for which $\theta \equiv \eta$, is called the *canonical link*. If X is the model matrix, so $\eta = X\beta$, it is easy to see that with φ known, $A_i \equiv 1$ and the canonical link $X^T y$ is a minimal sufficient statistic for β. Also using (7.3) the score equations for the regression parameters β reduce to

$$X^T y = X^T \widehat{\mu} \tag{7.4}$$

This relation is sometimes described by saying that the "observed and fitted values have the same marginal totals". Equation (7.4) is the basis for certain simple fitting procedures, for example Stevens' algorithm for fitting an additive model to a non-orthogonal two-way layout by alternate row and column sweeps, (Stevens, 1948), and the Deming-Stephan (1940) or *iterative proportional scaling* algorithm for fitting log-linear models to frequency tables (see Darroch & Ratcliff, 1972).

Table 7.2 shows the canonical links and variance functions. The canonical link function is the default link for the families catered for by the S software (except for the inverse Gaussian, where the factor of 2 is dropped).

Table 7.2: Canonical (default) links and variance functions.

Family	Canonical link	Name	Variance	Name
binomial	$\log(\mu/(1-\mu))$	logit	$\mu(1-\mu)$	mu(1-mu)
Gamma	$-1/\mu$	inverse	μ^2	mu^2
gaussian	μ	identity	1	constant
inverse.gaussian	$-2/\mu^2$	1/mu^2	μ^3	mu^3
poisson	$\log\mu$	log	μ	mu

Iterative estimation procedures

Since explicit expressions for the maximum likelihood estimators are not usually available estimates must be calculated iteratively. It is convenient to give an outline of the iterative procedure here; for a more complete description the reader is referred to McCullagh & Nelder (1989, §2.5, pp. 40ff) or Firth (1991, §3.4).

An initial estimate to the linear predictor is found by some guarded version of $\widehat{\eta}_0 = \ell(y)$. (Guarding is necessary to prevent problems such as taking logarithms of zero.) Define *working weights* W and *working values* z by

$$W = \frac{A}{V}\left(\frac{\mathrm{d}\mu}{\mathrm{d}\eta}\right)^2, \qquad z = \eta + \frac{y-\mu}{\mathrm{d}\mu/\mathrm{d}\eta}$$

Initial values for W_0 and z_0 can be calculated from the initial linear predictor.

At iteration k a new approximation to the estimate of β is found by a weighted regression of the working values z_k on X with weights W_k. This provides a new linear predictor and hence new working weights and values which allow the iterative process to continue. The difference between this process and a Newton-Raphson scheme is that the Hessian matrix is replaced by its expectation. This statistical simplification was apparently first used in Fisher (1925), and is often called *Fisher scoring*. The estimate of the large sample variance of $\widehat{\beta}$ is $\widehat{\varphi}(X^T\widehat{W}X)^{-1}$, which is available as a by-product of the iterative process. It is easy to see that the iteration scheme for $\widehat{\beta}$ does not depend on the scale parameter φ.

This iterative scheme depends on the response distribution only through its mean and variance functions. This has led to ideas of *quasi-likelihood*, implemented in the family `quasi`.

The analysis of deviance

A *saturated model* is one in which the parameters θ_i, or almost equivalently the linear predictors η_i, are unconstrained. It is then clear from (7.3) that the maximum likelihood estimator of θ_i is obtained by $y_i = \gamma'(\widehat{\theta_i}) = \widehat{\mu_i}$ which is also a special case of (7.4). Denote the saturated model by S.

Assume temporarily that the scale parameter φ is known and has value $\varphi = 1$. Let M be a model involving $p < n$ regression parameters, and let $\widehat{\theta_i}$ be the maximum likelihood estimate of θ_i under M. Twice the log-likelihood ratio statistic for testing M within S is given by

$$D_M = 2\sum_{i=1}^{n} A_i \left[\left\{ y_i\theta(y_i) - \gamma\left(\theta(y_i)\right) \right\} - \left\{ y_i\widehat{\theta_i} - \gamma(\widehat{\theta_i}) \right\} \right] \qquad (7.5)$$

The quantity D_M is called the *deviance* of model M, even when the scale parameter is unknown or is known to have a value other than one. In the latter case D_M/φ, the difference in twice the log-likelihood, is known as the *scaled deviance*. (Confusingly, sometimes either is called the *residual deviance*, for example McCullagh & Nelder, p. 119.)

For a Gaussian family with identity link the scale parameter φ is the variance and D_M is the residual sum of squares. Hence in this case the scaled deviance has distribution

$$D_M/\varphi \sim \chi^2_{n-p} \qquad (7.6)$$

leading to the customary unbiased estimator

$$\widehat{\varphi} = \frac{D_M}{n-p} \qquad (7.7)$$

In other cases the distribution (7.6) for the deviance under M may be approximately correct suggesting $\widehat{\varphi}$ as an approximately unbiased estimator of φ. It should be noted that sufficient (if not always necessary) conditions under which (7.6) becomes approximately true are that the individual distributions for the components y_i should become closer to normal form and the link effectively closer to an identity link. The approximation will often *not* improve as the sample size n increases since the number of parameters under S also increases and the usual likelihood ratio approximation argument does not apply. Nevertheless, (7.6) may sometimes be a good approximation, for example in a binomial GLM with large values of n_i. Firth (1991, p. 69) discusses this approximation, including the extreme case of a binomial GLM with only one observation per case.

Let $M_0 \subset M$ be a submodel with $q < p$ regression parameters and consider testing M_0 within M. If φ is known, by the usual likelihood ratio argument under M_0 we have a test given by

$$\frac{D_{M_0} - D_M}{\varphi} \mathrel{\dot\sim} \chi^2_{p-q} \tag{7.8}$$

where $\mathrel{\dot\sim}$ denotes 'is approximately distributed as'. The distribution is exact only in the Gaussian family with identity link. If φ is not known, by analogy with the Gaussian case it is customary to use the approximate result

$$\frac{(D_{M_0} - D_M)}{\widehat\varphi (p - q)} \mathrel{\dot\sim} F_{p-q,n-p} \tag{7.9}$$

although this must be used with some caution in non-Gaussian cases.

Some of the fitting functions use a quantity AIC defined by

$$D_M + 2p\widehat\varphi \tag{7.10}$$

(Chambers & Hastie, 1992, p. 234). If the scale parameter is one, this corresponds up to an additive constant to the commonly used definition of AIC (Akaike, 1974) of

$$AIC = -2 \text{ maximized log likelihood} + 2 \# \text{parameters}$$

and if φ is known but not one it also differs by a factor. In both cases choosing models to minimize AIC will give the same result. However, as we saw in Section 6.6, the concepts differ when φ is estimated. It is important when using (7.10) to hold $\widehat\varphi$ constant when comparing models.

7.1 Functions for generalized linear modelling

The linear predictor part of a generalized linear model may be specified by a model formula using the same notation and conventions as linear models. Generalized linear models also require the family to be specified, that is the response distribution, the link function and perhaps the variance function for `quasi` models.

The fitting function is `glm` for which main arguments are

```
glm(formula, family, data, weights, control)
```

The `family` argument is usually given as the name of one of the standard family functions listed under "Family Name" in Table 7.1. Where there is a choice of links, the name of the link may also be supplied in parentheses as a parameter, for example `binomial(link=probit)`. (The variance function for the `quasi` family may also be specified in this way.) For user-defined families (such as our `neg.bin` discussed in Section 7.4) other arguments to the family function may be allowed or even required.

Prior weights A_i may be specified using the `weights` argument. For binomial models these are implied and should not be specified separately.

The iterative process can be controlled by many parameters. The only ones which are at all likely to need altering are `maxit`, which controls the maximum number of iterations and the default value of 10 is occasionally too small, and `trace` which will often be set as `trace=T` to trace the iterative process.

Generic functions with methods for `glm` objects include `coef`, `resid`, `print`, `summary` and `deviance`. It is useful to have a `glm` method function to extract the variance-covariance matrix of the estimates. This can be done using part of the result of `summary`:

```
vcov.glm <- function(obj)
{
    so <- summary(obj)
    so$dispersion * so$cov.unscaled
}
```

Our library contains the generic function and methods for class `lm` and `nls` objects as well as `glm`.

For `glm` fitted model objects the `anova` function allows an additional argument `test` to specify which test is to be used. Two possible choices are `test="Chisq"` for chi-squared tests using (7.8) and `test="F"` for F-tests using (7.9). The default is `test="Chisq"` for the binomial and Poisson families, otherwise `test="F"`.

Prediction and residuals

The `predict` method function for `glm` has a `type` argument to specify what is to be predicted. The default is `type="link"` which produces predictions of the linear predictor η. Predictions on the scale of the mean μ (for example, the fitted values $\widehat{\mu}_i$) are specified by `type="response"`.

For `glm` models there are four types of residual that may be requested, known as *deviance, working, Pearson* and *response* residuals. The response residuals are simply $y_i - \widehat{\mu}_i$. The Pearson residuals are a standardized version of the response residuals, $(y_i - \widehat{\mu}_i)/\sqrt{\widehat{V}_i}$. The working residuals come from the last stage of the iterative process, $(y_i - \widehat{\mu}_i) \, / \, \mathrm{d}\mu_i/\mathrm{d}\eta_i$. The deviance residuals d_i are defined as the signed square roots of the summands of the deviance (7.5) taking the same sign as $y_i - \widehat{\mu}_i$.

For Gaussian families all four types of residual are identical. For binomial and Poisson GLMs the sum of the Pearson residuals is the Pearson chi-squared statistic, which often approximates the deviance, and the deviance and Pearson residuals are normally then very similar.

Method functions for the resid function have an argument type which defaults to type="deviance" for objects of class glm. Other values are "response", "pearson" or "working"; these may be abbreviated to the initial letter. Deviance residuals are the most useful for diagnostic purposes.

Concepts of leverage and its effect on the fit are as important for GLMs as they are in linear regression, and are discussed in some detail by Davison & Snell (1991) and extensively for binomial GLMs by Collett (1991). On the other hand, they seem less often used, as GLMs are most often used either for simple regressions or for contingency tables where, as in designed experiments, high-leverage can not occur.

Robust versions of GLM models are discussed in Chapter 8 on page 215.

The default Gaussian family

A call to glm with the default family, gaussian, achieves the same purpose as a call to lm, less efficiently. The gaussian family is not provided with a choice of links, so no argument is allowed. If a problem requires a Gaussian family with a non-standard link, this can usually be handled using the quasi family. Although the gaussian family is the default, it is virtually never used in practice since lm or aov are to be preferred.

7.2 Binomial data

Consider first a small example. Collett (1991, p. 75) reports an experiment on the toxicity of the tobacco budworm *Heliothis virescens* to doses of the pyrethroid *trans*-cypermethrin to which the moths were beginning to show resistance. Batches of twenty moths of each sex were exposed for 3 days to the pyrethroid and the number in each batch which were dead or knocked down was recorded. The results were

		dose				
sex	1	2	4	8	16	32
Male	1	4	9	13	18	20
Female	0	2	6	10	12	16

The doses were in μg. We fit a logistic regression model using $\log_2$ (dose) since the doses are powers of two. To do so we must specify the vector of n_i. This is done using glm with the binomial family in one of two ways:

1. If the response is a vector it is assumed to hold binary data and so must be a vector of 0s and 1s. In this case it is assumed that $n_i \equiv 1$.

2. If the response is a two-column matrix it is assumed that the first column holds the number of successes and the second holds the number of failures for each trial.

In both cases the response is y_i/n_i, so the means μ_i are the probabilities p_i. Hence fitted yields probabilities, not binomial means.

Since we have binomial data we use the second possibility:

```
> options(contrasts=c("contr.treatment", "contr.poly"))
> ldose <- rep(0:5, 2)
> numdead <- c(1, 4, 9, 13, 18, 20, 0, 2, 6, 10, 12, 16)
> sex <- factor(rep(c("M", "F"), c(6, 6)))
> SF <- cbind(numdead, numalive=20-numdead)
> budworm.lg <- glm(SF ~ sex*ldose, family=binomial)
> summary(budworm.lg)
    ....
Coefficients:
                Value Std. Error  t value
(Intercept) -2.99354    0.55253 -5.41789
        sex  0.17499    0.77816  0.22487
      ldose  0.90604    0.16706  5.42349
  sex:ldose  0.35291    0.26994  1.30735
    ....
    Null Deviance: 124.88 on 11 degrees of freedom
Residual Deviance: 4.9937 on 8 degrees of freedom
    ....
```

This shows some evidence of a difference in slope between the sexes. Note that we use treatment contrasts to make interpretation easier. Since female is the first level of sex (they are in alphabetical order) the parameter for sex:ldose represents the increase in slope for males. We can plot the data and the fitted curves by

```
plot(c(1,32), c(0,1), type="n", xlab="dose",
   ylab="prob", log="x")
text(2^ldose, numdead/20,as.character(sex))
ld <- seq(0, 5, 0.1)
lines(2^ld, predict(budworm.lg, data.frame(ldose=ld,
   sex=factor(rep("M", length(ld)), levels=levels(sex))),
   type="response"))
lines(2^ld, predict(budworm.lg, data.frame(ldose=ld,
   sex=factor(rep("F", length(ld)), levels=levels(sex))),
   type="response"))
```

see Figure 7.1. Note that when we set up a factor for the new data we must specify all the levels or both lines would refer to level one of sex. (Of course, here we could have predicted for both sexes and plotted separately, but it helps to understand the general mechanism needed.)

The apparently non-significant sex effect in this analysis has to be interpreted carefully. Since we are fitting separate lines for each sex, it tests the (uninteresting) hypothesis that the lines do not differ at zero log-dose. If we re-parameterize to include the intercept at dose 8 we find

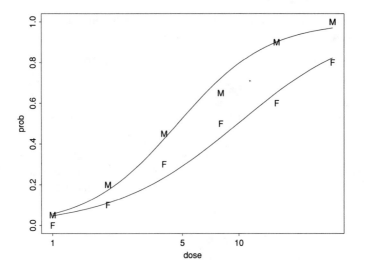

Figure 7.1: Tobacco budworm destruction *versus* dosage of *trans*-cypermethrin by sex.

```
> budworm.lgA <- glm(SF ~ sex*I(ldose-3), family=binomial)
> summary(budworm.lgA)$coefficients
                     Value Std. Error t value
   (Intercept) -0.27543    0.23049 -1.1950
           sex  1.23373    0.37694  3.2730
    I(ldose - 3)  0.90604    0.16706  5.4235
sex:I(ldose - 3)  0.35291    0.26994  1.3074
```

which shows a significant difference between the sexes at dose 8. The model fits
very well (4.9937 is a small value for a χ^2_8 variate, and the estimated probabilities
are based on a reasonable number of trials) so there is no suspicion of non-linearity.
We can confirm this by an analysis of deviance:

```
> anova(update(budworm.lg, . ~ sex + ldose + factor(ldose)),
      test="Chisq")

Terms added sequentially (first to last)
            Df Deviance Resid. Df Resid. Dev Pr(Chi)
       NULL                  11      124.88
       sex  1     6.08       10      118.80 0.01370
     ldose  1   112.04        9        6.76 0.00000
factor(ldose)  4     1.74        5        5.01 0.78267
```

which shows no evidence of a better fit using a separate probability for each
dose. Note how anova when applied to a glm fitted object produces a sequential
analysis of deviance, which will always be order-dependent. The additional
argument test="Chisq" to the anova method may be used to specify tests
using equation (7.8). The default corresponds to test="none". The other
possible choices are "F" and "Cp", neither of which is appropriate here.

Our analysis so far suggests a model with parallel lines (on logit scale) for each sex. We re-parameterize them as separate lines:[1]

```
> budworm.lg0 <- glm(SF ~ sex + ldose - 1, family=binomial)
> summary(budworm.lg0)$coefficients
        Value Std. Error t value
 sexF -3.4732    0.46829 -7.4166
 sexM -2.3724    0.38539 -6.1559
ldose  1.0642    0.13101  8.1230
```

We now set out to estimate for each sex the dose at which the probability of the moth being knocked down or dead is 50%, usually called the LD50. More generally consider estimating x_p, the log-dose for which the probability is p:

$$x_p = (\ell(p) - \beta_0)/\beta_1$$

where β_0 and β_1 are the slope and intercept. Our library contains functions to calculate and print $\widehat{x}_p$ and its asymptotic standard error:

```
dose.p <- function(obj, cf = 1:2, p = 0.5)
{
    eta <- family(obj)$link(p)
    b <- coef(obj)[cf]
    dose <- (eta - b[1])/b[2]
    names(dose) <- paste("p = ", format(p), ":", sep = "")
    pd <-  - cbind(1, dose)/b[2]
    vr <- ((pd %*% vcov(obj)[cf, cf]) * pd) %*% c(1, 1)
    attr(dose, "SE") <- sqrt(vr)
    attr(dose, "p") <- p
    class(dose) <- "glm.dose"
    dose
}
print.glm.dose <- function(x, ...)
{
    M <- cbind(x, attr(x, "SE"))
    dimnames(M) <- list(names(x), c("Dose", "SE"))
    x <- M
    NextMethod("print")
}
```

Notice that the `family` method function is used to extract the link function as an S function. Consider estimating x_p for the budworm data for females at three quartiles:

```
> xp.F <- dose.p(budworm.lg0, c(1,3), (1:3)/4)
> xp.F
              Dose      SE
p = 0.25: 2.2313 0.24983
p = 0.50: 3.2636 0.22971
p = 0.75: 4.2959 0.27462
```

---

[1] S-PLUS 3.2 rel 1 for Sun Sparc and for Windows incorrectly label the coefficients in this model; a correction is in the `fixes` directory on the disk.

Note that this x_p is on a $\log_2$ scale, so we need to convert to the original scale. Consider approximate 95% confidence intervals using two standard errors on either side of the estimate:

```
> u.F <- xp.F + 2*attr(xp.F, "SE")
> l.F <- xp.F - 2*attr(xp.F, "SE")
> 2^cbind(l.F, xp.F, u.F)
                      xp.F
p = 0.25:  3.3210  4.6955  6.6388
p = 0.50:  6.9846  9.6037 13.2049
p = 0.75: 13.4232 19.6426 28.7434
```

In biological assays such as this example it used to be conventional to perform a *probit* analysis, using the probit link in a binomial family. The logistic and standard normal distributions can approximate each other very well at least between the 10th and 90th percentiles by a simple scale change in the abscissa. (See for example Cox & Snell 1989, p. 21.) Unless the estimated probabilities are concentrated in the tails, probit and logit models would be expected to give equivalent results. We can confirm this for the budworm data by fitting a probit model and looking at, for example, the estimated x_p's for females:

```
> budworm.pb0 <- update(budworm.lg0,
      family=binomial(link=probit))
> xp.F <- dose.p(budworm.pb0, c(1,3), (1:3)/4)
> u.F <- xp.F + 2*attr(xp.F, "SE")
> l.F <- xp.F - 2*attr(xp.F, "SE")
> 2^cbind(l.F, xp.F, u.F)
                      xp.F
p = 0.25:  3.2816  4.5669  6.3556
p = 0.50:  7.0109  9.5646 13.0484
p = 0.75: 13.8372 20.0313 28.9983
```

A binary data example: Low birth weight in infants

Hosmer & Lemeshow (1989) give a dataset on 189 births at a US hospital, with the main interest being in low birth weight. The following variables are available in our data frame `birthwt`:

low	birth weight less than 2.5kg (0/1)
age	age of mother in years
lwt	weight of mother (lbs) at last menstrual period
race	white / black / other
smoke	smoking status during pregnancy (0/1)
ptl	number of previous premature labours
ht	history of hypertension (0/1)
ui	has uterine irritability (0/1)
ftv	number of physician visits in the first trimester
bwt	actual birth weight (grams)

Although the actual birth weights are available, we shall be concentrating on predicting if the birth weight is low from the remaining variables. The dataset contains a small number of pairs of rows which are identical apart from the ID; it is possible that these refer to twins but identical birth weights seem unlikely.

We will use a logistic regression with a binomial (in fact 0/1) response. It is worth considering carefully how to use the variables. It is unreasonable to expect a linear response with `ptl`. Since the numbers with values greater than one are so small we reduce it to a indicator of past history. Similarly, `ftv` can be reduced to three levels. With non-Gaussian GLMs it is usual to use treatment contrasts, as these are easier to interpret.

```
> options(contrasts=c("contr.treatment", "contr.poly"))
> attach(birthwt)
> race <- factor(race, labels=c("white", "black", "other"))
> table(ptl)
   0   1 2 3
 159 24 5 1
> ptd <- factor(ptl > 0)
> table(ftv)
   0   1  2 3 4 6
 100 47 30 7 4 1
> ftv <- factor(ftv)
> levels(ftv)[-(1:2)] <- "2+"
> table(ftv)   # as a check
   0   1 2+
 100 47 42
> bwt <- data.frame(low=factor(low), age, lwt, race,
     smoke=(smoke>0), ptd, ht=(ht>0), ui=(ui>0), ftv)
> detach("birthwt"); rm(race, ptd, ftv)
```

We can then fit a full logistic regression, and omit the rather large correlation matrix from the summary.

```
> birthwt.glm <- glm(low ~ ., family=binomial, data=bwt)
> summary(birthwt.glm, correlation=F)
    ....
Coefficients:
                Value Std. Error  t value
(Intercept)  0.823013 1.2440732  0.66155
        age -0.037234 0.0386777 -0.96267
        lwt -0.015653 0.0070759 -2.21214
  raceblack  1.192409 0.5357458  2.22570
  raceother  0.740681 0.4614609  1.60508
      smoke  0.755525 0.4247645  1.77869
        ptd  1.343761 0.4804445  2.79691
         ht  1.913162 0.7204344  2.65557
         ui  0.680195 0.4642156  1.46526
       ftv1 -0.436379 0.4791611 -0.91071
      ftv2+  0.179007 0.4562090  0.39238
```

```
        Null Deviance: 234.67 on 188 degrees of freedom
    Residual Deviance: 195.48 on 178 degrees of freedom
```

Since the data are binary, even if the model is correct there is no guarantee that the deviance will have even an approximately chi-squared distribution. Since the value is about in line with its degrees of freedom there seems no serious reason to question the fit. Rather than select a series of sub-models by hand, we make use of the `step` function. This uses (an approximation to) AIC to select models. We first allow this to drop variables, then to choose pairwise interactions. By default argument `trace` is true and produces voluminous output.

```
> birthwt.step <- step(birthwt.glm, trace=F)
> birthwt.step$anova
Initial Model:
low ~ age + lwt + race + smoke + ptd + ht + ui + ftv
Final Model:
low ~ lwt + race + smoke + ptd + ht + ui

    Step Df Deviance Resid. Df Resid. Dev    AIC
1                          178     195.48 217.48
2 - ftv  2   1.3582       180     196.83 214.83
3 - age  1   1.0179       181     197.85 213.85
> birthwt.step2 <- step(birthwt.glm, ~ .^2, trace=F)
> birthwt.step2$anova
Initial Model:
low ~ age + lwt + race + smoke + ptl + ht + ui + ftv
Final Model:
low ~ age + lwt + smoke + ptd + ht + ui + ftv + age:ftv
     + smoke:ui

          Step Df Deviance Resid. Df Resid. Dev    AIC
1                              178     195.48 217.48
2  + age:ftv -2  -12.475       176     183.00 209.00
3 + smoke:ui -1   -3.057       175     179.94 207.94
4     - race  2    3.130       177     183.07 207.07

> summary(birthwt.step2)$coef
                Value Std. Error  t value
(Intercept) -0.582520   1.418729 -0.41059
        age  0.075535   0.053880  1.40190
        lwt -0.020370   0.007465 -2.72878
      smoke  0.780057   0.419249  1.86061
        ptd  1.560205   0.495741  3.14722
         ht  2.065549   0.747204  2.76437
         ui  1.818252   0.664906  2.73460
       ftv1  2.920800   2.278843  1.28170
      ftv2+  9.241693   2.631548  3.51188
    ageftv1 -0.161809   0.096472 -1.67726
   ageftv2+ -0.410873   0.117548 -3.49537
   smoke:ui -1.916401   0.970786 -1.97407
```

```
> table(bwt$low, predict(birthwt.step2) > 0)
   FALSE TRUE
0    116   14
1     28   31
```

Note that although both `age` and `ftv` were dropped previously, their interaction is included, the slopes on `age` differing considerably within the three `ftv` groups. The AIC criterion penalizes terms less severely than an F-test would, and so although `smoke:ui` reduces the AIC it is only just significant at the 5% level. We also considered three-way interactions, but none were chosen.

Residuals are not easy to examine with zero-one observations. None are particularly large here.

The next step of the analysis is to examine the linearity in age and mother's weight. To do so it is easiest to use generalized additive models and include smooth terms. This will be considered in Section 10.1 after generalized additive models have been discussed, but no serious evidence of non-linearity on the logistic scale arises, suggesting the model chosen here is acceptable.

An alternative approach is to predict the actual live birth weight and later threshold at 2.5 kilograms. This is left as an exercise for the reader; surprisingly it produces somewhat worse predictions with around 52 errors. We also consider a tree-based prediction rule in Section 13.3, with a smaller AIC but similar error rate.

7.3   Poisson models

The canonical link for the Poisson family is `log`, and the major use of this family is to fit surrogate Poisson log-linear models to what is actually multinomial frequency data. Such log-linear models have a large literature. (For example, Plackett, 1974; Bishop, Fienberg & Holland, 1975; Haberman, 1978, 1979; Goodman, 1978; Whittaker, 1990). The surrogate Poisson models approach we adopt here is in line with that of McCullagh & Nelder (1989).

It is convenient to divide the factors classifying a multi-way frequency table into *response* and *stimulus* factors. Stimulus factors have their marginal totals fixed in advance (or for the purposes of inference). The main interest lies in the conditional probabilities of the response factor given the stimulus factors.

It is well known that the conditional distribution of a set of independent Poisson random variables given their sum is multinomial with probabilities given by the ratios of the Poisson means to their total. This result is applied to the counts for the multi-way response within each combination of stimulus factor levels. This allows models for multinomial data with a multiplicative probability specification to be fitted and tested using Poisson log-linear models.

Identifying the multinomial model corresponding to any surrogate Poisson model is straightforward. Suppose A, B, ... are factors classifying a frequency table. The *minimum model* is the interaction of all stimulus factors, and must be included for the analysis to respect the totals over response factors. This model is usually of no interest, corresponding to a uniform distribution over response factors independent of the stimulus factors. Interactions between response and stimulus factors indicate interesting structure. For large models is often helpful to use a graphical structure to represent the conditional independencies (Whittaker, 1990).

A four-way contingency table

As an example of a surrogate Poisson model analysis consider the data in Table 7.3. This shows the result of a detergent brand preference study where the respondents are also classified according to the temperature and softness of their washing water and whether or not they were previous users of brand M. The dataset was reported by Ries & Smith (1963) and analysed by Cox & Snell (1989).

Table 7.3: Numbers preferring brand X of a detergent to brand M.

M user?	No				Yes			
Temperature	Low		High		Low		High	
Preference	X	M	X	M	X	M	X	M
Water softness								
Hard	68	42	42	30	37	52	24	43
Medium	66	50	33	23	47	55	23	47
Soft	63	53	29	27	57	49	19	29

We can enter this dataset by

```
detg <- cbind(expand.grid(Brand=c("X","M"),
  Temp=c("Low","High"), M.user=c("N","Y"),
  Soft=c("Hard","Medium","Soft")),
  Fr= c(68,42,42,30,37,52,24,43,  66,50,33,23,47,55,23,47,
        63,53,29,27,57,49,19,29))
detg$Soft <- ordered(detg$Soft,
  levels=c("Soft","Medium","Hard"))
```

The levels of Soft are specified to ensure their ordering is correct.

We study how the proportions of users preferring one or other brand varies with the other factors. In our terminology Brand is the only response factor and M.user, Temp and Soft are stimulus factors. The minimum model is M.user*Temp*Soft. The aim is to arrive at the simplest model not contradicted by the data. The best way to achieve this is to be guided by any insights the research worker can provide and to test a sequence of models suggested by the context. In the absence of such detailed information, and partly for illustration, we note here what the automatic stepwise procedure suggests as a final model. Note that we include the minimum model within the lower bound on the scope of step, and start with the simplest interesting model, which has probability of choice independent of the stimulus factors.

```
> detg.m0 <- glm(Fr ~ M.user*Temp*Soft + Brand,
                 family=poisson, data=detg)
> detg.m0
    . . . .
Degrees of Freedom: 24 Total; 11 Residual
Residual Deviance: 32.826
> detg.step <- step(detg.m0, list(lower=formula(detg.m0),
    upper= ~ .^3), scale=1, trace=F)
> detg.step$anova
Initial Model:
Fr ~ M.user * Temp * Soft + Brand

Final Model:
Fr ~ M.user + Temp + Soft + Brand + M.user:Temp + M.user:Soft +
    Temp:Soft + M.user:Brand + Temp:Brand + M.user:Temp:Soft +
    Brand:M.user:Temp
```

	Step Df	Deviance	Resid. Df	Resid. Dev	AIC
1			11	32.826	58.826
2	+ M.user:Brand -1	-20.581	10	12.244	40.244
3	+ Temp:Brand -1	-3.800	9	8.444	38.444
4	+ Brand:M.user:Temp -1	-2.788	8	5.656	37.656

The effect of M.user is overwhelmingly significant, but there is a suggestion of an effect of Temp and a M.user × Temp interaction.

It is helpful to re-fit the model keeping together terms associated with the minimum model, which we do using keep.order and omit these in our display:

```
> detg.mod <- glm(terms(Fr ~ M.user*Temp*Soft +
                  Brand*M.user*Temp, keep.order=T),
                  family=poisson, data=detg)
> summary(detg.mod, correlation=F)
Coefficients:
                      Value Std. Error    t value
    . . . .
          Brand -0.306470   0.109420   -2.80087
    Brand:M.user  0.407566   0.159608    2.55354
     Brand:Temp  0.044106    0.18463    0.23889
Brand:M.user:Temp  0.444267   0.26673    1.66560

    Null Deviance: 118.63 on 23 degrees of freedom
Residual Deviance: 5.656 on 8 degrees of freedom
```

From the sign of the M.user term previous users of brand M are less likely to prefer brand X. The interaction term, though non-significant, suggests that for M users this proportion differs for those who wash at low and high temperatures.

Equivalence to a binomial logit analysis

When there is just one response factor with two levels, as here, it is also possible to treat it as binomial data. It is important to note that if we do so using the logit link the results are exactly equivalent to the log-linear analysis:

```
> attach(detg)
> deterg <- cbind(detg[Brand=="X", -4],
        M = detg[Brand=="M","Fr"])
> detach()
> detg.lg <- glm(cbind(M, Fr) ~ M.user*Temp,
        family=binomial, data=deterg)
> summary(detg.lg, correlation=F)
Coefficients:
                 Value Std. Error t value
(Intercept) -0.306470    0.10941 -2.8011
     M.user  0.407566    0.15960  2.5537
       Temp  0.044106    0.18462  0.2389
M.user:Temp  0.444267    0.26670  1.6658

    Null Deviance: 32.826 on 11 degrees of freedom
Residual Deviance: 5.656 on 8 degrees of freedom
```

The null deviance is for a model with just an intercept, which corresponds to the surrogate Poisson model `detg.m0`.

Fitting by iterative proportional scaling

The function `loglin` fits log-linear models by iterative proportional scaling. This starts with an array of fitted values that has the correct multiplicative structure, (for example with all values equal to 1) and makes multiplicative adjustments so that the observed and fitted values come to have the same marginal totals, in accordance with equation (7.4). (See Darroch & Ratcliff, 1972.) This is usually very much faster than GLM fitting but is less flexible.

To use `loglin` we need to form the frequencies into an array. A simple way to do this is to construct a matrix subscript from the factors:

```
attach(detg)
detg.tab <- table(M.user, Temp, Soft, Brand)
detg.tab[cbind(M.user, Temp, Soft, Brand)] <- Fr
detg.ips <- loglin(detg.tab,  margin=list(c(1,2,3), c(1,2,4)),
                   fit=T)
c(detg.ips$df, detg.ips$lrt, detg.ips$pearson)
[1] 8.000 5.656 5.650
```

The model is specified by giving the margins to be fitted, which correspond to the two third-order interactions in the final `glm` model. The fitted values are returned as a table in component `fit`; there are no parameter values in this approach.

7.4   A negative binomial family

It is possible to define a new GLM family. We illustrate the procedure using a negative binomial family with known shape parameter, a restriction which will be relaxed later.

Using the negative binomial distribution in modelling is important in its own right and has a long history. An excellent modern reference is Lawless (1987). The variance is greater than the mean, suggesting its use be considered for frequency data where this is a prominent feature, often loosely called "overdispersed Poisson data".

The negative binomial can arise from a two-stage model for the distribution of a discrete variable Y. We suppose there is an unobserved random variable E having a gamma distribution $\text{gamma}(\theta)/\theta$, that is with mean 1 and variance $1/\theta$. Then the model postulates that conditionally on E, Y is Poisson with mean μE. Thus:

$$Y \mid E \sim \text{Poisson}(\mu E), \qquad \theta E \sim \text{gamma}(\theta)$$

The marginal distribution of Y is then negative binomial with probability function, mean and variance given by

$$\text{E}(Y) = \mu, \quad \text{var}(Y) = \mu + \mu^2/\theta, \quad f_Y(y; \theta, \mu) = \frac{\Gamma(\theta + y)}{\Gamma(\theta)\, y!} \frac{\mu^y\, \theta^\theta}{(\mu + \theta)^{\theta + y}}$$

If θ is known, as for the present we assume it is, this distribution has the general form (7.1).

A function `make.family` is available to piece together the information required for the `family` argument of the `glm` fitting function. It requires three main arguments

name   a character string giving the name of the family,

link   a list supplying information about the link function, its inverse and derivative and an initialization expression,

variance   a list specifying the variance and deviance functions.

The two datasets `glm.links` and `glm.variances` are matrices of lists with examples that can be used as templates. We have in mind a negative binomial model for the Quine data introduced in Section 6.6, so we consider only a `log` link. The following function provides the family, passing the parameter θ as a required argument.

```
neg.bin <- function(theta = stop("theta must be given"))
{
   nb.lnk <- list( names = "Log: log(mu)",
        link = function(mu) log(mu),
        inverse = function(eta) exp(eta),
        deriv = function(mu) 1/mu,
        initialize = expression(mu <- y + (y == 0)/6) )
   nb.var <- list(
```

```
        names = "mu + mu^2/theta",
        variance = substitute( function(mu, th = .Theta)
            mu * (1 + mu/th), list(.Theta = theta)),
        deviance = substitute(
            function(mu, y, A, residuals = F, th = .Theta)
            {
                devi <- 2 * A * (y * log(pmax(1, y)/mu) -
                    (y + th) * log((y + th)/(mu + th)))
                if(residuals) sign(y - mu) * sqrt(abs(devi))
                else sum(devi)
            }, list(.Theta = theta) )
        )
    make.family("Negative Binomial", link = nb.lnk,
        variance = nb.var)
}
```

A negative binomial model for the Quine data

A Poisson model for the Quine data has an excessively large deviance:

```
> glm(Days ~ .^4, family=poisson, data=quine)
    ....
Degrees of Freedom: 146 Total; 118 Residual
Residual Deviance: 1173.9
```

Inspection of the mean–variance relationship in Figure 6.5 suggests a negative binomial model with $\theta \approx 2$ might be appropriate. We will assume at first that $\theta = 2$ is known. A negative binomial model may be fitted by

```
quine.nb <- glm(Days ~ .^4, family=neg.bin(2), data=quine)
```

The standard generic functions may now be used to fit sub-models, produce analysis of variance tables, and so on. For example let us check the final model found in Chapter 6. (The output has been edited.)

```
> quine.nb0 <- update(quine.nb, . ~ Sex/(Age + Eth*Lrn))
> anova(quine.nb0, quine.nb, test="Chi")
  Resid. Df Resid. Dev   Test  Df Deviance  Pr(Chi)
1       132     198.51
2       118     171.98 1 vs. 2  14   26.527 0.022166
```

which suggests that model to be an over-simplification in the fixed-θ negative binomial setting.

Consider now what happens when θ is estimated rather than held fixed. The function negative.binomial supplied with our library is similar to neg.bin defined above, but allows more links. We have also included a function glm.nb, a modification of glm which incorporates maximum likelihood estimation of θ. This has summary and anova methods; the latter produces likelihood ratio tests for the sequence of fitted models. (For deviance tests to be applicable the θ parameter has to be held constant for all fitted models.)

The following models summarize the results of a manual selection (step does not work with glm.nb), and indicate model quine.nb2 .

```
> quine.nb1 <- glm.nb(Days ~ Sex/(Age + Eth*Lrn), data=quine)
> quine.nb2 <- update(quine.nb1, . ~ . + Sex:Age:Lrn)
> quine.nb3 <- update(quine.nb2, Days ~ .^4)
> anova(quine.nb1, quine.nb2, quine.nb3)
Likelihood ratio tests of Negative Binomial Models
   ....
     theta Resid. df  2 x log-lik.   Test   df LR stat.  Pr(Chi)
1 1.5980       132         10254
2 1.6869       128         10262 1 vs 2    4   7.6272 0.10623
3 1.9284       118         10278 2 vs 3   10  16.0738 0.09754
```

Testing the first model within the third is still borderline significant, so model 2, which is equivalent to `Sex*Lrn/(Age + Eth)`, should be retained.

The estimate of θ and its standard error are available from the summary or as components of the fitted model object:

```
> c(theta=quine.nb2$theta, SE=quine.nb2$SE)
  theta      SE
1.6869 0.22677
```

We can perform some diagnostic checks by examining the deviance residuals:

```
rs <- resid(quine.nb2, type="deviance")
plot(predict(quine.nb2), rs, xlab="Linear predictors",
    ylab="Deviance residuals")
abline(h=0, lty=2)
qqnorm(rs, ylab="Deviance residuals")
qqline(rs)
```

The result is shown in Figure 7.2. There is perhaps a little skewness, but no indication of serious violations of the model assumptions.

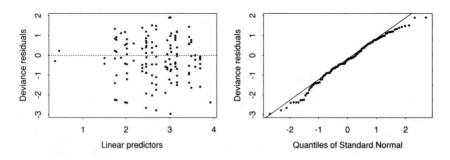

Figure 7.2: Two residual plots for the deviance residuals of a negative binomial model for the Quine data.

Chapter 8

Robust Statistics

Outliers are sample values which cause surprise in relation to the majority of the sample. This is not a pejorative term; outliers may be correct, but they should always be checked for transcription errors. They can play havoc with standard statistical methods, and many *robust* and *resistant* methods have been developed since 1960 to be less sensitive to outliers.

The sample mean $\bar{y}$ can be upset completely by a single outlier; if any data value $y_i \rightarrow \pm\infty$, then $\bar{y} \rightarrow \pm\infty$. This contrasts with the sample median, which is little affected by moving any single value to $\pm\infty$. We say that the median is *resistant* to *gross errors* whereas the mean is not. In fact the median will tolerate up to 50% gross errors before it can be made arbitrarily large; we say its *breakdown point* is 50% whereas that for the mean is 0%. Although the mean is the optimal estimator of the location of the normal distribution, it can be substantially sub-optimal for distributions close to the normal. Robust methods aim to have high efficiency in a neighbourhood of the assumed statistical model.

There are now a number of books on robust statistics. Huber (1981) is rather theoretical, Hampel *et al.* (1986) and Staudte & Sheather (1990) less so. Rousseeuw & Leroy (1987) is principally concerned with regression, but is very practical. A series of books edited by Hoaglin, Mosteller and Tukey (Hoaglin *et al.*, 1983, 1985, 1991) provide a wealth of angles on modern developments in resistant statistics. The chapters by Goodall (1983) and Iglewicz (1983) provide concise introductions to univariate robust statistics. Marazzi (1993) documents a set of Fortran routines for robust statistics that have a (rather complex) interface to S-PLUS. [1]

The earlier techniques of outlier rejection and their relation to modern methods are discussed by Barnett & Lewis (1985).

Why will it not suffice to screen data and remove outliers? There are several aspects to consider:

1. Users, even expert statisticians, do not always screen the data. One of us met this 15 years ago with an example of multiple regression. The data had originally been entered on punched cards, and for one of the regressors about

---

[1] These routines are available from `statlib` (see Appendix D) and include code for the resistant methods discussed in Section 8.4.

half the observations were two columns out and so were read as 100 times their true value. Several cohorts of students and some eminent teachers had used the example without reporting the problem.

2. The sharp decision to keep or reject an observation is wasteful. We can do better by down-weighting dubious observations than by rejecting them, although we may wish to reject completely wrong observations.

3. It can be difficult or even impossible to spot outliers in multivariate or highly structured data.

4. Rejecting outliers affects the distribution theory, which ought to be adjusted. In particular, variances will be under-estimated from the 'cleaned' data.

Robust and resistant methods have been provided by S and S-PLUS for some years, and a bewildering variety of functions are available. We consider first univariate problems, then regression and finally robust covariance estimation for use in multivariate techniques. Robust time-series methods are considered briefly in Chapter 14.

8.1   Univariate samples

For a fixed underlying distribution, we define the *relative efficiency* of an estimator $\tilde{\theta}$ relative to another estimator $\hat{\theta}$ by

$$RE(\tilde{\theta}; \hat{\theta}) = \frac{\text{variance of } \hat{\theta}}{\text{variance of } \tilde{\theta}}$$

since $\hat{\theta}$ needs only RE times as many observations as $\tilde{\theta}$ for the same precision, approximately. The asymptotic relative efficiency (ARE) is the limit of the RE as the sample size $n \to \infty$. (It may be defined more widely via asymptotic variances.) If $\hat{\theta}$ is not mentioned, it is assumed to be the optimal estimator. There is a difficulty with biased estimators whose variance can be small or zero. One solution is to use the mean square-error, another to rescale by $\theta/E(\hat{\theta})$. Iglewicz (1983) suggests using $\mathrm{var}\,(\log \hat{\theta})$ (which is scale-free) for estimators of scale.

We can apply the concept of ARE to the mean and median. At the normal distribution $ARE(\text{median}; \text{mean}) = 2/\pi \approx 64\%$. For longer-tailed distributions the median does better; for the t distribution with 5 degrees of freedom (which is often a better model of error distributions than the normal) $ARE(\text{median}; \text{mean}) \approx 96\%$.

The following example from Tukey (1960) is more dramatic. Suppose we have n observations $Y_i \sim N(\mu, \sigma^2), i = 1, \ldots, n$ and we want to estimate σ^2. Consider $\hat{\sigma}^2 = s^2$ and $\tilde{\sigma}^2 = d^2\pi/2$ where

$$d = \frac{1}{n} \sum_i |Y_i - \overline{Y}|$$

and the constant is chosen since for the normal $d \to \sqrt{2/\pi}\,\sigma$. The $ARE(\tilde{\sigma}^2; s^2)$ = 0.876. Now suppose that each Y_i is from $N(\mu, \sigma^2)$ with probability $1 - \epsilon$ and from $N(\mu, 9\sigma^2)$ with probability ϵ. (Note that both the overall variance and the variance of the uncontaminated observations are proportional to σ^2.) We have

ϵ (%)	$ARE(\tilde{\sigma}^2; s^2)$
0	0.876
0.1	0.948
0.2	1.016
1	1.44
5	2.04

Since the mixture distribution with $\epsilon = 1\%$ is indistinguishable from normality for all practical purposes, the optimality of s^2 is very fragile. We say it lacks *robustness of efficiency*.

There are better estimators of σ than $d\sqrt{\pi/2}$ (which has breakdown point 0%). Two alternatives are proportional to

$$IQR = X_{(3n/4)} - X_{(n/4)}$$
$$MAD = \underset{i}{\text{median}}\,\{|Y_i - \underset{j}{\text{median}}\,(Y_j)|\}$$

(Order statistics are linearly interpolated where necessary.) At the normal,

$$MAD \to \text{median}\,\{|Y - \mu|\} \approx 0.6745\sigma$$
$$IQR \to \sigma\big[\Phi^{-1}(0.75) - \Phi^{-1}(0.25)\big] \approx 1.35\sigma$$

(We will now refer to $MAD/0.6745$ as the MAD estimator.) Both are not very efficient but are very resistant to outliers in the data. The MAD estimator has ARE 37% at the normal (Staudte & Sheather, 1990, p. 123).

The S-PLUS function mad calculates the rescaled MAD.[2] A centre other than the median can be specified. There is a choice of median if n is even. By default mad uses the central median (the linear interpolant) but by specifying the parameter low=T it will use the low median, the smaller of the two order statistics which the span the 50% point. (By analogy the high median would be the larger of the two.)

Consider n independent observations Y_i from a location family with pdf $f(y - \mu)$ for a function f symmetric about zero, so it is clear that μ is the centre (median, mean if it exists) of the distribution of Y_i. We also think of the distribution as being not too far from the normal. There are a number of obvious estimators of μ, including the sample mean, the sample median, and the MLE.

The *trimmed mean* is the mean of the central $1 - 2\alpha$ part of the distribution, so αn observations are removed from each end. This is implemented by the function mean with the argument trim specifying α. Obviously, trim=0 gives

---

[2] For S users we provide a simpler mad function.

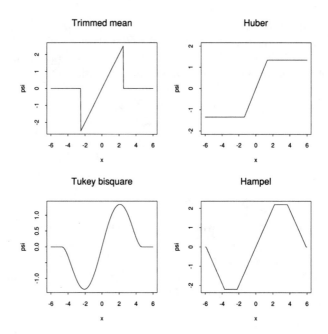

Figure 8.1: The ψ-functions for four common M-estimators.

the mean and `trim=0.5` gives the median (although it is easier to use the function `median`). (If αn is not an integer, the integer part is used.)

Most of the location estimators we consider are *M-estimators*. The name derives from 'MLE-like' estimators. If we have density f, we can define $\rho = -\log f$. Then the MLE would solve

$$\min_{\mu} \sum_i -\log f(y_i - \mu) = \min_{\mu} \sum_i \rho(y_i - \mu)$$

and this makes sense for functions ρ not corresponding to pdfs. Let $\psi = \rho'$ if this exists. Then we will have $\sum_i \psi(y_i - \hat{\mu}) = 0$ or $\sum_i w_i (y_i - \hat{\mu}) = 0$ where $w_i = \psi(y_i - \hat{\mu})/(y_i - \hat{\mu})$. This suggests an iterative method of solution similar to that for GLMs.

Examples of M-estimators

The mean corresponds to $\rho(x) = x^2$, and the median to $\rho(x) = |x|$. (For even n any median will solve the problem). The function

$$\psi(x) = \begin{cases} x & |x| < c \\ 0 & \text{otherwise} \end{cases}$$

corresponds to *metric trimming* and large outliers have no influence at all. The function

$$\psi(x) = \begin{cases} -c & x < -c \\ x & |x| < c \\ c & x > c \end{cases}$$

is known as *metric Winsorizing*[3] and brings in extreme observations to $\mu \pm c$. The corresponding $-\log f$ is

$$\rho(x) = \begin{cases} x^2 & \text{if } |x| < c \\ c(2|x| - c) & \text{otherwise} \end{cases}$$

and corresponds to a density with a Gaussian centre and double-exponential tails. This estimator is due to Huber. Note that its limit as $c \to 0$ is the median, and as $c \to \infty$ the limit is the mean. The value $c = 1.345$ gives 95% efficiency at the normal.

Tukey's *biweight* has

$$\psi(t) = t \left[1 - \left(\frac{t}{R} \right)^2 \right]_+^2$$

where $[\ ]_+$ denotes the positive part of. This implements 'soft' trimming. The value $R = 4.685$ gives 95% efficiency at the normal.

Hampel's ψ has several linear pieces,

$$\psi(x) = \operatorname{sgn}(x) \begin{cases} |x| & 0 < |x| < a \\ a & a < |x| < b \\ a(c - |x|)/(c - b)a & b < |x| < c \\ 0 & c < |x| \end{cases}$$

for example with $a = 2.2s, b = 3.7s, c = 5.9s$. Figure 8.1 illustrates these functions.

There is a scaling problem with the last four choices, since they depend on a scale factor (c, R, or s). We can apply the estimator to rescaled results, that is

$$\min_{\mu} \sum_i \rho \left(\frac{y_i - \mu}{s} \right)$$

for a scale factor s, for example the MAD estimator.

Alternatively, we can estimate s in a similar way. The MLE for density $s^{-1} f((x - \mu)/s)$ gives rise to the equation

$$\sum_i \psi \left(\frac{y_i - \mu}{s} \right) \left(\frac{y_i - \mu}{s} \right) = n$$

which is not resistant (and is biased at the normal). We modify this to

$$\sum_i \chi \left(\frac{y_i - \mu}{s} \right) = (n - 1)\gamma$$

for bounded χ, where γ is chosen for consistency at the normal distribution, so $\gamma = E\chi(N)$. The main example is "Huber's proposal 2" with

$$\chi(x) = \psi(x)^2 = \min(|x|, c)^2 \tag{8.1}$$

---

[3] A term attributed by Dixon (1960) to Charles P. Winsor

In very small samples we need to take account of the variability of $\hat{\mu}$ in performing the Winsorizing. As $Y_i - \hat{\mu}$ has approximate scale factor $s\sqrt{(1 - 1/n)}$, we might replace c by $c\sqrt{(1 - 1/n)}$. (This is not done internally in the software provided.)

If the location μ is known we can apply these estimators with $n - 1$ replaced by n to estimate the scale s alone.

Another approach to scale estimation is the A-estimators described by Iglewicz (1983). The bi-weight A-estimator is

$$s_{bi} = \frac{\sqrt{n}\left[\sum (x_i - \hat{\mu})^2 (1 - u_i^2)_+^4\right]^{1/2}}{[\sum (1 - u_i^2)_+ (1 - 5u_i^2)]}$$

for $u_i = (x_i - \hat{\mu})/(c \times MAD)$, which will be re-scaled to be correct at the normal. For $c = 6$ this has efficiency over 80% for a wide range of distributions.

Implementation

The current S-PLUS function for location M-estimation is `location.m` which has parameters including

```
location.m(x, location, scale, weights, na.rm=F,
    psi.fun="bisquare", parameters)
```

The initial values specified by `location` defaults to the low median, and `scale` is by default fixed as MAD (in fact with the high median in version 3.2, the low median in 3.1). The `psi.fun` can be Tukey's bi-square, "huber" or a user-supplied function. The `parameters` are passed to the `psi.fun` and default to 5 for bi-square and 1.45 for Huber, for efficiency at the normal of about 96%. Hampel's ψ function can be implemented by the example (using C) given on the help page.

For scale estimation there are the functions `mad` described above, `scale.a` and `scale.tau`. All are centred at the median by default. The function `scale.tau` implements Huber's scale based on (8.1) with parameter 1.95 chosen for 80% efficiency at the normal. The function `scale.a` uses Tukey's bisquare A-estimate of scale with parameter 3.85, again chosen for 80% efficiency at the normal. (Both have an optional argument `tuning` to set the parameter.) The author of the `scale.tau` function seems unaware that for the Huber function integration by parts gives

$$E(\chi(N)) = 2c^2 + 2(1 - c^2)\Phi(c) - 2c\,\phi(c) - 1$$

and so the function uses numerical integration. (See, for example, Marazzi, 1993, p. 105.)

Examples

We give two datasets taken from analytical chemistry (Abbey, 1988; Analytical
Methods Committee, 1989a, b). The dataset `abbey` contains 31 determinations
of nickel content ($\mu g\, g^{-1}$) in SY-3, a Canadian syenite rock, and `chem` contains
24 determinations of copper ($\mu g\, g^{-1}$) in wholemeal flour. These data are part of
a larger study which suggests $\mu = 3.68$.

We also supply S functions `huber` and `hubers` for the Huber M-estimator
with MAD and "proposal 2" scale respectively, with default $c = 1.5$.

```
> sort(chem)
 [1]   2.20  2.20  2.40  2.40  2.50  2.70  2.80  2.90  3.03
[10]   3.03  3.10  3.37  3.40  3.40  3.40  3.50  3.60  3.70
[19]   3.70  3.70  3.70  3.77  5.28 28.95
> mean(chem)
[1] 4.2804
> median(chem)
[1] 3.385
> location.m(chem)
[1] 3.1452
      ....
> location.m(chem, psi.fun="huber")
[1] 3.2132
      ....
> mad(chem)
[1] 0.52632
> scale.tau(chem)
[1] 0.639
> scale.a(chem)
[1] 0.64411
> scale.tau(chem, 3.68)
[1] 0.91578
> scale.a(chem, 3.68)
[1] 0.85875
> unlist(huber(chem))
      mu       s
  3.2067 0.52632
> unlist(hubers(chem))
      mu       s
  3.2055 0.67365
```

The sample is clearly highly asymmetric with one value than appears to be out by
a factor of 10. It was checked and reported as correct by the laboratory.

```
> sort(abbey)
 [1]    5.2   6.5   6.9   7.0   7.0   7.0   7.4   8.0   8.0
[10]    8.0   8.0   8.5   9.0   9.0  10.0  11.0  11.0  12.0
[19]   12.0  13.7  14.0  14.0  14.0  16.0  17.0  17.0  18.0
[28]   24.0  28.0  34.0 125.0
> mean(abbey)
```

```
[1] 16.006
> median(abbey)
[1] 11
> location.m(abbey)
[1] 10.804
> location.m(abbey, psi.fun="huber")
[1] 11.517
> unlist(hubers(abbey))
      mu       s
  11.732 5.2585
> unlist(hubers(abbey, 2))
      mu       s
  12.351 6.1052
> unlist(hubers(abbey, 1))
      mu       s
  11.365 5.5673
```

Note how reducing the constant reduces the estimate of location, as this sample (like many in analytical chemistry) has a long right tail.

8.2   Median polish

Consider a two-way layout. The additive model is

$$\hat{y}_{ij} = \mu + \alpha_i + \beta_j, \qquad \alpha. = \beta. = 0$$

The least squares fit corresponds to choosing the parameters μ, α_i and β_j so that the row and column sums of the residuals are zero.

Means are not resistant. Suppose we use medians instead. That is, we seek a fit of the same form, but with $\text{median}(\alpha_i) = \text{median}(\beta_j) = 0$ and $\text{median}_i(e_{ij}) = \text{median}_j(e_{ij}) = 0$. This is no longer a set of linear restrictions, so there may be many solutions. The median polish algorithm (Mosteller & Tukey, 1977; Emerson & Hoaglin, 1983) is to augment the table with row and column effects as

$$
\begin{array}{cccc}
e_{11} & \cdots & e_{1c} & a_1 \\
\vdots & \ddots & \vdots & \vdots \\
e_{r1} & \cdots & e_{rc} & a_r \\
b_1 & \cdots & b_r & m
\end{array}
$$

where initially $e_{ij} = y_{ij}$, $a_i = b_j = m = 0$. At all times we maintain

$$y_{ij} = m + a_i + b_j + e_{ij}$$

In a *row sweep* for each row we subtract the median of columns $1, \ldots, c$ from those columns and add it to the last column. For a *column sweep* for each column we subtract the median of rows $1, \ldots, r$ from those rows and add it to the bottom row.

Median polish operates by alternating row and column sweeps until the changes made become small or zero (or the human computer gets tired!). (Often just two pairs of sweeps are recommended.) The answer may depend on whether rows or columns are tried first and is very resistant to outliers. Using means rather medians will give the least-squares decomposition without iteration.

An example

The table below gives specific volume (*cc/gm*) of rubber at four temperatures ($^\circ C$) and six pressures (*kg/cm^2 above atmo*). These data were published by Wood & Martin (1964, p. 260), and used by Mandel (1969) and Emerson & Wong (1985).

	\multicolumn{6}{c}{Pressure}					
Temperature	500	400	300	200	100	0
0	1.0637	1.0678	1.0719	1.0763	1.0807	1.0857
10	1.0697	1.0739	1.0782	1.0828	1.0876	1.0927
20	1.0756	1.0801	1.0846	1.0894	1.0944	1.0998
25	1.0786	1.0830	1.0877	1.0926	1.0977	1.1032

The default `trim=0.5` option of `twoway` performs median polish.   We have, after multiplying by 10^4,

Temperature	\multicolumn{6}{c}{Pressure}						a_i
	500	400	300	200	100	0	
0	7.0	4.5	1.5	-1.5	-6.5	-9.0	-96.5
10	3.0	1.5	0.5	-0.5	-1.5	-3.0	-32.5
20	-3.0	-1.5	-0.5	0.5	1.5	3.0	32.5
25	-4.5	-4.0	-1.0	1.0	3.0	5.5	64.0
b_j	-111.0	-67.5	-23.5	23.5	72.5	125.0	$m = 10837.5$

This is interpreted as

$$y_{ij} = m + a_i + b_j + e_{ij}$$

and the body of the table contains the residuals e_{ij}. These have both row medians and column medians zero. Originally the value for temperature 0, pressure 400 was entered as 1.0768; the only change was to increase the residual to 94.5×10^{-4} which was easily spotted.

Note the pattern of residuals in the table; this suggests a need for transformation. Note also how linear the row and column effects are in the factor levels. Emerson & Wong (1985) fit Tukey's 'one degree of freedom for non-additivity' model

$$y_{ij} = m + a_i + b_j + e_{ij} + k a_i b_j \qquad (8.2)$$

by plotting the residuals against $a_i b_j / m$ and estimating a power transformation y^λ with $\lambda = 1 - mk$ estimated as -6.81. As this is such an awkward power, they thought it better to retain the model (8.2).

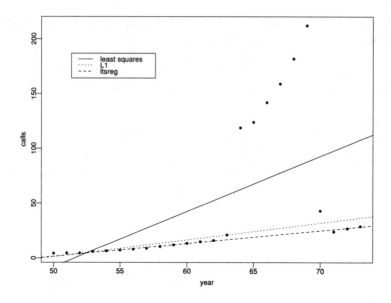

Figure 8.2: Millions of phone calls in Belgium, 1950-73, from Rousseeuw & Leroy (1987), with three fitted lines.

8.3 Robust regression

There are a number of ways to perform robust regression in S and S-PLUS, but all have drawbacks. First consider an example. Rousseeuw & Leroy (1987) give data on annual numbers of Belgian 'phone calls, given in our dataset `phones`.

```
attach(phones)
phones.lm <- lm(calls ~ year)
plot(year, calls)
abline(phones.lm$coef)
abline(l1fit(year, calls), lty=2)
abline(ltsreg(year, calls), lty=3)
legend(locator(1), legend=c("least squares", "L1", "ltsreg"),
    lty=1:3)
```

Figure 8.2 shows the least squares line, the L_1 regression and the least trimmed squares regression (Section 8.4). The `ltsreg` line is $-56.45 + 1.16\,\text{year}$. Rousseeuw & Leroy's investigations showed that for 1964–9 the total length of calls (in minutes) had been recorded rather than the number, with each system being used during 1963 and 1970.

Next some theory. In a regression problem there are two possible sources of errors, the observations y_i and the corresponding row vector of p regressors, $\mathbf{x}_i$. Most robust methods in regression only consider the first, and in some cases (designed experiments?) errors in the regressors can be ignored. This is the case for M-estimators, the only ones we shall consider in this section.

Consider a regression problem with n cases $(y_i, \mathbf{x}_i)$ from the model

$$y = \mathbf{x}\beta + \epsilon$$

for a p-variate row vector $\mathbf{x}$.

L_1 regression

The median is the L_1 estimator in the location problem. A natural generalization to regression is:

$$\min_b \sum_{i=1}^n |y_i - \mathbf{x}_i b|$$

The solution is not necessarily unique (as with the median). There is a solution which fits exactly at p observations.

To compute the solutions, we may reduce the problem to one of linear programming. Define non-negative variables e_i^+ and e_i^- and set

$$e_i^+ - e_i^- = e_i = y_i - \mathbf{x}_i b$$

Then we minimize $\sum_i (e_i^+ + e_i^-)$ subject to $e_i^+, e_i^- \geqslant 0$ and the equation above. Clearly either e_i^+ or e_i^- will be zero, and so $(e_i^+ + e_i^-) = |e_i|$.

Like the median, L_1 regression is not very efficient. In fact, asymptotically we have

$$\text{var}\,(b) \approx \frac{1}{4f(0)^2}(X^T X)^{-1}$$

(Bassett & Koenker, 1978) so the ARE is the same as that of the median.

Bloomfield & Steiger (1983) is a good general reference for L_1 regression; Narula & Wellington (1982) provide a concise survey for (then) known results.

The S function `l1fit` implements L_1 regression. Its call is

```
l1fit(x, y, intercept=T, print=T)
```

where x is the design matrix, y the vector of dependent observations and an intercept is added to x by default. The parameter `print` refers to warnings and diagnostics.

M-estimators

If we assume a scaled pdf $f(e/s)/s$ for ϵ and set $\rho = -\log f$, the maximum likelihood estimator minimizes

$$\sum_{i=1}^n \rho\left(\frac{y_i - \mathbf{x}_i b}{s}\right) + n \log s \tag{8.3}$$

Suppose for now that s is known. Let $\psi = \rho'$. Then the MLE b of β solves

$$\sum_{i=1}^n \mathbf{x}_i \psi\left(\frac{y_i - \mathbf{x}_i b}{s}\right) = 0 \tag{8.4}$$

Let $r_i = y_i - \mathbf{x}_i b$ denote the residuals.

The solution to equation (8.4) or to minimising over (8.3) can be used to define an M-estimator of β. Note that L_1 regression corresponds to the Laplace or double-exponential distribution.

There are a number of versions of asymptotic theory for M-estimators. Several of the most precise forms are given by Huber (1981, §7.6). Let X as usual be the design matrix and let

$$\kappa = 1 + \frac{p}{n} \frac{\text{var}(\psi')}{(E\psi')^2}$$

evaluated at the distribution of ϵ (and in practice estimated from the distribution of residuals). Then Huber gives three forms of estimators of the variance-covariance matrix of b, of which we use

$$\kappa^2 \frac{[1/(n-p)] \sum \psi(r_i)^2}{[(1/n) \sum \psi'(r_i)]^2} (X^T X)^{-1} \tag{8.5}$$

A simplification occurs with Huber's ψ function, for which ψ' picks out the un-Winsorized observations. Let m denote the proportion of these, and S the variance (with divisor $n - p$) of the Winsorized residuals. Then the simplified forms are

$$\kappa = 1 + \frac{p(1-m)}{nm}, \qquad \kappa^2 \frac{S}{m^2} (X^T X)^{-1}$$

A common way to solve (8.4) is by iterated re-weighted least squares (IWLS or IRLS), with weights

$$w_i = \psi\left(\frac{y_i - \mathbf{x}_i b}{s}\right) \bigg/ \left(\frac{y_i - \mathbf{x}_i b}{s}\right) \tag{8.6}$$

The iteration is only guaranteed to converge for *convex* ρ functions, and for redescending functions (such as those of Tukey and Hampel), equation (8.4) may have multiple roots. In such cases it is usual to choose a good starting point (such as a fully-iterated Huber estimator) and perform a small and fixed number of iterations.

Unknown scale

Of course, in practice the scale s is not known. A simple and very resistant scale estimator is the MAD about some centre. This is applied to the residuals about zero, either to the current residuals within the loop or to the residuals from a prior very resistant fit (e.g. L_1 or those of Section 8.4).

Alternatively, we can estimate s in an MLE-like way. Finding a stationary point of (8.3) with respect to s gives

$$\sum_i \psi\left(\frac{y_i - \mathbf{x}_i b}{s}\right)\left(\frac{y_i - \mathbf{x}_i b}{s}\right) = n$$

which is not resistant (and is biased at the normal). As in the univariate case we modify this to

$$\sum_i \chi\left(\frac{y_i - \mathbf{x}_i b}{s}\right) = (n - p)\gamma \tag{8.7}$$

The function rreg

The oldest robust regression function is rreg which was re-designed for the 1990 release of S (Heiberger & Becker, 1992). Its usage is, for example,

```
rreg(stack.x, stack.loss, iter=20,
    method=function(x)wt.huber(x, c=1.5))
```

This function does not use model formulae, and no functions are provided to analyze its output. In particular there is no way to calculate standard errors for the parameter estimates. It uses the iterated MAD scale estimate, and a wide range of ψ functions are provided via S functions. The default is to use a converged Huber estimator followed by the Tukey bisquare, both with constants (which can not be changed) chosen to give 95% efficiency at the normal.

There is a simplified version, rbiwt , for use with a single regressor.

robust family generator for glm

This uses the iterative re-weighted least squares mechanism of glm to fit Huber M-regressions with the weights given by (8.6) and default tuning parameter $c = 1.345$. Occasionally glm fails to converge in the default 10 iterations, and the iteration control mechanism is not explained well. The form needed is

```
attach(phones)
glm(calls ~ year, robust(maxit=50))
```

The maxit and control parameters to glm do not change the iteration limit. The tuning parameter is set by the argument k= to robust .

The standard errors in the summary functions are given by

$$\text{var}\,(b) = s^2 (X^T W X)^{-1}$$

where W is a diagonal matrix of the current weights. This is often a reasonable estimate, if less accurate than (8.5), but can be poor as our example below and Street, Carroll and Ruppert (1988) show.

The scale estimation is by $MAD/0.67$ *(sic)* of the residuals from the least squares fit. Because it is not iterated, it can seriously over-estimate the scale. The summary and print functions for glm do not know much about robust and so are inappropriate. In particular, they base the scale estimate on the usual GLM method of

$$\sum_i \rho\left(\frac{y_i - \mathbf{x}_i b}{s}\right) = (n - p) \tag{8.8}$$

Note that this is not resistant (has breakdown point zero) as ρ is unbounded for the Huber function, and can be a serious over-estimate.

Our class "rlm"

An alternative method is provided in our main library. This introduces a new class rlm and model-fitting function rlm, building on lm. The syntax in general follows lm, and the current version of the S code is based on that for the lm class and of rreg. Huber's M-estimator is used, and the summary function computes standard errors from (8.5). The default tuning parameter is $c = 1.345$. By default the scale s is estimated by iterated MAD, but at the iteration number of the sw parameter it switches to solving (8.7). (Typically sw = 3 works well if this scale estimator is desired.) The default iteration limit is 20, but this may be increased by the maxit parameter.

To show that these differences do matter, consider again the Belgian 'phones data.

```
> attach(phones)
> summary(lm(calls ~ year))
             Value Std. Error  t value Pr(>|t|)
(Intercept) -260.059  102.607    -2.535   0.019
       year    5.041    1.658     3.041   0.006
Residual standard error: 56.2 on 22 degrees of freedom
> summary(rlm(calls ~ year, maxit=50))
             Value Std. Error  t value
(Intercept) -102.812   26.680    -3.853
       year    2.045    0.431     4.743
Residual standard error: 9.08 on 22 degrees of freedom
> summary(rlm(calls ~ year, sw=3))
             Value Std. Error  t value
(Intercept) -227.828  101.783    -2.238
       year    4.451    1.645     2.707
Residual standard error: 57.2 on 22 degrees of freedom
> summary(glm(calls ~ year, robust))
             Value Std. Error t value
(Intercept) -216.0149   92.7864 -2.3281
       year    4.2308    1.5077  2.8061
(Dispersion Parameter for Robust ... 2487.1)
Residual Deviance: 64349 on 22 degrees of freedom
> summary(rlm(calls ~ year, k=0.25, maxit=50))
             Value Std. Error t value
(Intercept) -74.057   15.277    -4.848
       year   1.499    0.247     6.071
Residual standard error: 3.96 on 22 degrees of freedom
> summary(glm(calls ~ year, robust(k=0.25, maxit=20)))
             Value Std. Error t value
(Intercept) -105.5421   45.90532 -2.2991
       year    2.0959    0.76789  2.7294
(Dispersion Parameter for Robust ... 478.73)
Residual Deviance: 19211 on 22 degrees of freedom
> rreg(year, calls, method=wt.hampel)
$coef:
```

```
(Intercept)      x
    -248.77 4.8362
> rreg(year, calls, init=l1fit(year,calls)$coef,
       method=wt.hampel)
$coef:
 (Intercept)      x
    -52.382 1.1006
```

Note how the robust GLM method inflates the quoted standard errors for severe Winsorizing. The dispersion parameter is also over-estimated because the scale is not iterated.

As Figure 8.2 shows, in this example there is a batch of outliers from a different population in the late 1960's, and these should probably be rejected, which the Huber M-estimators do not. The final two fits show how crucially the iterated re-descending Hampel estimator depends on the initial scale.

8.4   Resistant regression

As an attempt to get a very resistant regression fit, Rousseeuw suggested

$$\min_b \operatorname*{median}_i |y_i - \mathbf{x}_i b|^2$$

called the *least median of squares* (LMS) estimator. The square is necessary if n is even, when the central median is taken.

This fit is very resistant, and needs no scale estimate. Note that unlike the robust regression it can reject values that fit badly because their $\mathbf{x}_i$ are outliers. It is however very inefficient, converging at rate $1/\sqrt[3]{n}$. Further, it displays marked sensitivity to central data values: see Hettmansperger & Sheather (1992) and Davies (1993, §2.3).

Rousseeuw later suggested least trimmed squares (LTS) regression:

$$\min_b \sum_{i=1}^{q} |y_i - \mathbf{x}_i b|_{(i)}^2$$

as this is more efficient, but shares the same extreme resistance. The sum is over the smallest $q = \lfloor n/2 \rfloor + \lfloor (p+1)/2 \rfloor$ squared residuals. (Other authors differ; for example Marazzi (1993, p. 200) has $q = \lfloor n/2 \rfloor + 1$ and Rousseeuw & Leroy (1987, p.132) mention both choices.)

LMS and LTS are very suitable methods for finding a starting point for rejecting outliers or for a few steps of a M-estimator; LTS is preferable to LMS.

S-PLUS implementation

Both methods are implemented in S-PLUS, but `lmsreg` is slower and is now deprecated. The essential parts of their calls are

```
lmsreg(x, y, intercept=T)
ltsreg(x, y, intercept=T)
```

where `x` is the design matrix, `y` the dependent variable and by default a column of 1's is added to `x`. Both algorithms involve an approximate random search, so it would be more accurate to say that they approximate LMS and LTS. As a consequence, the results given here will not be reproducible.

Brownlee's stack loss data

We consider Brownlee's (1965) much-studied stack loss data, given in the S datasets `stack.x` and `stack.loss`. The data are from the operation of a plant for the oxidation of ammonia to nitric acid, measured on 21 consecutive days. There are 3 explanatory variables (air flow to the plant, cooling water inlet temperature, and acid concentration) and the response, 10 times the percentage of ammonia lost.

```
> summary(lm(stack.loss ~ stack.x))
Residuals:
   Min    1Q Median    3Q Max
  -7.24 -1.71 -0.455  2.36 5.7

Coefficients:
                     Value Std. Error t value Pr(>|t|)
      (Intercept) -39.920    11.896    -3.356   0.004
    stack.xAir Flow  0.716     0.135     5.307   0.000
  stack.xWater Temp  1.295     0.368     3.520   0.003
  stack.xAcid Conc.  -0.152     0.156    -0.973   0.344

Residual standard error: 3.24 on 17 degrees of freedom

> lmsreg(stack.x, stack.loss)
$coef:
 Intercept Air Flow Water Temp  Acid Conc.
     -34.5  0.71429     0.35714 -3.3255e-17

    Min.   1st Qu. Median  Mean 3rd Qu.   Max.
  -7.643 -1.776e-15 0.3571 1.327   1.643  9.714

$wt:
 [1] 0 0 0 0 1 1 1 1 1 1 1 1 1 1 1 1 1 1 1 1 0
> ltsreg(stack.x, stack.loss)
$coefficients:
 (Intercept) Air Flow Water Temp Acid Conc.
     -35.435  0.76316    0.32619  -0.011591
```

```
    Min. 1st Qu. Median   Mean 3rd Qu.  Max.
  -8.269  -0.361 0.2133 0.9869   1.283 9.313
```

The weights returned by lmsreg indicate that five points have large residuals and should be considered as possible outliers. For each of lmsreg and ltsreg we give the results of summary applied to their residuals. Now consider M-estimators:

```
> l1fit(stack.x, stack.loss)
$coefficients:
 Intercept Air Flow Water Temp Acid Conc.
    -39.69  0.83188     0.57391   -0.06087


    Min. 1st Qu. Median   Mean 3rd Qu.   Max.
  -9.481  -1.217       0 0.08944  0.5275 7.635
> stack.rl <- rlm(stack.loss ~ stack.x)
> summary(stack.rl)
Residuals:
   Min    1Q Median    3Q Max
  -8.92 -1.73 0.0617 1.54 6.5

Coefficients:
                    Value Std. Error t value
       (Intercept) -41.027     9.793   -4.189
    stack.xAir Flow   0.829     0.111    7.470
  stack.xWater Temp   0.926     0.303    3.057
  stack.xAcid Conc.  -0.128     0.129   -0.994

Residual standard error: 2.44 on 17 degrees of freedom
> stack.rl$w:
 [1] 1.000 1.000 0.786 0.505 1.000 1.000 1.000
 [8] 1.000 1.000 1.000 1.000 1.000 1.000 1.000
[15] 1.000 1.000 1.000 1.000 1.000 1.000 0.368
```

The L_1 method fits observations 2, 8, 16 and 18 exactly. The component w returned by rlm contains the final weights in (8.6). Although all methods seem to agree about observation 21, they differ in their view of the early observations. Atkinson (1985, pp. 129–136, 267–8) discusses this example in some detail, as well as the analyses performed by Daniel & Wood (1971). They argue for a logarithmic transformation, dropping acid concentration and fitting interactions or products of the remaining two regressors. However, the question of outliers and change of model are linked, since most of the evidence for changing the model comes from the possible outliers.

Rather than fit a parametric model we examine the points in the air flow – water temp space, using the robust fitting option of loess (discussed briefly in Chapter 10); see Figure 8.3.

```
x1 <- stack.x[,1]; x2 <- stack.x[,2]
stack.loess <- loess(log(stack.loss) ~ x1*x2, span=0.5,
    family="symmetric")
stack.mar <- list(x1 = seq(50,80,2.5), x2=seq(17, 27, 1))
stack.lop <- predict(stack.loess, expand.grid(stack.mar))
contour(stack.mar$x1,stack.mar$x2, stack.lop, levels=0:5,
    xlab="Air flow", ylab="Water temp")
points(x1, x2)
identify(x1, x2)
```

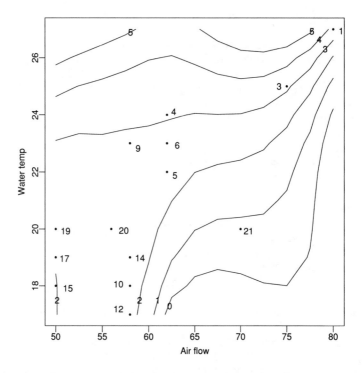

Figure 8.3: Fitted surface for Brownlee's stack loss data on log scale using `loess`.

This shows clearly that the 'outliers' are also outliers in this space. (There are duplicate points; in particular points 1 and 2 are coincident.)

Scottish hill races revisited

We return to the data on Scottish hill races studied in the introduction and Section 6.3. There we saw one gross outlier and a number of other extreme observations. Many of the robust and resistant methods do not allow weighted fits.

```
> hills.lm
Coefficients:
  (Intercept)  dist    climb
      -8.992  6.218  0.011048
```

```
Degrees of freedom: 35 total; 32 residual
Residual standard error: 14.676
> hills1.lm # omitting Knock Hill
Coefficients:
 (Intercept)    dist    climb
      -13.53  6.3646 0.011855

Degrees of freedom: 34 total; 31 residual
Residual standard error: 8.8035
> rlm(time ~ dist + climb, hills)
Coefficients:
 (Intercept)    dist    climb
     -9.6067  6.5507 0.0082959

Degrees of freedom: 35 total; 32 residual
Scale estimate: 5.21
> summary(rlm(time ~ dist + climb, hills, weights=1/dist^2))
Coefficients:
              Value Std. Error t value
(Intercept) -1.270     1.843    -0.689
       dist  5.389     0.257    20.929
      climb  0.007     0.001     7.922

Residual standard error: 4.24 on 32 degrees of freedom

> attach(hills)
> rreg(cbind(dist, climb), time)$coef
 (Intercept)    dist    climb
     -8.2671  6.6358 0.006631
> rreg(cbind(dist, climb), time, wx=1/dist^2)$coef
 (Intercept)    dist    climb
     -1.3783  5.1361 0.007578
> ltsreg(cbind(dist, climb), time)$coef
 (Intercept)    dist    climb
     -1.2426  4.8894 0.0083503
```

Notice that the intercept is no longer significant in the robust weighted fit.

   If we move to the model for inverse speed:

```
> summary(hills2.lm) # omitting Knock Hill
Coefficients:
              Value Std. Error t value Pr(>|t|)
(Intercept)  4.900     0.474    10.344    0.000
       grad  0.008     0.002     5.022    0.000

Residual standard error: 1.28 on 32 degrees of freedom
> ltsreg(grad, ispeed)$coef
 (Intercept)       x
      4.7488 0.0080721
```

```
> rreg(grad, ispeed)$coef
 (Intercept)          x
     5.1522 0.0072142
> summary(rlm(ispeed ~ grad, hills))
Coefficients:
            Value Std. Error t value
(Intercept) 5.176  0.381       13.593
       grad 0.007  0.001        5.431

Residual standard error: 0.869 on 33 degrees of freedom
```

The results are in close agreement with the least-squares results removing
Knock Hill.

8.5 Multivariate location and scale

Somewhat counter-intuitively, it does not suffice to apply a robust location estimator to each component of a multivariate mean (Rousseeuw & Leroy, 1987, p. 250),
and it is easier to consider the estimation of mean and variance simultaneously.

Multivariate variances are very sensitive to outliers. One method for robust
covariance estimation is available via the S-PLUS function cov.mve. Let there
be n observations of p variables. The method seeks an ellipsoid containing
$\lfloor (n+p+1)/2 \rfloor$ points of minimum volume. Having found such an ellipsoid by a
random search, it returns a (product-moment) covariance estimate of those points
whose Mahalanobis distance from the mean computed via the ellipsoid covariance
is not too large (specifically within the 97.5% point). This estimator was proposed
by Rousseeuw & van Zomeren (1990). Note that the centre of the ellipsoid and the
mean of the 'cleaned' data provide reasonably robust estimates of the population
mean. Inspecting the code shows that cov.mve returns the mean of the 'cleaned'
data as its component center.

Chapter 9

Non-linear Regression Models

In linear regression the mean surface in sample space is a plane. In non-linear regression the mean surface may be an arbitrary curved surface but in other respects the models are similar. In practice the mean surface in most non-linear regression models will be approximately planar in the region(s) of high likelihood allowing good approximations based on linear regression techniques to be used. Non-linear regression models can still present tricky computational and inferential problems. (Indeed, the examples here exceeded the capacity of S-PLUS for Windows 3.1.)

A thorough treatment of non-linear regression is given in Bates & Watts (1988). Another encyclopaedic reference is Seber & Wild (1989), and the books of Ratkowsky (1983, 1990) and Ross (1990) also offer some practical statistical advice. The S software is described by Bates & Chambers (1992) who state that its methods are based on those described in Bates & Watts (1988).

We start with a simple example. Obese patients on a weight rehabilitation programme tend to lose adipose tissue at a diminishing rate as the treatment progresses. Our dataset wtloss has kindly been supplied by Dr T. Davies (personal communication). The two variables are Days, the time (in days) since start of the programme, and Weight, weight in kilograms measured under standard conditions. The dataset pertains to a male patient, aged 48, height 193 cm (6' 4") with a large body frame, and is illustrated in Figure 9.1, produced by

```
attach(wtloss)
# alter margin 4; others are default
par(mar=c(5.1, 4.1, 4.1, 4.1))
plot(Days, Weight, type="p",ylab="Weight (kg)")
Wt.lbs <- pretty(range(Weight*2.205))
axis(side=4, at=Wt.lbs/2.205, lab=Wt.lbs, srt=90)
mtext("Weight (lb)", side=4, line=3)
```

Although polynomial regression models may describe such data very well within the time range they can fail spectacularly outside this range (see Figure 9.2 on page 232). A more useful model with some theoretical and empirical support has an exponential form. (For notational convenience we denote Days by x and Weight by y.)

$$y = \beta_0 + \beta_1 2^{-x/\theta} + \epsilon \tag{9.1}$$

Notice that all three parameters have a ready interpretation:

223

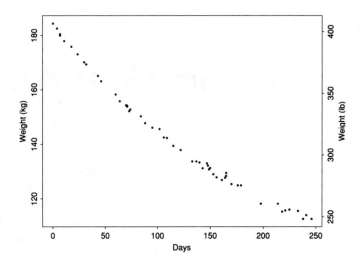

Figure 9.1: Weight loss from an obese patient.

β_0 is the notionally stable lean weight, or asymptote,
β_1 is the total weight to be lost and
θ is the 'half life': the time taken to lose half of the remaining weight
to be lost.

The parameters β_0 and β_1 are called *linear parameters* since the second partial derivative of the model function with respect to them is identically zero. Other parameters for which this is not the case, such as θ in this example, are called *non-linear parameters*.

9.1 Fitting non-linear regression models

The general form of a non-linear regression model is

$$y = \eta(\mathbf{x}, \boldsymbol{\beta}) + \epsilon \qquad (9.2)$$

where $\mathbf{x}$ is a vector of covariates, $\boldsymbol{\beta}$ is a p–component vector of unknown parameters and ϵ is a $N(0, \sigma^2)$ error term. In the weight loss example the parameter vector is $\boldsymbol{\beta} = (\beta_0, \beta_1, \theta)^T$.

Suppose $\mathbf{y}$ is a sample vector of size n and $\boldsymbol{\eta}(\boldsymbol{\beta})$ is its mean vector. It is easy to show that the maximum likelihood estimate of $\boldsymbol{\beta}$ is a least squares estimate, that is a minimizer of $\|\mathbf{y} - \boldsymbol{\eta}(\boldsymbol{\beta})\|^2$. The variance parameter, σ^2, is then estimated by the residual mean square as in linear regression.

For varying $\boldsymbol{\beta}$ the vector $\boldsymbol{\eta}(\boldsymbol{\beta})$ traces out a p–dimensional surface in $\mathbb{R}^n$ that we refer to as the *solution locus*. The parameters $\boldsymbol{\beta}$ define a coordinate system within the solution locus. From this point of view a linear regression model is one for which the solution locus is a plane through the origin and the coordinate system within it defined by the regression coefficient vector has no curvature. The estimation problem is to find the point on the solution locus closest to the sample vector **y** in the sense of Euclidean distance.

The process of fitting non-linear regression models in S is similar to that for fitting linear models, but note two important differences:

1. There is no explicit formula for the estimates, so iterative procedures are required, for which initial values will be needed.

2. Linear regression model formulae are not adequate to specify non-linear regression models. A more flexible protocol is needed.

The primary S function for fitting a non-linear regression model is `nls`. To fit the weight loss model we can use:

```
> wtloss.fm <- nls(Weight ~ b0 + b1*2^(-Days/th),
    data = wtloss,
    start = list(b0=90, b1=95, th=120),
    trace = T)
67.5435 : 90 95 120
40.1808 : 82.7263 101.305 138.714
39.2449 : 81.3987 102.658 141.859
39.2447 : 81.3738 102.684 141.911
> wtloss.fm
Residual sum of squares : 39.245
parameters:
     b0     b1     th
 81.374 102.68 141.91
formula: Weight ~ b0 + b1 * 2^( - Days/th)
52 observations
```

The arguments to `nls` are:

formula A non-linear model formula expressing the model. The formula operator, ~, is still used as in a linear model but the right hand side is an ordinary expression where the operators have their usual arithmetic meaning.

data An optional data frame as the primary reference point for variables in the model.

start A list or vector specifying the starting values for the parameters in the model. The names for the components of `start` specify which of the variables occurring on the right hand side of the model formula are parameters.

control An optional argument allowing some features of the iterative procedure to be specified.

algorithm An optional character string argument allowing a particular iterative procedure to be specified. The default procedure is denoted by `"default"`.

trace An argument allowing tracing information from the iterative procedure to be printed. By default no information is printed.

In our example the names of the parameters were specified as b0, b1 and th. The initial values of 90, 95 and 120 were found by inspection of Figure 9.1. The procedure converged in three iterations.

9.2   Parametrized data frames

The arguments data and start to nls can be combined by specifying a data frame that also has the starting information list included with it as a special attribute, thus making it a *parametrized data frame*. In the weight loss data example we can use:

```
> parameters(wtloss) <- list(b0=90, b1=95, th=120)
> wtloss.fm <- nls(Weight ~ b0 + b1*2^(-Days/th),
      data = wtloss, trace=T)
   ....
```

The parameter list attached to the data frame gives it an additional class, pframe. If a parametrized data frame is attached to the search list the parameters become available as well as the columns of the data frame itself.

The functions parameters and param operate in the same way as the functions attributes and attr. Thus to change a single parameter one can use the assignment form of param as follows:

```
param(wtloss, "b0") <- 80
```

From now on we will assume that wtloss is a parametrized data frame.

9.3   Using function derivative information

Most non-linear regression fitting algorithms operate in outline as follows:

1. Near an initial point $\eta(\beta^{(0)})$ the solution locus is approximated by its tangent plane.

2. The observation vector y is projected onto the tangent by linear regression to give a new coefficient vector $\beta^{(1)}$.

3. The tangent plane is calculated at $\eta(\beta^{(1)})$ and the procedure continues until either convergence or abnormal termination.

The process can be made more explicit using a Taylor series expansion. To first order

$$\eta_k(\beta) \approx \eta_k(\beta^{(0)}) + \sum_{j=1}^{p}(\beta_j - \beta_j^{(0)}) \left.\frac{\partial \eta_k}{\partial \beta_j}\right|_{\beta=\beta^{(0)}}$$

In vector terms we have

$$\eta(\beta) \approx \omega^{(0)} + Z^{(0)}\beta \qquad\qquad (9.3)$$

where

$$Z_{kj}^{(0)} = \left.\frac{\partial \eta_k}{\partial \beta_j}\right|_{\beta=\beta^{(0)}} \qquad \text{and} \qquad \omega_k^{(0)} = \eta_k(\beta^{(0)}) - \sum_{j=1}^{p}\beta_j^{(0)} Z_{kj}^{(0)}$$

Equation (9.3) defines the tangent plane to the surface at the coordinate point $\beta = \beta^{(0)}$. For a linear regression the *offset vector* $\boldsymbol{\omega}^{(0)}$ is $\mathbf{0}$ and the matrix $Z^{(0)}$ is the model matrix X, a constant matrix. In the non-linear case the tangent plane at $\beta^{(0)}$ passes through the offset vector $\boldsymbol{\omega}^{(0)}$ and is spanned by the columns of the *gradient matrix* $Z^{(0)}$. Hence the new coefficient vector is

$$\beta^{(1)} = \left(Z^{(0)\,T} Z^{(0)} \right)^{-1} Z^{(0)\,T} \left(\mathbf{y} - \boldsymbol{\omega}^{(0)} \right)$$

The default algorithm for nls is actually a group of related algorithms, and which used depends on what information is supplied.

1. If only the model function $\eta(\beta)$ is supplied the tangent plane must be approximated numerically.

2. If the model function and its first derivatives are supplied the tangent plane is used.

3. If both the first and the second derivatives are supplied, the second derivatives can be used in the iterative process.

To supply first derivative information, the right hand side of the model formula must deliver the value of the model function, to which is attached the first derivative matrix, Z, as a gradient attribute. To so we write an S function for the right-hand-side of the model formula.

For our simple example the three derivatives are

$$\frac{\partial \eta}{\partial \beta_1} = 1, \qquad \frac{\partial \eta}{\partial \beta_2} = 2^{-x/\theta}, \qquad \frac{\partial \eta}{\partial \theta} = \frac{\log(2)\,\beta_1 x 2^{-x/\theta}}{\theta^2}$$

so an S function to specify these derivatives is

```
expn <- function(b0, b1, th, x)
{
    temp <- 2^(-x/th)
    model.func <- b0 + b1 * temp
    Z <- cbind(1, temp, (b1 * x * temp * log(2))/th^2)
    dimnames(Z) <- list(NULL, c("b0","b1","th"))
    attr(model.func, "gradient") <- Z
    model.func
}
```

Note that the gradient matrix must have column names matching those of the corresponding parameters.

We can fit our model again using first derivative information:

```
> wtloss.gr <- nls(Weight ~ expn(b0, b1, th, Days),
    data = wtloss, trace = T)
67.5435 : 90 95 120
40.1808 : 82.7263 101.305 138.714
39.2449 : 81.3987 102.658 141.859
39.2447 : 81.3738 102.684 141.911
```

This appears to make no difference to the speed of convergence, but tracing the function `expn` shows that only 6 function evaluations are required with the explicit derivatives whereas 21 function evaluations are needed if no derivative information is supplied.

Functions such as `expn` can often be generated automatically. The function `deriv` will in most cases carry through the symbolic differentiation and produce as output an S function similar to ours. It is called with three arguments:

(a) the model formula (where the left hand side may be left vacant),

(b) a character vector giving the names of the parameters for which derivatives are sought, and

(c) a NULL function with an argument specification as required for the result function.

An example makes the process clearer. For the weight loss data with the exponential model, we can use:

```
expn1 <- deriv(y ~ b0 + b1 * 2^(-x/th), c("b0", "b1", "th"),
               function(b0, b1, th, x) NULL)
```

The result is the function

```
expn1 <- function(b0, b1, th, x)
{
    .expr3 <- 2^(( - x)/th)
    .value <- b0 + (b1 * .expr3)
    .grad <- array(0, c(length(.value), 3),
                  list(NULL, c("b0", "b1", "th")))
    .grad[, "b0"] <- 1
    .grad[, "b1"] <- .expr3
    .grad[, "th"] <- b1 *
        (.expr3 * (0.693147180559945 * (x/(th^2))))
    attr(.value, "gradient") <- .grad
    .value
}
```

If second derivatives are used, they must be supplied as an $n \times p \times p$ array, as a `hessian` attribute of the model function. Each $p \times p$ layer of this array is then the (symmetric) matrix of second derivatives of the model function for that observation. There are no tools in S-PLUS to build functions to supply both first and second derivatives, but a function `deriv3` written by David Smith, is provided in our library. (For complicated functions or cases with a large number of parameters `deriv3` can be slow and hit memory limits.)

Supplying second-derivative information usually does not offer any computational advantage for non-linear regression problems fitted by `nls`. However, in general model-fitting problems using `ms` second-derivative information can be computationally very important, as we will see in Section 9.8.

Second derivatives are also needed to assess the extent of the curvature in the model surface and hence indirectly to give some indication whether methods based on the planar approximation are accurate. We will explore this a little further in Section 9.7.

9.4   Non-linear fitted model objects and method functions

The result of a call to `nls` is an object of class `nls`. The standard method functions are available, the most important of which are

`print` to print an abbreviated synopsis (usually called implicitly),

`summary` to print a summary of the fitting process results,

`coef` to extract the estimated mean parameters,

`fitted` to extract the fitted value,

`residuals` to extract residuals,

`profile` to explore the contours of the least squares surface in the vicinity of the least squares estimate.

The summary function gives for our example:

```
> summary(wtloss.gr)

Formula: Weight ~ expn1(b0, b1, th, Days)
Parameters:
      Value Std. Error t value
b0  81.374    2.2690  35.863
b1 102.684    2.0828  49.302
th 141.911    5.2945  26.803
Residual standard error: 0.894937 on 49 degrees of freedom
Correlation of Parameter Estimates:
          b0      b1
b1 -0.989
th -0.986   0.956
```

It is perhaps surprising that no deviance method function exists, and the default function does not work since no `deviance` component is supplied as part of the fitted object. The function is easy to supply, though as a one-line function:

```
> deviance.nls <- function(object) sum(object$residuals^2)
> deviance(wtloss.gr)
[1] 39.245
```

It is also useful to have available a generic function that will extract the full estimated variance matrix of the mean parameters, since this is often a necessary ingredient in subsequent calculations. Again this is very simple to do, since most of the calculation is done already by `summary.nls`:

```
> vcov.nls <- function(fm) {
      sm <- summary(fm)
      sm$cov.unscaled * sm$sigma^2
}
> vcov(wtloss.gr)
          b0       b1       th
b0    5.1484  -4.6745 -11.841
b1   -4.6745   4.3379  10.543
th  -11.8414  10.5432  28.032
```

It is convenient to work with a generic function and a specific method for objects of class `nls` since it may be useful to add later further method functions for, say, objects of class `lm`, `glm`, `aov`, and so on (see page 234).

Another surprising omission is a `predict` method for `nls` objects.

9.5   Taking advantage of linear parameters

For models with both linear and non-linear parameters, if the non-linear parameters were known the model would become linear and standard linear regression methods could be used for fitting. This simple idea lies behind the `plinear` algorithm.

With the `default` algorithm the non-linear model formula specifies the mean vector as the right-hand-side expression. With the `plinear` algorithm the right-hand-side expression specifies a matrix whose columns are functions of the non-linear parameters. The linear parameters are then implicitly specified as regression coefficients for the columns of the matrix, that is, by the same method as for linear regression. Initial values are needed only for the non-linear parameters.

In the weight loss example there are two linear parameters, β_0 and β_1. To fit the model using the `plinear` algorithm we use:

```
> parameters(wtloss) <- list(th=120)
> wtloss.pl <- nls(Weight ~ cbind(1, 2^(-Days/th)),
               data=wtloss, algorithm="plinear", trace=T)
58.0817 : 120
39.538 : 138.778
39.2448 : 141.855
39.2447 : 141.911
> summary(wtloss.pl)

Formula: Weight ~ cbind(1, 2^( - Days/th))
Parameters:
     Value Std. Error t value
th 141.911    5.2945  26.803
    81.374    2.2690  35.863
   102.684    2.0828  49.302
Residual standard error: 0.894937 on 49 degrees of freedom
Correlation of Parameter Estimates:
       th
  -0.986
   0.956 -0.989
```

Notice that the linear parameters are unnamed.

There are two advantages in using the partially linear algorithm. It can be much more stable than methods that do not take account of linear parameters; it requires initial estimates for fewer parameters and it can often converge from poor starting positions where other procedures fail. In cases where there are very many linear parameters it may be possible to generate the matrix involved fairly easily but to specify it in full as a standard non-linear regression model formula may be very tedious.

9.6 Examples

The weight loss data, continued

For the weight loss data the high correlations between parameter estimates give a warning that the example is very ill conditioned, but it is the only dataset we have. (In interim analyses before the weight rise at 160 days the correlations were even higher.) The estimated asymptotic lean stable weight is lower than originally anticipated, so the half-life parameter is higher.

For the individual concerned, a crucial question is how long he expects to be on the programme before achieving any particular goal weight. If x_0 is the time to achieve a mean weight of y_0, then solving the equation $y_0 = \beta_0 + \beta_1 2^{-x_0/\theta}$ gives

$$x_0 = -\theta \log_2\{(y_0 - \beta_0)/\beta_1\}$$

The estimated (large sample) variance of the estimate, $\widehat{x}_0$, is then

$$\mathrm{Var}\big[\widehat{x}_0\big] = \boldsymbol{\lambda}^T \mathrm{Var}\big[\widehat{\beta}\big]\boldsymbol{\lambda}, \qquad \boldsymbol{\lambda} = \partial x_0/\partial \boldsymbol{\beta_0}$$

Rather than explicitly carry out the differentiation we can use `deriv` to supply a function for calculating both times and derivatives:

```
time.pds <- deriv(~ -th * log((y0 - b0)/b1)/log(2),
     c("b0","b1","th"), function(y0, b0, b1, th) NULL)
```

(Note that we do not use `log(x, 2)` as this is not differentiated correctly.)

Since several different potential goal weights may be involved, we wrote an S function for the remaining calculations:

```
est.time <- function(y0, obj) {
    b <- coef(obj)
    tmp <- time.pds(y0, b["b0"], b["b1"], b["th"])
    x0 <- as.vector(tmp)
    lam <- attr(tmp, "gradient")
    v <- (lam %*% vcov(obj) * lam) %*% matrix(1, 3, 1)
    a <- cbind(x0, sqrt(v))
    dimnames(a) <- list(paste(y0, "kg: "), c("x0", "SE"))
    a
}
```

We estimated the time to three goal weights, 100, 90 and 85 kilograms.

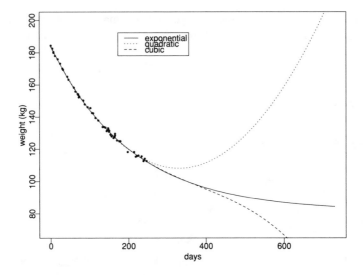

Figure 9.2: Weight loss data extrapolated to two years: exponential, quadratic and cubic polynomial fitted models.

```
> est.time(c(100, 90, 85), wtloss.gr)
                x0      SE
100 kg:     349.50  8.1754
 90 kg:     507.09 31.2179
 85 kg:     684.52 98.9002
```

Now consider a comparison of the non-linear model with polynomial regression models:

```
plot(Days, Weight, xlab= "days", ylab ="weight (kg)",
    xlim=c(0,730), ylim=c(70, 200))
xx <- seq(0, 730, 10)
lines(xx, 81.37+ 102.68 * 2^(-xx/141.91))
wtloss.quad <- lm(Weight ~ poly(Days, 2))
lines(xx, predict.gam(wtloss.quad, data.frame(Days=xx)), lty=2)
wtloss.cub <- lm(Weight ~ poly(Days, 3))
lines(xx, predict.gam(wtloss.cub, data.frame(Days=xx)), lty=3)
legend(locator(1), c("exponential", "quadratic", "cubic"),
    lty=1:3)
```

The plot of the fits in Figure 9.2 shows that the basis for extrapolation is weak. The exponential model gives a credible estimate for the asymptotic lean weight and fairly realistic estimates for the time to achieve some reasonable goal weights. This is in contrast to polynomial models which account for the data almost equally as well in the observed time range, but break down outside this range.

Note the use of predict.gam with the polynomial models to get the correct predictions.

A connection with generalized linear models

The following example comes from Williams (1959). The Stormer viscometer measures the viscosity of a fluid by measuring the time taken for an inner cylinder in the mechanism to perform a fixed number of revolutions in response to an actuating weight. The viscometer is calibrated by measuring the time taken with varying weights while the mechanism is suspended in fluids of accurately known viscosity. The dataset comes from such a calibration, and theoretical considerations suggest a non-linear relationship between time T, weight w and viscosity V of the form

$$T = \frac{\beta_1 v}{w - \beta_2} + \epsilon$$

where β_1 and β_2 are unknown parameters to be estimated. Note that β_1 is a linear parameter and β_2 is non-linear. The dataset is given in Table 9.1.

Table 9.1: The Stormer viscometer calibration data. The body of the table shows the times in seconds.

Weight	Viscosity (poise)								
(grams)	14.7	27.5	42.0	75.7	89.7	146.6	158.3	161.1	298.3
20	35.6	54.3	75.6	121.2	150.8	229.0	270.0		
50	17.6	24.3	31.4	47.2	58.3	85.6	101.1	92.2	187.2
100				24.6	30.0	41.7	50.3	45.1	89.0,86.5

Williams suggested that a suitable initial value may be obtained by writing the regression model in the form

$$wT = \beta_1 v + \beta_2 T + (w - \beta_2)\epsilon$$

and regressing wT on v and T using ordinary linear regression. With the data available in a data frame `stormer` with variables `Viscosity`, `Wt` and `Time`, we used:

```
> fm0 <- lm(Wt*Time ~ Viscosity + Time - 1, data=stormer)
> b0 <- as.list(coef(fm0))
> names(b0) <- paste("b",1:2,sep="")
> unlist(b0)
      b1      b2
 28.876 2.8437
> parameters(stormer) <- b0
> storm.fm <- nls(Time ~ b1*Viscosity/(Wt-b2), data=stormer,
            trace=T)
885.365 : 28.8755 2.84373
825.110 : 29.3935 2.23328
825.051 : 29.4013 2.21823
```

With a different parametrization this particular model may also be expressed as a generalized linear model. To do this we write

$$\frac{\beta_1 v}{w - \beta_2} = \frac{1}{\gamma_1 z_1 + \gamma_2 z_2}$$

where $\gamma_1 = 1/\beta_1$, $\gamma_2 = \beta_2/\beta_1$, $z_1 = w/v$ and $z_2 = -1/v$. This has the form of generalized linear model with inverse link, and we can use

```
> attach(stormer,1)
> z1 <- Wt/Viscosity
> z2 <- -1/Viscosity
> detach(1,save="stormer")
> attach(stormer)
> storm.gm <- glm(Time ~ z1 + z2 - 1,
      family=quasi(link=inverse, variance=constant),
      data=stormer, trace=T)
GLM    linear loop 1: deviance = 860.92
GLM    linear loop 2: deviance = 825.06
GLM    linear loop 3: deviance = 825.05
```

The deviance agrees with the residual sum of squares from `nls`. We extract the coefficients, back transform, and compare them with the `nls` estimates:

```
> g <- coef(storm.gm)
> b <- coef(storm.fm)
> b0 <- c(1/g[1], g[2]/g[1])
> cbind(b,b0)
          b      b0
z1 29.4013 29.4013
z2  2.2182  2.2183
```

Finding the standard errors is a little more tedious. We need to find the variance matrix from the generalized linear model. The objects from `glm` fits inherit from `lm`, and the two classes of object are similar enough to make the one method function apply to both.

```
vcov.lm <- function(obj) {
    so <- summary(obj)
    so$deviance/so$df[2] * so$cov.unscaled
}
```

The large sample variance matrix for $\widehat{\beta}$ is related to that for $\widehat{\gamma}$ by $\mathrm{Var}\big[\widehat{\beta}\big] = J\mathrm{Var}\big[\widehat{\gamma}\big]J^T$ where

$$J = \begin{bmatrix} \partial\beta_1/\partial\gamma_1 & \partial\beta_1/\partial\gamma_2 \\ \partial\beta_2/\partial\gamma_1 & \partial\beta_2/\partial\gamma_2 \end{bmatrix} = \begin{bmatrix} -1/\gamma_1^2 & 0 \\ -\gamma_2/\gamma_1^2 & 1/\gamma_1 \end{bmatrix}$$

Here J is the Jacobian matrix of the inverse of the parameter transformation, calculated by

```
> J <- matrix(c(-1/g[1]^2, -g[2]/g[1]^2, 0, 1/g[1]), 2, 2)
> J %*% vcov(storm.gm) %*% t(J)
          [,1]      [,2]
[1,]   0.83763 -0.56018
[2,]  -0.56018  0.44265
> vcov(storm.fm)
            beta     theta
  beta   0.83820 -0.56055
  theta -0.56055  0.44292
```

The agreement with `nls` is certainly adequate for statistical purposes.

The advantage of the generalized linear model approach is that starting values can be generated automatically and the iteration process can be rather more stable. The disadvantage is that the estimates obtained may not be of the parameters of most interest.

Since there are just two parameters we can display a confidence region for the regression parameters as a contour map. To this end put:

```
bc <- coef(storm.fm)
se <- sqrt(diag(vcov(storm.fm)))
dv <- deviance(storm.fm)
```

Define $d(\beta_1, \beta_2)$ as the sum of squares function:

$$d(\beta_1, \beta_2) = \sum_{i=1}^{23} \left(T_i - \frac{\beta_1 v_i}{w_i - \beta_2} \right)^2$$

Then `dv` contains the minimum value, $d_0 = d(\widehat{\beta}_1, \widehat{\beta}_2)$, the residual sum of squares or model deviance.

If β_1 and β_2 are the true parameter values the statistic

$$F(\beta_1, \beta_2) = \frac{(d(\beta_1, \beta_2) - d_0)/2}{d_0/21}$$

is approximately distributed as $F_{2,21}$. An approximate confidence set contains those values in parameter space for which $F(\beta_1, \beta_2)$ is less than the 95% point of the $F_{2,21}$ distribution. We will construct a contour plot of the $F(\beta_1, \beta_2)$ function and mark off the confidence region. The result is shown in Figure 9.3.

A suitable region for the contour plot is three standard errors either side of the least squares estimates in each parameter. Since these ranges are equal in their respective standard error units it is useful to make the plotting region square.

```
par(pty = "s")
b1 <- bc[1] + seq(-3*se[1], 3*se[1], length = 51)
b2 <- bc[2] + seq(-3*se[2], 3*se[2], length = 51)
bv <- expand.grid(b1, b2)
```

The simplest way to calculate the sum of squares function is to use `apply` function:

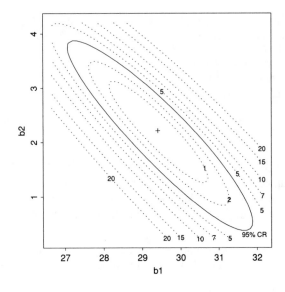

Figure 9.3: The Stormer data. The F–statistic surface and a confidence region for the regression parameters.

```
ssq <- function(b)
       sum((Time - b[1] * Viscosity/(Wt-b[2]))^2)
dbeta <- apply(bv, 1, ssq)
```

However using a function such as `outer` makes better use of the vectorizing facilities of S, and in S-PLUS 3.2 a direct calculation is the most efficient:

```
cc <- matrix(Time - rep(bv[,1],rep(23,2601)) *
       Viscosity/(Wt - rep(bv[,2], rep(23,2601))), 23)
dbeta <- matrix(drop(rep(1, 23) %*% cc^2),51)
```

The F–statistic array is then:

```
fstat <- matrix( ((dbeta - dv)/2) / (dv/21), 51, 51)
```

We can now produce a contour map of the F–statistic, taking care that the contours occur at relatively interesting levels of the surface. Note that the confidence region contour is at about 3.5:

```
> qf(0.95, 2, 21)
[1] 3.4668
```

Rather than use `contour` to set up the plot directly, we begin with a call to `plot`:

```
plot(b1, b2, type="n")
lev <- c(1,2,5,7,10,15,20)
contour(b1, b2, fstat, levels=lev, labex=0.75, lty=2, add=T)
contour(b1, b2, fstat, levels=qf(0.95,2,21), add=T, labex=0)
text(31.6,0.3,"95% CR", adj=0, cex=0.75)
points(bc[1], bc[2], pch=3, mkh=0.1)
```

Since the likelihood function has the same contours as the F–statistic the near elliptical shape of the contours is an indication that the approximate theory based on normal linear regression is probably accurate, although more than this is needed to be confident. See the discussion of profiles in Section 9.7. Given the way the axis scales have been chosen, the elongated shape of the contours shows that the estimates $\widehat{\beta}_1$ and $\widehat{\beta}_2$ are highly (negatively) correlated.

9.7 Assessing the linear approximation: profiles

It is convenient to separate two sources of curvature, that of the solution locus itself, the *intrinsic curvature*, and that of the coordinate system within the solution locus, the *parameter-effects curvature*. The intrinsic curvature is fixed by the data and solution locus, but the parameter-effects curvature also depends upon the parametrization.

Summary measures for both kinds of *relative* curvature were proposed by Beale (1960) and elegantly interpreted by Bates & Watts (1980, 1988 §7.3). (The measures are relative to the estimated standard error of y and hence scale free.) The two measures are denoted by c^θ and c^ι for the parameter-effects and intrinsic root-mean-square curvatures respectively. If F is the $F_{p,n-p}$ critical value, Bates & Watts suggest that a value of $c\sqrt{F} > 0.3$ should be regarded as indicating unacceptably high curvature of either kind. Readers are referred to Bates & Watts (1988) or Seber & Wild (1989, §4.3) for further details.

The function `rms.curv` supplied with our library can be used to calculate and display $c^\theta\sqrt{F}$ and $c^\iota\sqrt{F}$. The only required argument is an `nls` fitted model object, provided the model function has both a `gradient` and a `hessian` attribute. Consider our weight loss example. The function `enlist` is a simple utility that coerces a vector to a list of its components retaining any names attribute, a companion to `unlist`.

```
> enlist <- function(vec)
    {
        x <- as.list(vec)
        names(x) <- names(vec)

        x
    }
> plist <- enlist(coef(wtloss.gr))
> expn2 <- deriv3(~ b0 + b1*2^(-x/th), c("b0","b1","th"),
        function(x, b0, b1, th) NULL)
> wtloss.he <- nls(Weight ~ expn2(Days, b0, b1, th),
        start=plist, data=wtloss)
> rms.curv(wtloss.he)
Parameter effects: c^theta x sqrt(F) = 0.1679
          Intrinsic: c^iota  x sqrt(F) = 0.0101
```

Although this result is acceptable, a lower parameter-effects curvature would be preferable.

Stable parameters

Ross (1970) has suggested using *stable parameters* for non-linear regression, mainly to achieve estimates which are as near to uncorrelated as possible. It turns out that in many cases stable parameters also define a coordinate system within the solution locus with a small curvature.

The idea is to use the means at p well separated points in sample space as the parameters. Since the range of the Days variable is 0 to 246 days, in this case we could choose as parameters the mean values at, say, 40, 120 and 200 days. These parameters will then mainly be determined by the first, middle and last third of the data points. Hence we set

$$\mu_0 = \beta_0 + \beta_1 2^{-40/\theta}, \quad \mu_1 = \beta_0 + \beta_1 2^{-120/\theta}, \quad \mu_2 = \beta_0 + \beta_1 2^{-200/\theta}$$

Writing the regression function in terms of the stable parameters is often intractable but here it is possible:

$$\eta = \frac{\mu_0 \mu_2 - \mu_1^2}{\mu_0 - 2\mu_1 + \mu_2} + \frac{(\mu_0 - \mu_1)^2}{\mu_0 - 2\mu_1 + \mu_2} \left(\frac{\mu_1 - \mu_2}{\mu_0 - \mu_1} \right)^{(x-40)/80}$$

We can now fit the model using the stable parametrization. Good initial values are always easy to find by estimating the mean at the required points by an approximating linear model.

```
> stab <- deriv3(~ ((u0*u2-u1^2) +
    (u0-u1)^2 *((u1-u2)/(u0-u1))^((x-40)/80))/(u0-2*u1+u2),
    c("u0","u1","u2"), function(x, u0, u1, u2) NULL)
> mu <- predict(lm(Weight~Days+Days^2, data=wtloss),
    newdata=data.frame(Days=c(40,120,200)))
> names(mu) <- paste("u", 0:2, sep="")
> wtloss.st <- nls(Weight ~ stab(Days, u0, u1, u2),
    start=mu, data=wtloss, trace=T)
43.3655 : 166.18 138.526 119.742
39.2447 : 165.834 138.515 120.033
> rms.curv(wtloss.st)
Parameter effects: c^theta x sqrt(F) = 0.0101
        Intrinsic: c^iota  x sqrt(F) = 0.0101
```

The intrinsic curvature remains unchanged (as it should). Since any coordinate system on a curved solution locus must itself have some curvature, the result is clearly very good. The parameter estimates are still somewhat correlated, but much less so than with the previous parametrization:

```
> summary(wtloss.st)$correlation
        u0        u1        u2
u0  1.00000  0.43675 -0.11960
u1  0.43675  1.00000  0.25806
u2 -0.11960  0.25806  1.00000
```

Profiles

Another way of assessing the accuracy of the approximating linear model is to look at *profiles* of the residual sum of squares function and check if they are close to quadratic, as they would be if the model were linear.

For a single component, β_i, of the parameter vector let $S(\beta_i)$ denote the residual sum of squares, $\|y - \eta(\beta)\|^2$, minimized with respect to all other components. The statistic

$$\tau(\beta_i) = \left(\text{sign}(\beta_i - \hat{\beta}_i)\sqrt{S(\beta_i) - S(\hat{\beta}_i)} \right) / s$$

has an interpretation as a non-linear t–statistic. For coordinate directions along which the approximate linear methods are accurate a plot of $\tau(\beta_i)$ against β_i over several standard deviations on either side of the maximum likelihood estimate should be straight. Non-linearity serves as a warning that the linear approximation may be misleading in that direction.

The function `profile` is available to automate these calculations. It takes a fitted model object and returns a list of data frames with the required information. In most cases there is sufficient information in the fitted model object to allow the default values of the remaining arguments to be used. Appropriate plot methods are also available. As an example consider both parametrizations of the weight loss data.

```
par(mfrow=c(2,3), pty="s")
plot(profile(wtloss.gr), ask=F)
plot(profile(wtloss.st))
```

The result is shown in Figure 9.4.

Profiles can also be defined for two or more parameters simultaneously. These present much greater difficulties of both computation and visual assessment. An example is given in the previous section with the two parameters of the stormer data model. In Figure 9.3 an assessment of the linear approximation requires checking both that the contours are approximately elliptical and that each one-dimensional cross section through the minimum is approximately quadratic. In one dimension it is much easier visually to check the linearity of the signed square root than the quadratic nature of the sum of squares itself.

9.8   General minimization and maximum likelihood estimation

The `ms` function is a general facility for minimizing quantities that can be written as sums of one or more terms. Sums of squares functions are one example, so `ms` can be also used to fit non-linear regressions, but there may be advantages in using `nls` for this special case.

The call to `ms` is very similar to that for `nls`, but the interpretation of the formula is slightly different. It has nothing to the left of the $\tilde{}$ operator, and the quantity on the right hand side specifies the entire sum to be minimized. Hence the parameter estimates from the call

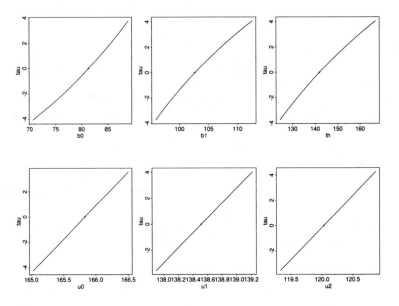

Figure 9.4: Profile plots for two parametrizations of the weight loss model.

```
fm <- nls(y ~ fn(a,b,c,x), start=p0, data=dat)
```

should be the same, in principle, as those from

```
gm <- ms( ~(y - fn(a,b,c,x))^2, start=p0, data=dat)
```

The resulting object is of class `ms`, for which very few method functions are available. The four main ones are

`print`   for basic printing (mostly used implicitly),

`coef`   for the parameter estimates,

`summary`   for a succinct report of the fitted results,

`profile`   for an exploration of the region near the minimum.

The fitted object has additional components, two of independent interest being

`gm$value`   the minimum value of the sum, and

`gm$pieces`   the vector of summands in the minimum sum.

Note that `fitted` and `resid` are not available since in this general context they have no unique meaning.

The most common statistical use of `ms` is to minimize negative log-likelihood functions, thus finding maximum likelihood estimates. The fitting algorithm can make use of first and second derivative information, the first and second derivative arrays of the summands rather than of a model function. In the case of a negative log-likelihood function, these are arrays whose "row sums" are the negative score vector and observed information matrix respectively.

The convergence criterion for ms is that *one of* the relative changes in the parameter estimates are small, the relative change in the achieved value is small *or* that the optimized function value is close to zero. The latter is only appropriate for deviance-like quantities, and can be eliminated by setting the argument

```
control = ms.control(f.tolerance=-1)
```

Fitting a mixture model

The waiting times between eruptions in the data faithful are strongly bimodal. Figure 9.5 shows a histogram with a density estimate superimposed (with bandwidth chosen by the cross-validation methods of Chapter 5).

```
attach(faithful)
hist(waiting, prob=T, xlim=c(35,105), ylim=c(0,0.045),
    style.bar="old")
wait.dns <- density(waiting, 200, width=10)
lines(wait.dns, lty=2)
```

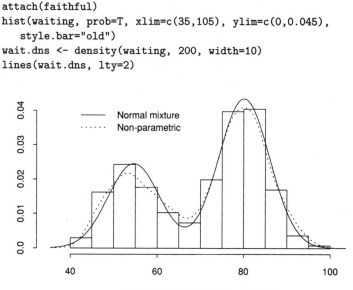

Waiting time between eruptions (min)

Figure 9.5: Histogram for the waiting times between successive eruptions for the "Old Faithful" geyser, with non-parametric and parametric estimated densities superimposed.

From inspection of this figure a mixture of two normal distributions would seem to be a reasonable descriptive model for the marginal distribution of waiting times. We now consider how to fit this by maximum likelihood. The observations are not independent since successive waiting times are strongly negatively correlated. In this section we propose to ignore this, both for simplicity and because ignoring this aspect is not likely seriously to misrepresent the information in the sample on the marginal distribution in question.

Useful references for mixture models include Everitt & Hand (1981), Titterington, Smith & Makov (1985) and McLachlan & Basford (1988). Everitt & Hand describe the EM algorithm for fitting a mixture model, which is simple (and provides an interesting S programming exercise), but we shall consider here a more direct function minimization method.

If y_i, $i = 1, 2, \ldots, n$ is a sample waiting time the log-likelihood function for a mixture of two normal components is

$$L(\pi, \mu_1, \sigma_1, \mu_2, \sigma_2) = \sum_{i=1}^{n} \log \left[\frac{\pi}{\sigma_1} \phi \left(\frac{y_i - \mu_1}{\sigma_1} \right) + \frac{1 - \pi}{\sigma_2} \phi \left(\frac{y_i - \mu_2}{\sigma_2} \right) \right]$$

We will estimate the parameters by minimizing $-L$.

It will be important in this example to use both first and second derivative information and we would like to use deriv3 to produce a model function for us. Both deriv and deriv3 use a system function, D, to generate symbolic derivatives but as it stands it does not differentiate functions involving the normal distribution or density functions. Our first task will be to extend D and its ancillary function make.call so that they can do so. (Extended versions are supplied in our library, which must be attached with first=T to make them effective.)

The function D is a collection of nested switch statements and to extend it we need only insert two additional branches in the appropriate switch that declare how the standard normal distribution and density functions are differentiated. The changes are:

```
    ....
    sqrt = D(call("^", expr[[2]], 0.5), name),
    pnorm = make.call("*", make.call("dnorm",      # Extension
            expr[[2]]), D(expr[[2]], name)),
    dnorm = make.call("*", make.call("*",          # Extension
            make.call("-", expr[[2]]), expr),
            D(expr[[2]], name)),
    stop(paste("Function", expr[[1]],
            "is not listed in the derivatives table"
            ))),
    ....
```

Similarly make.call requires two additional branches in its major switch:

```
    ....
    tanh = call("tanh", arg1),
    pnorm = call("pnorm", arg1),     # Extension
    dnorm = call("dnorm", arg1),     # Extension
    NA)
    ....
```

Only the second of these is needed in our example but the first is included for completeness. Note that these extensions only allow calls to pnorm and dnorm with one argument. Thus calls such as pnorm(x,u,s) must be written as pnorm((x-u)/s).

We can now generate a function which calculates the summands of $-L$ and also returns first and second derivative information for each summand by:

```
lmix2 <- deriv3(
        ~ -log(p*dnorm((x-u1)/s1)/s1 + (1-p)*dnorm((x-u2)/s2)/s2),
        c("p", "u1", "s1", "u2", "s2"),
        function(x, p, u1, s1, u2, s2) NULL)
```

which will take minutes.

Initial values for the parameters could be obtained by the method of moments described in Everitt & Hand (1985, page 31*ff*) but for well–separated components as we have here, simply choosing initial values by reference to the plot suffices. For the initial value of π we take the proportion of the sample below the density low point at 70.

```
p0 <- list(p=sum(waiting < 70)/length(waiting),
            u1=50, s1=5, u2=80, s2=5)
unlist(p0)
         p u1 s1 u2 s2
 0.37868 50  5 80  5
```

We will trace the iterative process, using a trace function that produces just one line of output per iteration:

```
tr.ms <- function(info, theta, grad, scale, flags, fit.pars)
{
    cat(signif(info[3]), ":", signif(theta), "\n")
    invisible(list())
}
```

Note that the trace function must have an argument specification the same as the standard trace function, `trace.ms`.

We can now fit the mixture model:

```
> wait.mix2 <- ms(~ lmix2(waiting, p, u1, s1, u2, s2),
    start=p0, data = faithful, trace = tr.ms)
1077.34 : 0.378676 50 5 80 5
1073.09 : 0.342766 56.2689 3.6252 79.8122 5.7471
1043.70 : 0.340921 54.6143 4.15463 79.6473 6.24173
1036.19 : 0.349106 54.5222 4.92496 79.9016 6.08841
1034.17 : 0.359554 54.6176 5.59894 80.0729 5.87407
1034.00 : 0.360617 54.6104 5.83976 80.0865 5.87119
1034.00 : 0.360882 54.6148 5.87079 80.0910 5.86778
1034.00 : 0.360886 54.6149 5.87122 80.0911 5.86773
```

Without any derivative information the procedure converges from this starting point in 32 iterations whereas with first derivatives only it requires 35 iterations.

We can now add the parametric density estimate to our original histogram plot.

```
dmix2 <- function(x, p, u1, s1, u2, s2)
              p * dnorm(x, u1, s1) + (1-p) * dnorm(x, u2, s2)
cf <- enlist(coef(wait.mix2))
wait.fdns <- list(x = wait.dns$x, y =
    eval(expression(dmix2(wait.dns$x, p, u1, s1, u2, s2)),
```

```
        local = cf))
    lines(wait.fdns)
    par(usr = c(0,1,0,1))
    legend(0.1, 0.9, c("Normal mixture", "Nonparametric"),
        lty = c(1,2), bty = "n")
```

The computations for Figure 9.5 are now complete.

The parametric and nonparametric density estimates are in fair agreement on the right component, but there is a suggestion of disparity on the left. We can informally check the adequacy of the parametric model by a Q-Q plot. First we solve for the quantiles using a Newton method:

```
    pmix2 <- deriv(~ p*pnorm((x-u1)/s1) + (1-p)*pnorm((x-u2)/s2),
        "x", function(x, p, u1, s1, u2, s2) NULL)
    pr0 <- (1:272 - 0.5)/272
    x0 <- x1 <- sort(waiting)
    for(i in 1:10) {
        pr <- eval(expression(pmix2(x0, p, u1, s1, u2, s2)),
                   local = cf)
        del <- (pr - pr0)/attr(pr, "gradient")
        x0 <- x0 - del
        print(del <- max(abs(del)))
        if(del < 0.0005) break
    }
    [1] 1.9953
    [1] 1.1259
    [1] 0.35148
    [1] 0.029407
    [1] 0.00019061
```

The plot is shown in Figure 9.6 and confirms the adequacy of the mixture model.

```
    par(pty = "s")
    plot(x0, x1, xlim = range(x0, x1), ylim = range(x0, x1),
        xlab = "Model quantiles", ylab = "Waiting time")
    abline(0,1)
```

Another advantage of using second derivative information is that it provides an estimate of the (large sample) variance matrix of the parameters. The fitted model object has a component, hessian, giving the sum of the second derivative array over the observations and since the function we minimized is the negative of the log-likelihood, this component is the observed information matrix, calculated (incorrectly before S-PLUS 3.2, and still misleadingly labelled) by summary.ms

```
    vmat <- wait.mix2$hessian          # lower triangle only
    vmat <- solve(vmat + t(vmat) - diag(diag(vmat)))
    # S-Plus 3.2 users can use the inverse information
    # vmat <- summary(wait.mix2)$Information
    cbind(coef(wait.mix2), sqrt(diag(vmat)))
             [,1]       [,2]
```

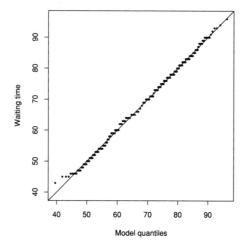

Figure 9.6: Sorted waiting times against normal mixture model quantiles for the "Old Faithful" eruptions data.

```
 p   0.36089 0.031165
u1 54.61486 0.699675
s1   5.87122 0.537322
u2 80.09107 0.504595
s2   5.86773 0.400962
```

It appears that a simpler model of two normal components with equal variances would have been adequate. We can easily check this via a score test or a likelihood ratio test:

```
> (cf$s1 - cf$s2) /sqrt(c(0,0,1,0,1) %*% vmat %*% c(0,0,1,0,1))
[1,] 0.0061413
> wait.mix2e <- ms(~ -log(p * dnorm(waiting, u1, s1) +
    (1-p) * dnorm(waiting, u2, s1)), start = cf[-5],
    data = faithful)

> coef(wait.mix2e)
        p      u1     s1     u2
 0.36085 54.614 5.8691 80.09
> 2*(wait.mix2e$value - wait.mix2$value)
> [1] 2.107e-05
```

The S-PLUS dataset `geyser` has data of the same type from the same source. A normal mixture model also fits this dataset well, but the variances of the two components are significantly different, which suggests the agreement here is fortuitous.

Chapter 10

Modern Regression

S-PLUS has a 'Modern Regression Module' which contains functions for a number of regression methods. These are not necessarily non-linear in the sense of Chapter 9, which refers to a non-linear parametrization, but they do allow non-linear functions of the independent variables to be chosen by the procedures. The methods are all fairly computer-intensive, and so are only feasible in the era of plentiful computing power (and hence are 'modern'). Some of these methods are part of the S modelling language, and others have been added by S-PLUS. As the latter predate the modelling language and have not been updated, the functions of this chapter do not have a consistent style and user interface.

There are few texts covering this material. Although it does not cover all our topics in equal detail, for what it does cover Hastie & Tibshirani (1990) is an excellent reference. Chambers & Hastie (1992) cover gam and loess in considerably greater detail than this chapter. Thisted (1988) gives a brief computationally-oriented introduction to many of the methods of this chapter.

10.1 Additive models and scatterplot smoothers

For linear regression we have a dependent variable Y and a set of predictor variables $X_1, \ldots, X_p$, and model

$$Y = \alpha + \sum_{j=1}^{p} \beta_j X_j + \epsilon$$

Additive models replace the linear function $\beta_j X_j$ by a non-linear function, to get

$$Y = \alpha + \sum_{j=1}^{p} f_j(X_j) + \epsilon \tag{10.1}$$

Since the functions f_j are rather general, they can subsume the α. Of course, it will not be useful to allow an arbitrary function f_j, and it will help to think of it as a *smooth* function.

The classical way to introduce non-linear functions of dependent variables is to add a limited range of transformed variables to the model, for example quadratic and cubic terms, or to split the range of the variable and use a piecewise constant function. (In S these can be achieved by the functions `poly` and `cut`.) A 'modern' alternative is to use *spline* functions. We will only need cubic splines. Divide the real line by an ordered set of points $\{z_i\}$ known as *knots*. On the interval $[z_i, z_{i+1}]$ the spline is a cubic polynomial, and it is continuous and has continuous first and second derivatives, imposing 3 conditions at each knot. With n knots, $n+4$ parameters are needed to represent the cubic spline (from $4(n+1)$ for the cubic polynomials minus $3n$ continuity conditions). Of the many possible parametrizations, that of B-splines has desirable properties. The S function `bs` generates a matrix of B-splines, and so can be included in a linear-regression fit.

A restricted form of B-splines known as *natural splines* and implemented by the S function `ns`, is linear on $(-\infty, z_1]$ and $[z_n, \infty)$ and thus would have n parameters. However, `ns` adds an extra knot at each of the maximum and minimum of the data points, and so has $n+2$ parameters. The functions `bs` and `ns` may have the knots specified or be allowed to choose the knots as quantiles of the empirical distribution of the variable to be transformed, by specifying the number `df` of parameters.

Prediction from models including splines (and indeed the orthogonal polynomials generated by `poly`) needs care, as the basis for the functions depends on the observed values of the independent variable. If `predict.lm` is used, it will form a new set of basis functions and then erroneously apply the fitted coefficients. The function `predict.gam` will work more nearly correctly. (It uses both the old and new data to choose the knots.)

Splines are by no means the only way to fit smooth functions in (10.1), and indeed as we shall see, fitting as part of the regression is not the only way to use splines to find a smooth term f_j. To consider these approaches we need to digress to consider the simplest case of just one independent variable, so the relationship between x and y can be shown on a scatterplot. The single-variable methods are used as building-blocks in fitting multi-variable additive models.

Scatterplot smoothing

These methods depend on being able to estimate 'smooth' functions. They make use of scatterplot smoothers, which, given a scatterplot of (x, y) values, draw a smooth curve against x. It is not necessary to understand the exact details, and there are several alternative methods, including splines, running means and running lines. However, some users will want to understand the smoother being used.

Our running example is a set of data on 133 observations of acceleration against time for a simulated motorcycle accident, taken from Silverman (1985). A series of smoothers is shown in Figure 10.1, obtained by the following S-PLUS code.

```
attach(mcycle)
par(mfrow = c(3,2))
plot(times, accel, main="Polynomial regression")
```

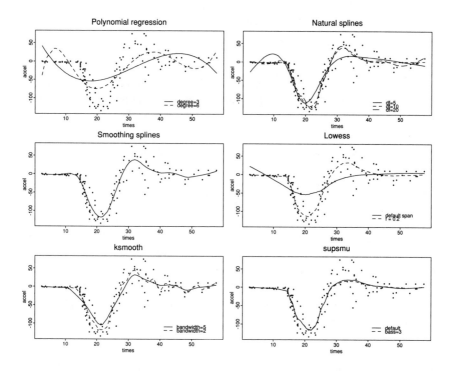

Figure 10.1: Scatterplot smoothers for the simulated motorcycle data given by Silverman (1985).

```
lines(times, fitted(lm(accel ~ poly(times, 3))))
lines(times, fitted(lm(accel ~ poly(times, 6))), lty=3)
legend(40, -100, c("degree=3", "degree=6"), lty=c(1,3),bty="n")
plot(times, accel, main="Natural splines")
lines(times, fitted(lm(accel ~ ns(times, df=5))))
lines(times, fitted(lm(accel ~ ns(times, df=10))), lty=3)
lines(times, fitted(lm(accel ~ ns(times, df=20))), lty=4)
legend(40, -100, c("df=5", "df=10", "df=20"), lty=c(1,3,4),
    bty="n")
plot(times, accel, main="Smoothing splines")
lines(smooth.spline(times, accel))
plot(times, accel, main="Lowess")
lines(lowess(times, accel))
lines(lowess(times, accel, 0.2), lty=3)
legend(40, -100, c("default span", "f = 0.2"), lty=c(1,3),
    bty="n")
plot(times, accel, main ="ksmooth")
lines(ksmooth(times, accel,"normal", bandwidth=5))
lines(ksmooth(times, accel,"normal", bandwidth=2), lty=3)
legend(40, -100, c("bandwidth=5", "bandwidth=2"), lty=c(1,3),
    bty="n")
plot(times, accel, main ="supsmu")
```

```
lines(supsmu(times, accel))
lines(supsmu(times, accel, bass=3), lty=3)
legend(40, -100, c("default", "bass=3"), lty=c(1,3), bty="n")
```

We have already discussed polynomials and regression splines, and Figure 10.1 shows how much better splines are than polynomials at adapting to general smooth curves. (The df parameter controls the number of terms in the regression spline.)

Suppose we have n pairs (x_i, y_i). A *smoothing spline* minimizes a compromise between the fit and the degree of smoothness of the form

$$\sum [y_i - f(x_i)]^2 + \lambda \int (f''(x))^2 \, dx$$

over all (measurably twice-differentiable) functions f. It is a cubic spline with knots at the x_i, but does not interpolate the data points for $\lambda > 0$ and the degree of fit is controlled by λ. The S function smooth.spline allows λ or the (equivalent) degrees of freedom to the specified, otherwise it will choose the degree of smoothness automatically by cross-validation (see page 139).

The algorithm used by lowess is quite complex; it uses robust locally linear fits. A window is placed about x; data points that lie inside the window are weighted so that nearby points get the most weight and a robust weighted regression is used to predict the value at x. The parameter f controls the window size and is the proportion of the data which is included. The default, f=2/3, is often too large for scatterplots with appreciable structure. The S function loess is an extension of the ideas of lowess which will work in one, two or more dimensions in a similar way. The function scatter.smooth plots a loess line on a scatter plot, using loess.smooth.

A *kernel smoother* is of the form

$$\hat{y}_i = \sum_{j=1}^{n} y_i K\left(\frac{x_i - x_j}{b}\right) \bigg/ \sum_{j=1}^{n} K\left(\frac{x_i - x_j}{b}\right) \tag{10.2}$$

where b is a bandwidth parameter, and K a kernel function, as in density estimation. In our example we use the S-PLUS function ksmooth and take K to be a standard normal density. The critical parameter is the the bandwidth b. Other S code (calling C code), including methods for bandwidth choice, is given by Härdle (1991) and in the library haerdle available from statlib. The function ksmooth seems rather slow, and the faster alternatives in the library haerdle are recommended for large problems.

The smoother supsmu is the one used by the S-PLUS functions for modern regression. It is based on a symmetric k-nearest neighbour linear least squares procedure. (That is, $k/2$ data points on each side of x are used in a linear regression to predict the value at x.) This is run for three values of k, $n/2$, $n/5$ and $n/20$, and cross-validation is used to choose a value of k for each x which is approximated by interpolation between these three. Larger values of the parameter bass (up to 10) encourage smooth functions.

Fitting additive models

As we have seen, smooth terms parametrized by regression splines can be fitted by `lm`. For smoothing splines it would be possible to set up a penalized least-squares problem and minimize that, but there would be computational difficulties in choosing the smoothing parameters simultaneously. Instead an iterative approach is used.

The *backfitting* algorithm fits the smooth functions f_j in (10.1) one at a time by taking the residuals

$$Y - \sum_{k \neq j} f_k(X_k)$$

and smoothing them against X_j using one of the scatterplot smoothers of the previous subsection. The process is repeated until it converges. Linear terms in the model (including any linear terms in the smoother) are fitted by least squares.

This procedure is implemented in S by the function `gam`. The model formulae are extended to allow the terms `s(x)` and `lo(x)` which respectively specify a smoothing spline and a loess smoother. These have similar parameters to the scatterplot smoothers; for `s()` the default degrees of freedom is 4, and for `lo()` the window width is controlled by `span` with default 0.5. There is a plot method, `plot.gam`, which shows the smooth function fitted for each term in the additive model.

Our dataset `rock` contains measurements on four cross-sections of each of 12 oil-bearing rocks; the aim is to predict permeability y (a property of fluid flow) from the other three measurements. As permeabilities vary greatly (6.3–1300), we use a log scale. The measurements are the end product of a complex image-analysis procedure and represent the total area, total perimeter and a measure of 'roundness' of the pores in the rock cross-section.

We first fit a linear model, then a full additive model. In this example convergence of the backfitting algorithm is unusually slow, so the `control` limits must be raised. The plots are shown in Figure 10.2.

```
attach(rock)
rock.lm <- lm(log(perm) ~ area + peri + shape)
summary(rock.lm)
rock.gam <- gam(log(perm) ~ s(area) + s(peri) + s(shape),
    control=gam.control(maxit=50, bf.maxit=50))
summary(rock.gam)
anova(rock.lm, rock.gam)
par(mfrow=c(2,2))
plot(rock.gam, se=T)
rock.gam1 <- gam(log(perm) ~ area + peri + s(shape))
par(mfrow=c(2,2))
plot(rock.gam1, se=T)
anova(rock.lm, rock.gam1, rock.gam)
```

It is worth showing the output from `summary.gam` and the analysis of variance table from `anova(rock.lm, rock.gam)`:

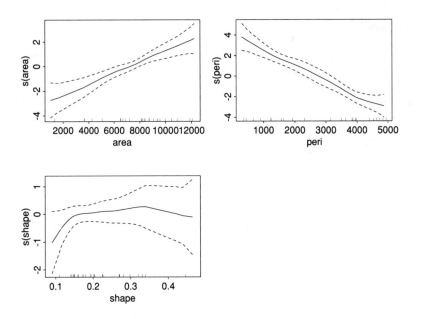

Figure 10.2: The results of fitting an additive model to the dataset `rock` with smooth functions of all three predictors. The dashed lines are approximate 95% pointwise confidence intervals. The tick marks show the locations of the observations on that variable.

```
Deviance Residuals:
     Min        1Q  Median       3Q      Max
 -1.6855 -0.46962 0.12531 0.54248  1.2963

(Dispersion Parameter ... 0.7446 )

    Null Deviance: 126.93 on 47 degrees of freedom
Residual Deviance: 26.065 on 35.006 degrees of freedom

Number of Local Scoring Iterations: 1

DF for Terms and F-values for Nonparametric Effects

            Df Npar Df Npar F   Pr(F)
(Intercept)  1
   s(area)   1      3 0.3417 0.79523
   s(peri)   1      3 0.9313 0.43583
  s(shape)   1      3 1.4331 0.24966

Analysis of Variance Table

                        Terms Resid. Df    RSS    Test
1             area + peri + shape   44.000 31.949
2 s(area) + s(peri) + s(shape)     35.006 26.065 1 vs. 2
```

```
        Df Sum of Sq F Value    Pr(F)
1
2 8.9943     5.8835  0.8785 0.55311
```

This shows that each smooth term has one linear and three non-linear degrees of freedom. The reduction of RSS from 31.95 (for the linear fit) to 26.07 is not significant with an extra 9 degrees of freedom, but Figure 10.2 shows that only one of the functions appears non-linear, and even that is not totally convincing. Although suggestive, the non-linear term for `peri` is not significant. With just that non-linear term (Figure 10.3) the RSS is 29.00.

We can also fit a smooth term for `shape` by regression splines (see Figure 10.3):

```
rock.ns <- lm(log(perm) ~ area + peri + ns(shape, df=4))
summary(rock.ns)
plot.gam(rock.ns, se=T)
```

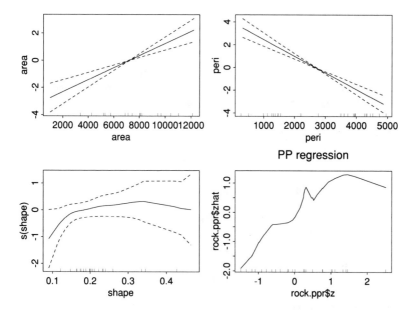

Figure 10.3: Additive model functions fitted to `rock` with linear `area`, linear `peri` and a smooth function of `shape`. The bottom right plot is of the smooth function fitted by a one-term projection pursuit regression.

The fit is very similar to that using smoothing splines. Note how `plot.gam` can still be used for regression splines fitted by `lm`.

Generalized additive models

These stand in the same relationship to additive models as generalized linear models do to regression models. Consider, for example, a logistic regression model. There are n observational units, each of which records a random variable Y_i with a Binomial (n_i, p_i) distribution, and (p_i) is determined by

$$\text{logit}\,(p) = \alpha + \sum_{j=1}^{p} \beta_j X_j \qquad (10.3)$$

This is readily generalized to the logistic additive model:

$$\text{logit}\,(p) = \alpha + \sum_{j=1}^{p} f_j(X_j) \qquad (10.4)$$

and the full form of a GAM is then obvious; replace the linear predictor in a GLM by an additive predictor.

Low birth weights

We return to the data on low birth weights in data frame `birthwt` studied in Section 7.2. We will examine the linearity in age and mother's weight, using generalized additive models and including smooth terms.

```
> attach(bwt)
> age1 <- age*(ftv=="1"); age2 <- age*(ftv=="2+")
> birthwt.gam <- gam(low ~ s(age) + s(lwt) + smoke + ptd +
      ht + ui + ftv + s(age1) + s(age2) + smoke:ui, binomial,
      bf.maxit=25)
> summary(birthwt.gam)

Residual Deviance: 170.35 on 165.18 degrees of freedom

DF for Terms and Chi-squares for Nonparametric Effects

         Df Npar Df Npar Chisq  P(Chi)
  s(age)  1     3.0      3.1089 0.37230
  s(lwt)  1     2.9      2.3392 0.48532
 s(age1)  1     3.0      3.2504 0.34655
 s(age2)  1     3.0      3.1472 0.36829

> table(low, predict(birthwt.gam) > 0)
    FALSE TRUE
0    115   15
1     28   31
> plot(birthwt.gam, ask=T, se=T)
```

Creating the variables `age1` and `age2` allows us to fit smooth terms for the *difference* in having one or more visits in the first trimester. Both the summary and the plots show no evidence of non-linearity. Note that the convergence of the fitting algorithm is slow in this example, so we increased the control parameter `bf.maxit` from 10 to 25. The parameter `ask=T` allows us to choose plots from a menu. Our choice of plots is shown in Figure 10.4.

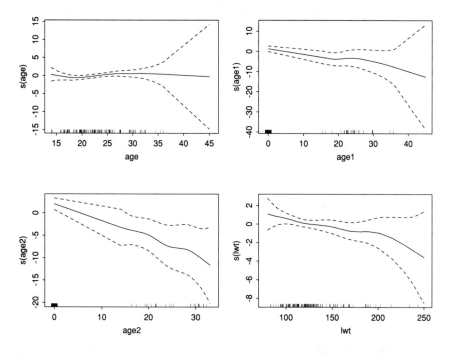

Figure 10.4: Plots of smooth terms in a generalized additive model for the data on low birth weight. The dashed lines indicate plus and minus two pointwise standard deviations.

10.2   Projection-pursuit regression

Now suppose that the explanatory vector $\mathbf{X} = (X_1, \dots, X_p)$ is of high dimension. The additive model (10.1) may be too flexible as it allows a few degrees of freedom per X_j, yet it does not cover the effect of interactions between the independent variables. Projection pursuit regression (Friedman & Stuetzle, 1981) applies an additive model to projected variables. That is, it is of the form:

$$Y = \alpha_0 + \sum_{j=1}^{M} f_j(\boldsymbol{\alpha}_k^T \mathbf{X}) + \epsilon \tag{10.5}$$

for vectors $\boldsymbol{\alpha}_k$, and a dimension M to be chosen by the user. Thus it uses an additive model on predictor variables which are formed by projecting $\mathbf{X}$ in M carefully chosen directions. For large enough M such models can approximate (uniformly on compact sets and in many other senses) arbitrary continuous functions of X (e.g. Diaconis & Shahshahani, 1984). The terms of (10.5) are called ridge functions, since they are constant in all but one direction.

The S-PLUS function `ppreg` fits (10.5) by least squares, and constrains the vectors α_k to be of unit length. It first fits $M_{\max}$ terms sequentially, then prunes back to M by at each stage dropping the least effective term and re-fitting. The function returns the proportion of the variance explained by all the fits from $M, \ldots, M_{\max}$. Rather than a model formula, the arguments are an X matrix and a y vector (or matrix, since multiple dependent variables can be included). In many cases a suitable X matrix can be obtained by calling `model.matrix`, but care is needed not to duplicate intercept terms.

For the `rock` example we have:

```
> attach(rock)
> x <- cbind(area, peri, shape)
# or  model.matrix(~ -1 + area + peri + shape)
> rock.ppr <- ppreg(x, log(perm), 1, 5)
> SStot <- var(log(perm))*(length(perm)-1)
> SStot*rock.ppr$esq
[1] 17.4248 11.3637  7.1111  5.9324  3.1139
> rock.ppr$allalpha[1,,1] # first alpha for first fit
[1]  0.00041412 -0.00116210  0.99999928
> rock.ppr$allalpha[1,,1]*sqrt(diag(var(x)))
[1]  1.111428 -1.663737  0.083496
> plot(rock.ppr$z, rock.ppr$zhat, type="l")
```

This shows that even a single smooth projection term fits significantly better than any of the additive models. The combination appears at first sight to be dominated by shape, but the variables are on very different scales, have standard errors (2680, 1430, 0.084), and the variable appears to be a contrast between `area` and `peri`. The non-linear function is shown in Figure 10.3. (This fit is very sensitive to rounding error, and will vary from system to system.)

The fit improves steadily with extra terms, but there are only 48 observations, and the number of parameters is approximately $1 + \sum_{i=1}^{M}[2 + \mathrm{df}_i - 1]$ where df_i is the equivalent degrees of freedom for the ith smooth fit. (The projection can be rescaled, so accounts for 2 not 3 degrees of freedom, and each smooth fit can fit a constant term.) Unfortunately df_i is not specified, but the plot suggests $\mathrm{df}_i \approx 6\text{–}8$, so using more than two or three terms would be over-fitting.

It is also possible to use `gam` with `lo` to include smooth functions of two or more variables. (This again fits robustly a locally weighted linear surface.) We test out the linear term in `area-peri` space. We deliberately under-smooth (by choosing a small value of `span`) and obtain a RSS of 10.22 on roughly 24 degrees of freedom.

```
rock.gam2 <- gam(log(perm) ~ shape+lo(area, peri, span=0.2))
summary(rock.gam2)
anova(rock.lm, rock.gam2)
plot(rock.gam2, eye=c(-10000, 50000, 20), pty="s")
```

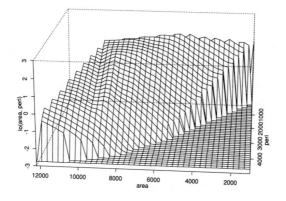

Figure 10.5: Perspective plot of `lo` term fitted to `rock` dataset.

The perspective plot for the `lo` term is shown in Figure 10.5. The plot confirms a broadly linear fitted surface, but with some suspicion of a curvature in the orthogonal direction, which is confirmed by a PP regression fit on just the first two variables.

```
> rock.ppr2 <- ppreg(x[, 1:2], log(perm), 1, 3)
> SStot*rock.ppr2$esq
[1] 20.064 15.309 14.145
> rock.ppr2$allalpha[1,,1]*sqrt(diag(var(x[, 1:2])))
[1]    932.37 -1342.49
> rock.ppr2$allalpha[1:2,,2]*sqrt(diag(var(x[, 1:2])))
          [,1]    [,2]
[1,]   960.55 -2506.1
[2,] -1012.56  1012.1
```

Note that dropping the third variable changes drastically the scaling of a unit-length combination α.

We can even include all pairwise combinations of variables. The very different scaling causes problems with the plots, so we re-scale.

```
area1 <- area/10000; peri1 <- peri/10000
rock.gam3 <- gam(log(perm) ~ lo(area1, peri1) +
  lo(area1, shape) + lo(peri1, shape),
  control=gam.control(maxit=50, bf.maxit=50))
summary(rock.gam3)
anova(rock.lm, rock.gam3)
par(mfrow=c(1,3))
plot(rock.gam3, pty="s")
```

Figure 10.6 looks like the sums of ridge functions, so it is not surprising that PP regression can fit very well. The RSS is 13.83 on about 25 degrees of freedom.

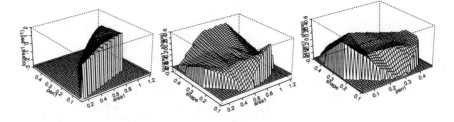

Figure 10.6: Perspective plots of the pairwise terms `lo` terms in the model for `rock`.

10.3   Response transformation models

If we want to predict Y, it may be better to transform Y as well, so we have

$$\theta(Y) = \alpha + \sum_{j=1}^{p} f_j(X_j) + \epsilon \qquad (10.6)$$

for an invertible smooth function $\theta()$, for example the log function we used for
the `rock` dataset.

The ACE (Alternating Conditional Expectation) algorithm of Breiman &
Friedman (1985) chooses the functions θ and $f_1, \ldots, f_j$ to maximize the correlation between the predictor $\alpha + \sum_{j=1}^{p} f_j(X_j)$ and $\theta(Y)$. Tibshirani's (1988)
procedure AVAS (additivity and variance stabilising transformation) fits the same
model (10.6), but with the aim of achieving constant variance of the residuals for
monotone θ.

The S-PLUS functions `ace` and `avas` fit (10.6). Both allow the functions
f_j and θ to be constrained to be monotone or linear, and the functions f_j to
be chosen for circular or categorical data. Thus these functions provide another
way to fit additive models (with linear θ) but they do not provide measures of
fit nor standard errors. They do, however, automatically choose the degree of
smoothness.

Our first example is data on body weights (*kg*) and brain weights (*g*) of 62
mammals, from Weisberg (1985, pp. 144–5), supplied in our data frame `mammals`.
For these variables we would expect log transformations will be appropriate. In
this example AVAS succeeds but ACE does not, as Figures 10.7 and 10.8 show:

```
attach(mammals)
a <- ace(body, brain)
o1<- order(body); o2 <- order(brain)
par(mfrow=c(2,2))
plot(body, brain, main="Original Data", log="xy")
plot(body[o1], a$tx[o1], main="Transformation of body wt",
    type="l", log="x")
plot(brain[o2], a$ty[o2], main="Transformation of brain wt",
    type="l", log="x")
plot(a$tx, a$ty, main="Transformed y vs  x")
```

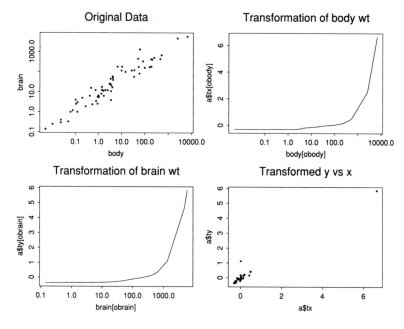

Figure 10.7: Plots of ACE fit of the `mammals` dataset.

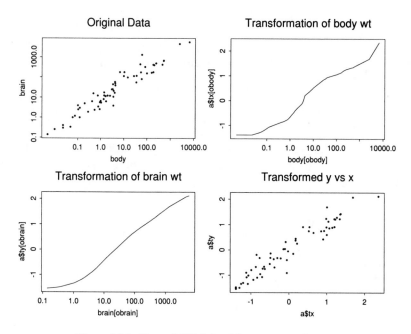

Figure 10.8: Plots of AVAS fit of the `mammals` dataset.

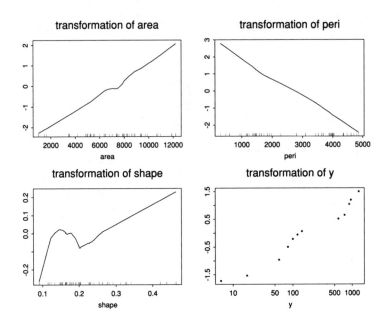

Figure 10.9: AVAS on permeabilities

```
a <- avas(body, brain)
par(mfrow=c(2,2))
plot(body, brain, main="Original Data", log="xy")
plot(body[o1], a$tx[o1], main="Transformation of body wt",
    type="l", log="x")
plot(brain[o2], a$ty[o2], main="Transformation of brain wt",
    type="l", log="x")
plot(a$tx, a$ty, main="Transformed y vs x")
```

For the rock dataset we can use the following S-PLUS code. The S function rug produces the ticks to indicate the locations of data points, as used by gam.plot.

```
attach(rock)
x <- cbind(area, peri, shape)
o1 <- order(area); o2 <- order(peri); o3 <- order(shape)
a <- avas(x, perm)
par(mfrow=c(2,2))
plot(area[o1], a$tx[o1,1], type="l")        see Figure 10.9
rug(area)
plot(peri[o2], a$tx[o2,2], type="l")
rug(peri)
plot(shape[o3], a$tx[o3,3], type="l")
rug(shape)
plot(perm, a$ty, log="x")                    note log scale
a <- avas(x, log(perm))                      looks like log(y)
```

```
a <- avas(x, log(perm), linear=0)          so force θ(y) = log(y)
# repeat plots
```

Here AVAS indicates a log transformation of permeabilities (expected on physical grounds) but little transformation of area or perimeter.

10.4 Neural networks

Feed-forward neural networks provide a flexible way to generalize linear regression functions.[1] General references are Hertz, Krogh & Palmer (1991), Ripley (1993) and White (1992).

We start with the simplest but most common form with one hidden layer as shown in Figure 10.10. The input units just provide a 'fan-out' and distribute the inputs to the 'hidden' units in the second layer. These units sum their inputs, add a constant (the 'bias') and take a fixed function ϕ_h of the result. The output units are of the same form, but with output function ϕ_o. Thus

$$y_k = \phi_o\left(\alpha_k + \sum_h w_{hk}\, \phi_h\left(\alpha_h + \sum_i w_{ih}\, x_i\right)\right) \tag{10.7}$$

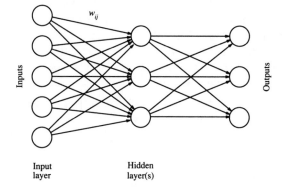

Figure 10.10: A generic feed-forward neural network.

The 'activation function' ϕ_h of the hidden layer units is almost always taken to be the logistic function

$$\ell(z) = \frac{\exp(z)}{1 + \exp(z)}$$

and the output units are linear, logistic or threshold units. (The latter have $\phi_o(x) = I(x > 0)$.) Note the similarity to projection pursuit regression (*cf* (10.5)), which has linear output units but general smooth hidden units. (However, arbitrary smooth functions can be approximated by sums of re-scaled logistics.)

---

[1] This software uses C.

The general definition allows more than one hidden layer, and also allows 'skip-layer' connections from input to output when we have

$$y_k = \phi_o \left(\alpha_k + \sum_{i \to k} w_{ik} x_i + \sum_{j \to k} w_{jk} \phi_h \left(\alpha_j + \sum_{i \to h} w_{ih} x_i \right) \right) \quad (10.8)$$

which allows the non-linear units to perturb a linear functional form.

We can eliminate the biases α_i by introducing an input unit 0 which is permanently at $+1$ and feeds every other unit. The regression function f is then parametrized by the set of weights w_{ij}, one for every link in the network (or zero for links which are absent).

The original biological motivation for such networks stems from McCulloch & Pitts (1943) who published a seminal model of a neuron as a binary thresholding device in discrete time, specifically that

$$n_i(t) = H \left(\sum_{j \to i} w_{ji} n_j(t-1) - \theta_i \right)$$

the sum being over neurons j connected to neuron i. Here H denotes the Heaviside or threshold function $H(x) = I(x > 0)$, $n_i(t)$ is the output of neuron i at time t, and $0 < w_{ij} < 1$ are attenuation weights. Thus the effect is to threshold a weighted sum of the inputs at value θ_i. Real neurons are now known to be more complicated; they have a graded response rather than the simple thresholding of the McCulloch–Pitts model, work in continuous time, and can perform more general non-linear functions of their inputs, for example logical functions. Nevertheless, the McCulloch–Pitts model has been extremely influential in the development of artificial neural networks.

Feed-forward neural networks can equally be seen as a way to parametrize a fairly general non-linear function. Such networks *are* rather general: Cybenko (1989), Funahashi (1989), Hornik, Stinchcombe and White (1989) and later authors have shown that neural networks with linear output units can approximate any continuous function f uniformly on compact sets, by increasing the size of the hidden layer.

The approximation results are nonconstructive, and in practice the weights have to be chosen to minimize some fitting criterion, for example least squares

$$E = \sum_p \| t^p - y^p \|^2$$

where t^p is the target and y^p the output for the pth example pattern. Other measures have been proposed, including for $y \in [0, 1]$ 'maximum likelihood' (in fact minus the logarithm of a conditional likelihood) or equivalently the Kullback–Leibler distance, which amount to minimizing

$$E = \sum_p \sum_k \left[t_k^p \log \frac{t_k^p}{y_k^p} + (1 - t_k^p) \log \frac{1 - t_k^p}{1 - y_k^p} \right] \quad (10.9)$$

This is half the deviance for a logistic model with linear predictor given by (10.7) or (10.8).

One way to ensure that f is smooth is to restrict the class of estimates, for example by using a limited number of spline knots. Another way is *regularization* in which the fit criterion is altered to

$$E + \lambda C(f)$$

for example, with a penalty C on the second derivatives of f such as

$$\int \sum_{i,o} \frac{\partial^2 y_o}{\partial x_i^2} \, dx \approx \frac{1}{P} \sum_{i,o,p} \frac{\partial^2 y_o}{\partial x_i^2}$$

for P patterns, as is used in the derivation of smoothing splines. *Weight decay*, specific to neural networks, uses as penalty the sum of squares of the weights w_{ij}. (This only makes sense if the inputs are rescaled to range about $[0, 1]$ to be comparable with the outputs of internal units.) The use of weight decay seems both to help the the optimization process and to avoid over-fitting. Arguments in Ripley (1993, 1994) based on a Bayesian interpretation suggest $\lambda \approx 10^{-4} - 10^{-2}$ depending on the degree of fit expected, for least-squares fitting to variables of range one and $\lambda \approx 0.01 - 0.1$ for the entropy fit.

Software to fit feed-forward neural networks with a single hidden layer but allowing skip-layer connections (as in (10.8)) is provided in our library nnet. The format of the call is

```
nnet(x, y, weights, size, Wts, linout=F, entropy=F, softmax=F,
    skip=F, rang=0.7, decay=0, maxit=100, trace=T)
```

The parameters are

x	matrix of dependent variables.
y	vector or matrix of independent variables.
weights	weights for training cases in E; default 1.
size	number of units in the hidden layer.
Wts	optional initial vector for w_{ij}.
linout	logical for linear output units.
entropy	logical for entropy rather than least-squares fit.
softmax	logical for log-probability models.
skip	logical for links from inputs to outputs.
rang	if Wts is missing, use random weights from runif(n,-rang, rang).
decay	parameter λ.
maxit	maximum of iterations for the optimizer.
trace	logical for output from the optimizer. Very reassuring!

There are predict, print and summary methods for neural networks, and a function nnet.Hess to compute the Hessian with respect to the weight parameters and so check if a secure local minimum has been found. For our rock example we have

```
> attach(rock)
> area1 <- area/10000; peri1 <- peri/10000
> rock.x <- cbind(area1, peri1, shape)
> set.seed(5555)
> rock.nn <- nnet(rock.x, log(perm), size=3, decay=1e-3,
        linout=T,skip=T,  maxit=1000)
# weights:  19
initial  value 1820.342503
      ....
final  value 12.757519
> summary(rock.nn)
a 3-3-1 network with 19 weights
options were - skip-layer connections  linear output units
   decay=0.001
    0->4   1->4   2->4   3->4   0->5   1->5   2->5   3->5   0->6
    4.47 -11.16  15.32  -8.78   9.14 -14.67  18.45 -22.92   1.24
    1->6   2->6   3->6   0->7   4->7   5->7   6->7   1->7   2->7
   -9.83   7.09  -3.79   8.77 -16.04   8.64   9.56  -1.98  -4.16
    3->7
    1.66
> sum((log(perm) - predict(rock.nn, rock.x))^2)
[1] 10.45
> eigen(nnet.Hess(rock.nn, rock.x, log(perm)), T)$values
 [1] 1.3510e+03 7.7394e+01 4.8628e+01 1.8258e+01 1.0199e+01
 [6] 4.0871e+00 1.2151e+00 8.5209e-01 5.3163e-01 3.5168e-01
[11] 1.3657e-01 7.6662e-02 2.8361e-02 1.3402e-02 1.0085e-02
[16] 7.0372e-03 6.0602e-03 4.0152e-03 3.2327e-03
```

The quoted values include the weight decay term. The eigenvalues of the Hessian suggest that a secure local minimum has been achieved.

A classification example

We consider an example from Section 12.4 on the classification by sex of crabs. The data are transformed to log scale there on physical grounds, which also scales them well for input to a neural network. As a neural network will give a nonlinear logistic discriminant, we try logistic discrimination first. The random subsets for cross-validation were the same as those used in Chapter 12.

```
> dcrabs <- log(crabs[,4:7])
> dfcrabs <- cbind(dcrabs, sex=crabs$sex)
> cr.glm <- glm(sex ~ ., data=dfcrabs, family=binomial,
        maxit=20)
> coef(cr.glm)
 (Intercept)      FL      RW      CL      CW
      -156.11 -51.489  -222.2   277.1 -26.376
> cr.pr <- predict(cr.glm, dcrabs, type="response")
> sum((crabs$sex=="M") != (cr.pr > 0.5))
[1] 2
> rand <- sample(10, dim(crabs)[1], replace = T)
```

```
> cnt <- 0; prsum <- 0
> for (i in unique(rand)) {
    t.glm <- glm(sex ~ ., data=dfcrabs, family=binomial,
                 subset=(rand!=i), maxit=50)
    t.pr <- predict(t.glm, dcrabs[rand == i,], type="response")
    cnt <- cnt + sum((crabs$sex=="M")[rand==i] != (t.pr > 0.5))
    prsum <- prsum + sum(pmax(1-t.pr, t.pr))
}
> c(cnt/200, 1-prsum/200)
[1] 0.030 0.0159
```

There are some warnings about extreme fitted probabilities, and convergence is slow.

We now use a neural net, and allow a small degree of re-training on each cross-validation fit.

```
> set.seed(777)
> cr.nn <- nnet(dcrabs, crabs$sex=="M", size=2, decay=1e-3,
    skip=T, entropy=T, maxit=500)
> eigen(nnet.Hess(cr.nn, dcrabs, crabs$sex=="M"), T)$values
> sum((crabs$sex=="M") != (predict(cr.nn, dcrabs) > 0.5))
[1] 2
> cnt <- 0; prsum <- 0
> for (i in unique(rand)) {
    sub <- rand!=i
    t.nn <- nnet(dcrabs[sub,], (crabs$sex=="M")[sub], size=2,
                 decay = 1e-3,  skip=T, entropy=T, maxit=500,
                 Wts=cr.nn$wts + runif(17, -0.1, 0.1), trace=F)
    t.nn <- predict(t.nn, dcrabs[rand == i,])
    cnt <- cnt + sum((crabs$sex=="M")[rand==i] != (t.nn > 0.5))
    prsum <- prsum + sum(pmax(1-t.nn, t.nn))
}
> c(cnt/200, 1-prsum/200)
[1] 0.02 0.0184
```

The cross-validation study suggests a small gain from a non-linear discriminant in this example.

10.5 Conclusions

We have considered a large, perhaps bewildering, variety of extensions to linear regression. These can be thought of as belonging to two classes, the 'black-box' fully automatic and maximally flexible routines represented by projection pursuit regression and neural networks, and the small steps under full user control of additive models. Although the latter gain in interpretation, as we saw in the rock example they may not be general enough.

There are at least two other, intermediate approaches. Tree-based models (Chapter 13) use piecewise-constant functions and can build interactions by successive partitions. Friedman's (1991) MARS explicitly has products of simple functions of each variable. (S software for MARS is available in the library `fda` from `statlib`.)

What is best for any particular problem depends on its aim, in particular whether prediction or interpretation is paramount. The methods of this chapter are powerful tools with very little distribution theory to support them so it is very easy to over-fit and over-explain features of the data. Be warned!

Chapter 11

Survival Analysis

Survival analysis is not part of S, but has been added to S-PLUS based on functions written by Terry Therneau (Mayo Foundation) and available as `survival2` code from `statlib` (see Appendix D for further information.) This has been superseded by `survival3` and then `survival4`. This is not part of S-PLUS 3.2, but is scheduled to be included in S-PLUS 3.3. As these functions are much easier to use and provide a higher capability, this chapter is based on their use. (This does mean that the methods are probably not accessible to Windows users of earlier versions, as the library uses C code. Section 11.6 sketches how to use `survival2`, for those who have no other choice.)

It is assumed throughout this chapter that `survival4` is available and has been attached with

```
library(survival, first=T)
```

or that the local version of S incorporates `survival4`. Note that the `first=T` is crucial, as some system functions are replaced.[1]

Survival analysis is concerned with the distribution of lifetimes, often of humans but also of components and machines. There are two distinct levels of mathematical treatment in the literature. Cox & Oakes (1984), Kalbfleisch & Prentice (1980), Lawless (1982) and Miller (1981) take a traditional and mathematically non-rigorous approach. The modern mathematical approach based on continuous-parameter martingales is given by Fleming & Harrington (1991) and Andersen *et al.* (1993). (The latter is needed here only to justify some of the distribution theory and for the concept of martingale residuals.)

Let T denote a lifetime random variable. It will take values in $(0, \infty)$, and its continuous distribution may be specified by a cumulative distribution function F with a density f. (Mixed distributions can be considered, but many of the formulae used by the software need modification.) For lifetimes it is more usual to work with the *survivor function* $S(t) = 1 - F(t) = P(T > t)$, the *hazard function* $h(t) = \lim_{\Delta t \to 0} P(t \leqslant T < t + \Delta t \mid T \geqslant t)/\Delta t$ and the *cumulative*

---

[1] To use our enhanced function `plot.survfit` this must override that in `library(survival)`. Either replace the function in `library(survival)` or load our library with `first=T` *after* `library(survival)`.

hazard function $H(t) = \int_0^t h(s)ds$. These are all related; we have

$$h(t) = \frac{f(t)}{S(t)}, \qquad H(t) = -\log S(t)$$

Common parametric distributions for lifetimes are (Kalbfleisch & Prentice, 1980) the exponential, with $S(t) = \exp -\lambda t$ and hazard λ, the Weibull with

$$S(t) = \exp -(\lambda t)^\alpha, \qquad h(t) = \lambda \alpha (\lambda t)^{\alpha-1}$$

the log-normal, the gamma and the log-logistic with

$$S(t) = \frac{1}{1 + (\lambda t)^\tau}, \qquad h(t) = \frac{\lambda \tau (\lambda t)^{\tau-1}}{1 + (\lambda t)^\tau}$$

The major distinguishing feature of survival analysis is *censoring*. An individual case may not be observed on the whole of its lifetime, so that, for example, we may only know that it survived to the end of the trial. More general patterns of censoring are possible, but all lead to data for each case of the form either of a precise lifetime or the information that the lifetime fell in some interval (possibly extending to infinity).

Clearly we must place some restrictions on the censoring mechanism, for if cases were removed from the trial just before death we would be misled. Consider right censoring, in which the case leaves the trial at time C_i, and we know either T_i if $T_i \leq C_i$ or that $T_i > C_i$. *Random censoring* assumes that T_i and C_i are independent random variables, and therefore in a strong sense that censoring is uninformative. This includes the special case of *type I* censoring, in which the censoring time is fixed in advance, as well as trials in which the patients enter at random times but the trial is reviewed at a fixed time. It excludes *type II* censoring in which the trial is concluded after a fixed number of failures. Most analyses (including all those based solely on likelihoods) are valid under a weaker assumption which Kalbfleisch & Prentice (1980, §5.2) call *independent* censoring in which the hazard at time t conditional on the whole history of the process only depends on the survival of that individual to time t. (Independent censoring does cover type II censoring.) Conventionally the time recorded is $\min(T_i, C_i)$ together with the indicator variable for observed death $\delta_i = I(T_i \leq C_i)$. Then under independent right censoring the likelihood for parameters in the lifetime distribution is

$$L = \prod_{\delta_i=0} f(t_i) \prod_{\delta_i=1} S(t_i) = \prod_{\delta_i=0} h(t_i)S(t_i) \prod_{\delta_i=1} S(t_i) = \prod_{i=1}^{n} h(t_i)^{\delta_i} S(t_i)$$

$$(11.1)$$

Usually we are not primarily interested in the lifetime distribution *per se*, but how it varies between groups (usually called *strata* in the survival context) or on measurements on the cases, called *covariates*. In the more complicated problems the hazard will depend on covariates which vary with time, such as blood-pressure measurements, or changes of treatments.

The function Surv(times, status) is used to describe the censored survival data to the routines in the survival3 library, and always appears on the left side of a model formula. In the simplest case of right censoring the variables are T_i and δ_i (logical or 0/1 or 1/2). Further forms allow left and interval censoring. The results of printing the object returned by Surv are the vector of the information available, either the lifetime or an interval.

We consider three small running examples. Uncensored data on survival times for leukaemia (Feigl & Zelen, 1965; Cox & Oakes, 1984, p. 9) are in data frame leuk. This has two covariates, the white blood count wbc, and ag a test result which returns 'present' or 'absent'. Two-sample data (Gehan, 1965; Cox & Oakes, 1984, p. 7) on remission times for leukaemia are given in data frame gehan. This trial has 42 individuals in matched pairs, and no covariates (other than the treatment group).[2] Data frame motors contains the results of an accelerated life test experiment with ten replicates at each of four temperatures reported by Nelson & Hahn (1972) and Kalbfleisch & Prentice (1980, pp. 4–5). The times are given in hours, but all but one is a multiple of 12, and only 14 values occur:

17 21 22 56 60 70 73.5 115.5 143.5 147.5833 157.5 202.5 216.5 227

in days, which suggests that observation was not continuous. Thus this is a good example to test the handling of ties.

11.1   Estimators of survivor curves

The estimate of the survivor curve for uncensored data is easy; just take one minus the empirical distribution function. For the leukaemia data we have

```
attach(leuk)
plot(survfit(Surv(time) ~ ag), lty=c(2,3))
legend(80, 0.8, c("ag absent", "ag present"), lty=c(2,3))
```

and confidence intervals are obtained easily from the binomial distribution of $\hat{S}(t)$. For example, the estimated variance is

$$\hat{S}(t)[1 - \hat{S}(t)]/n = r(t)[n - r(t)]/n^3 \tag{11.2}$$

when $r(t)$ is the number of cases still alive (and hence 'at risk') at time t.

This computation introduces the function survfit and its associated plot, print and summary methods. It takes a model formula, and if there are factors on the right-hand side, splits the data on those factors, and plots a survivor curve for each factor combination, here just presence or absence of ag. (Although the factors can be specified additively, the computation effectively uses their interaction.)

---

[2] Andersen *et al.* (1993, p. 22) indicate that this trial had a sequential stopping rule which invalidates most of the methods used here; it should be seen as illustrative only.

For censored data we have to allow for the decline in the number of cases 'at risk' over time. Let $r(t)$ be the number of cases at risk just before time t, that is those which are in the trial and not yet dead. If we consider a set of intervals $I_i = [t_i, t_{i+1})$ covering $[0, \infty)$, we can estimate the probability p_i of surviving interval I_i as $[r(t_i) - d_i]/r(t_i)$ where d_i is the number of deaths in interval I_i. Then the probability of surviving until t_i is

$$P(T > t_i) = S(t_i) \approx \prod_0^{i-1} p_j \approx \prod_0^{i-1} \frac{r(t_i) - d_i}{r(t_i)}$$

Now let us refine the grid of intervals. Non-unity terms in the product will only appear for intervals in which deaths occur, so the limit becomes

$$\hat{S}(t) = \prod \frac{r(t_i) - d_i}{r(t_i)}$$

the product being over times at which deaths occur before t (but they could occur simultaneously). This is the Kaplan-Meier estimator. Note that this becomes constant after the largest observed t_i, and for this reason the estimate is only plotted up to the largest t_i. However, the points at the right-hand end of the plot will be very variable, and it may be better to stop plotting when there are still a few individuals at risk.

We can apply similar reasoning to the cumulative hazard

$$H(t_i) \approx \sum_{j \leqslant i} h(t_j)(t_{j+1} - t_j) \approx \sum_{j \leqslant i} \frac{d_j}{r(t_j)}$$

with limit

$$\hat{H}(t) = \sum \frac{d_j}{r(t_j)} \tag{11.3}$$

again over times at which deaths occur before t. This is the Nelson estimator of the cumulative hazard, and leads to the Fleming-Harrington estimator of the survivor curve

$$\tilde{S}(t) = \exp{-\hat{H}(t)} \tag{11.4}$$

The two estimators are related by the approximation $\exp x \approx 1 - x$ for small x, so they will be nearly equal for large risk sets.

Similar arguments to those used to derive the two estimators lead to the standard error formula for the Kaplan-Meier estimator

$$\text{var}\left(\hat{S}(t)\right) = \hat{S}(t)^2 \sum \frac{d_j}{r(t_j)[r(t_j) - d_j]} \tag{11.5}$$

often called Greenwood's formula after its version for life tables, and

$$\text{var}\left(\hat{H}(t)\right) = \sum \frac{d_j}{r(t_j)[r(t_j) - d_j]} \tag{11.6}$$

We leave it to the reader to check that Greenwood's formula reduces to (11.2) in the absence of ties and censoring. Note that if censoring can occur, both the Kaplan-Meier and Nelson estimators are biased; the bias occurs from the inability to give a sensible estimate when the risk set is empty.

Tsiatis (1981) suggested the denominator $r(t_j)^2$ rather than $r(t_j)[r(t_j) - d_j]$ on asymptotic grounds. Both Fleming & Harrington (1991) and Andersen *et al.* (1993) give a rigorous derivation of these formulae (and corrected versions for mixed distributions), as well as calculations of bias and limit theorems which justify asymptotic normality. Klein (1991) discussed the bias and small-sample behaviour of the variance estimators; his conclusions for $\hat{H}(t)$ are that the bias is negligible and the Tsiatis form of the standard error is accurate (for practical use) provided the expected size of the risk set at t is at least 5. For the Kaplan-Meier estimator Greenwood's formula is preferred, and is accurate enough (but biased downwards) again provided the expected size of the risk set is at least 5.

We can use these formulae to indicate confidence intervals based on asymptotic normality, but we must decide on what scale to compute them. By default the function `survfit` computes confidence intervals on the log survivor (or cumulative hazard) scale, but linear and complementary log-log scales are also available (via the `conf.type` argument). These choices give

$$\hat{S}(t) \exp\left[\pm k_\alpha \text{ s.e.}(\hat{H}(t))\right]$$

$$\hat{S}(t) \left[1 \pm k_\alpha \text{ s.e.}(\hat{H}(t))\right]$$

$$\exp\left\{\hat{H}(t) \exp\left[\pm k_\alpha \frac{\text{s.e.}(\hat{H}(t))}{\hat{H}(t)}\right]\right\}$$

the last having the advantage of taking values in $(0, 1)$. Bie, Borgan & Liestøl (1987) and Borgan & Liestøl (1990) considered these and an arc-sine transformation; their results indicate that the complementary log-log interval is quite satisfactory for sample sizes as small as 25.

We will not distinguish clearly between log-survivor curves and cumulative hazards, which differ only by sign, yet the natural estimator of the first is the Kaplan-Meier estimator on log scale, and for the second it is the Nelson estimator. This is particularly true for confidence intervals, which we would expect to transform just by a change of sign. Fortunately, practical differences only emerge for very small risk sets, and are then swamped by the very large variability of the estimators.

The function `survfit` also handles censored data, and uses the Kaplan-Meier estimator by default. We can try it on the `gehan` data:

```
> attach(gehan)
> Surv(time, cens)
 [1]  1   10   22    7    3  32+ 12  23    8  22  17    6    2  16
[15] 11   34+   8  32+ 12  25+  2  11+   5  20+  4  19+ 15    6
[29]  8   17+ 23  35+   5    6  11  13    4    9+  1    6+  8  10+
> plot.factor(gehan)
```

```
> plot(log(time) ~ pair)
# product-limit estimators with Greenwood's formula for errors:
> gehan.surv <- survfit(Surv(time, cens) ~ treat,
      conf.type="log-log")
> summary(gehan.surv)
Call: survfit(formula = Surv(time, cens) ~ treat,
conf.type = "log-log")
```

```
                       treat=6-MP
 time n.risk n.event survival std.err lower 95% CI upper 95% CI
    6     21       3    0.857  0.0764        0.620        0.952
    7     17       1    0.807  0.0869        0.563        0.923
   10     15       1    0.753  0.0963        0.503        0.889
   13     12       1    0.690  0.1068        0.432        0.849
   16     11       1    0.627  0.1141        0.368        0.805
   22      7       1    0.538  0.1282        0.268        0.747
   23      6       1    0.448  0.1346        0.188        0.680
```

```
                     treat=control
 time n.risk n.event survival std.err lower 95% CI upper 95% CI
    1     21       2   0.9048  0.0641      0.67005        0.975
    2     19       2   0.8095  0.0857      0.56891        0.924
    3     17       1   0.7619  0.0929      0.51939        0.893
    4     16       2   0.6667  0.1029      0.42535        0.825
    5     14       2   0.5714  0.1080      0.33798        0.749
    8     12       4   0.3810  0.1060      0.18307        0.578
   11      8       2   0.2857  0.0986      0.11656        0.482
   12      6       2   0.1905  0.0857      0.05948        0.377
   15      4       1   0.1429  0.0764      0.03566        0.321
   17      3       1   0.0952  0.0641      0.01626        0.261
   22      2       1   0.0476  0.0465      0.00332        0.197
   23      1       1   0.0000      NA           NA           NA
```

```
> plot(gehan.surv, conf.int=T, lty=c(3,2), log=T,
      xlab="time of remission (weeks)", ylab="survival")
> plot(gehan.surv, lty=c(3,2), lwd=2, log=T, add=T)
> legend(25, 0.1 , c("control","6-MP"), lty=c(2, 3), lwd=2)
```

which calculates and plots (as shown in Figure 11.1) the product-limit estimators
for the two groups, giving standard errors calculated using Greenwood's formula.
(Confidence intervals are plotted automatically if there is only one group. The
add=T argument is from the version of plot.survfit in our library. Whereas
there is a lines method for survival fits, it is not as versatile.) Other options are
available, including error="tsiatis" and type="fleming-harrington"
(which can be abbreviated to the first character). Note that the plot method has
a log argument that plots $\hat{S}(t)$ on log scale, effectively showing the negative
cumulative hazard.

We can test for differences between the groups by

```
> survdiff(Surv(time, cens) ~ treat)
```

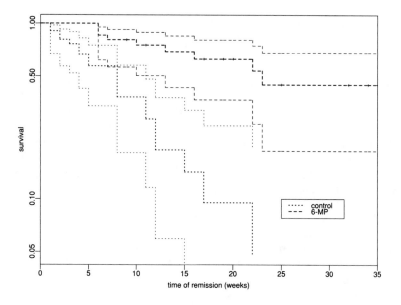

Figure 11.1: Survivor curves (on log scale) for the two groups of the gehan data. The crosses (on the 6-MP curve) represent censoring times.

```
                N Observed Expected (O-E)^2/E
    treat=6-MP  21        9    19.25      5.458
treat=control  21       21    10.75      9.775

Chisq= 16.8  on 1 degrees of freedom, p= 4.169e-05
```

This is one of a family of tests defined by Fleming and Harrington (1981) and controlled by a parameter ρ. The default $\rho = 0$ corresponds to the *log-rank test*. Suppose t_j are the observed death times. Then we can compute the mean and variance of the number of deaths in each group at time t_j from Fisher's exact test applied to the table of successes and failures by groups. If we sum over events, we obtain an expected number of deaths and a covariance matrix V of the numbers of deaths (assuming, incorrectly, independence of the tables). From this we compute a statistic $(O - E)^T V^{-1}(O - E)$ with an approximately chi-squared distribution.

The argument $\rho = 1$ of survdiff corresponds approximately to the Peto-Peto (1972) modification of the Wilcoxon test.

11.2   Parametric models

Some parametric models can be fitted via generalized linear models. Consider first the exponential distribution, and suppose we have covariate vector x for each case (including a constant if required). There are a number of ways to relate the rate λ_i of the exponential to the covariates. The 'natural' link is

$$\lambda_i = \boldsymbol{\beta}^T \mathbf{x}$$

but this need not be non-negative, and it is usually better to use a log-link:

$$\log \lambda_i = \boldsymbol{\beta}^T \mathbf{x}$$

These are tested on the leukaemia example by the following code. As usual with non-normal models we use treatment contrasts. (For the rest of this chapter we will assume that treatment contrasts are used.) With a log-link it seems appropriate to use `log(wbc)` and this does fit better.

```
attach(leuk)
plot(log(time) ~ ag + log(wbc))
# note use of variance-stabilising transform
options(contrasts=c("contr.treatment", "contr.poly"))
leuk.glm <- glm(time ~ ag*log(wbc), Gamma(log))
summary(leuk.glm, dispersion=1)
# force exponential data summary
anova(leuk.glm)
# drop interaction term
leuk.glm <- update(leuk.glm,  ~ .-ag:log(wbc))
summary(leuk.glm, dispersion=1)
# compare with natural link
leuk.glmi <- glm(time ~ ag*wbc, Gamma(inverse))
summary(leuk.glmi, dispersion=1)
anova(leuk.glmi)
detach("leuk")
```

The results for the means (the inverse of the rate) are given in Table 11.1:

Table 11.1: Exponential models for the `leuk` dataset.

	null	log link	log link	inverse link
(Intercept)		4.33 (1.97)	5.81 (1.29)	$5.63\ (1.89) \times 10^{-2}$
ag		4.15 (2.57)	1.02 (0.35)	$-4.78\ (1.93) \times 10^{-2}$
log(wbc)		−0.15 (0.20)	−0.30 (0.13)	$-0.23\ (4.24) \times 10^{-7}$
ag:log(wbc)		−0.33 (0.27)	—	$-5.93\ (5.23) \times 10^{-7}$
Deviance	58.18 (32)	38.55 (29)	40.32 (30)	39.73 (29)

We now consider other lifetime distributions. We can fit gamma distributions in the same way provided the shape parameter is known. The log-normal distribution is straightforward: regress log(time) on the covariates.

For the Weibull distribution with log link, the hazard function is

$$h(t) = \lambda^{\alpha} \alpha t^{\alpha-1} = \alpha t^{\alpha-1} \exp(\alpha \boldsymbol{\beta}^T \mathbf{x}) \tag{11.7}$$

and so we have the first appearance of the *proportional hazards* model

$$h(t) = h_0(t) \exp \boldsymbol{\beta}^T \mathbf{x} \qquad (11.8)$$

which we will consider again later. This identification suggests re-parametrising the Weibull by replacing λ^α by λ, but as this just re-scales the coefficients we can move easily from one parametrization to the other. Note that the proportional hazards form (11.8) provides a direct interpretation of the parameters $\boldsymbol{\beta}$.

The Weibull is also a member of the class of *accelerated life* models, which have survival time T such that $T \exp -\boldsymbol{\beta}^T \mathbf{x}$ has a fixed distribution; that is time is speeded up by the factor $\exp \boldsymbol{\beta}^T \mathbf{x}$ for an individual with covariate $\mathbf{x}$. This corresponds to replacing t in the survivor function and hazard by $t \exp \boldsymbol{\beta}^T \mathbf{x}$, and for models such as the exponential, Weibull and log-logistic with parametric dependence on λt, this corresponds to taking $\lambda = \exp \boldsymbol{\beta}^T \mathbf{x}$. For all accelerated-life models we will have

$$\log T = \log T_0 - \boldsymbol{\beta}^T \mathbf{x} \qquad (11.9)$$

for a random variable T_0 whose distribution does not depend on $\mathbf{x}$, so these are naturally considered as regression models.

For the Weibull the cumulative hazard is linear on a log-log plot, which provides a useful diagnostic aid. For example, for the gehan data (using our version of plot.survfit):

```
> plot(gehan.surv, lty=c(3,4), cloglog=T,
     xlab="time of remission (weeks)", ylab="H(t)")
> legend(2, 2 , c("control","6-MP"), lty=c(4, 3))
```

we see excellent agreement with the proportional hazards hypothesis and with a Weibull baseline (Figure 11.2).

It is possible to fit Weibull models using GLMs (Aitkin & Clayton, 1980) but is easier to use the function survreg from Therneau's library. This has arguments including

```
survreg(formula, data, subset, na.action,
   link=c("log", "identity"),
   dist=c("extreme", "logistic", "gaussian", "exponential"))
```

where the first-mentioned of the alternatives are the defaults. (The names can be abbreviated provided enough is given to uniquely identify them.) With the default log link, the extreme value option corresponds to the model

$$\log T \sim \boldsymbol{\beta}^T \mathbf{x} + \sigma \log E$$

for a standard exponential E whereas our Weibull parametrization corresponds to

$$\log T \sim -\log \lambda + \frac{1}{\alpha} \log E$$

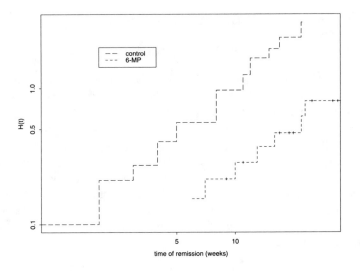

Figure 11.2: A log-log plot of cumulative hazard for the `gehan` dataset.

Thus `survreg` uses a log-linear Weibull model for $-\log \lambda$ and the scale factor estimates $1/\alpha$.

We repeat our exponential analyses, followed by Weibull and log-logistic regression analysis.

```
> attach(leuk)
> survreg(Surv(time) ~ ag*log(wbc), dist="exponential")
    ....
Coefficients:
 (Intercept)      ag log(wbc) ag:log(wbc)
      4.3433 4.1349 -0.15402     -0.32781
    ....
Degrees of Freedom: 33 Total; 29 Residual
Residual Deviance: 38.555
> summary(survreg(Surv(time) ~ ag + log(wbc), dist="exp"))
    ....
Coefficients:
             Value Std. Error z value        p
(Intercept)  5.815      1.263    4.60 4.15e-06
         ag  1.018      0.364    2.80 5.14e-03
   log(wbc) -0.304      0.124   -2.45 1.44e-02

> summary(survreg(Surv(time) ~ ag + log(wbc)))
    ....
Coefficients:
             Value Std. Error z value        p
(Intercept)  5.8524     1.323   4.425 9.66e-06
         ag  1.0206     0.378   2.699 6.95e-03
   log(wbc) -0.3103     0.131  -2.363 1.81e-02
```

```
Log(scale)  0.0399         0.139    0.287 7.74e-01

Extreme value distribution: Dispersion (scale) = 1.04
    ....
> summary(survreg(Surv(time) ~ ag + log(wbc), dist="log"))
    ....
Coefficients:
              Value Std. Error z value        p
(Intercept)   8.027      1.701    4.72 2.37e-06
         ag   1.155      0.431    2.68 7.30e-03
   log(wbc)  -0.609      0.176   -3.47 5.21e-04
 log(Scale)  -0.374      0.145   -2.59 9.74e-03

Logistic distribution: Dispersion (scale) est = 0.688
```

The Weibull analysis shows no support for non-exponential shape. The log-logistic, which is an accelerated life model but not a proportional hazards model (in our parametrization), gives a considerably more significant coefficient for log(wbc). Its scale parameter τ is estimated as $1/0.688 \approx 1.45$. Solomon (1984) considers fitting proportional-hazard models when an accelerated-life model is appropriate and vice-versa, and shows that to first order the coefficients β differ only by a factor of proportionality. Here there is definite evidence of non-proportionality.

Moving on the gehan dataset, which includes right-censoring, we find

```
> attach(gehan)
> survreg(Surv(time, cens) ~ factor(pair) + treat, dist="exp")
    ....
Degrees of Freedom: 42 Total; 20 Residual
Residual Deviance: 24.2
> summary(survreg(Surv(time, cens) ~ treat, dist="exp"))
Coefficients:
              Value Std. Error z value        p
(Intercept)   3.69       0.333   11.06 2.00e-28
      treat  -1.53       0.398   -3.83 1.27e-04

    Null Deviance: 54.5 on 41 degrees of freedom
Residual Deviance: 38.0 on 40 degrees of freedom  (LL= -49 )
> summary(survreg(Surv(time, cens) ~ treat))
Coefficients:
              Value Std. Error z value        p
(Intercept)   3.516      0.252   13.96 2.61e-44
      treat  -1.267      0.311   -4.08 4.51e-05
 log(Scale)  -0.312      0.147   -2.12 3.43e-02

Extreme value distribution: Dispersion (scale) = 0.732
```

There is no evidence of close matching of pairs. Note that survreg gives the reciprocal of the scale parameter for the Weibull in our notation. The sign of the coefficients is reversed, since this is a model for the mean and not the rate parameter.

The difference in log hazard between treatments is $-(-1.267)/0.732 = 1.731$ with a standard error of 0.425.

Finally, we consider the motors data, which are analysed by Kalbfleisch & Prentice (1980, §3.8.1). According to Nelson & Hahn (1972), the data were collected to assess survival at $130°C$, for which they found a median of 34 400 hours and a 10 percentile of 17 300 hours.

```
> attach(motors)
> plot(survfit(Surv(time, cens) ~ factor(temp)), conf.int=F)
> motor.wei <- survreg(Surv(time, cens) ~ temp)
> summary(motor.wei)
    ....
Coefficients:
              Value Std. Error z value         p
(Intercept) 16.3185    0.62296    26.2  3.03e-151
       temp -0.0453    0.00319   -14.2   6.74e-46
 Log(scale) -1.0956    0.21480    -5.1   3.38e-07

Extreme value distribution: Dispersion (scale) = 0.334
    ....
> unlist(predict(motor.wei, data.frame(temp=130), se.fit=T))
 fit.1 se.fit.1 residual.scale df
  10.4    0.659            3.2 37

> motor.wei$coef %*% c(1, 130)
         [,1]
[1,] 10.429
> cz <- c(1, 130,0) # compute s.e.
> sqrt(t(cz)%*% motor.wei$var %*% cz)
        [,1]
[1,] 0.222
```

At the time of writing `predict.survreg` was not yet written, and the predictions returned by `predict.glm` are not always appropriate (used as class `survreg` inherits from `glm`). The model for $\log T$ at $130°C$ is

$$\log T = 16.32 - 0.0453 \times \texttt{temp} + 0.334 \log E = 10.429(0.22) + 0.334 \log E$$

with a median for $\log T$ of

$$10.429(0.22) + 0.334 \times -0.3665 \approx 10.31(0.22)$$

or a median for T of 30 000 hours with 95% confidence interval (19 300, 46 600) and a 10 percentile for T of

$$10.429(0.22) + 0.334 \times -2.2503 \approx 9.68(0.22)$$

or as a 95% confidence interval for the 10 percentile for T, (10 300, 24 800) hours.

Nelson & Hahn worked with $z = 1000/(\texttt{temp} + 273.2)$. We leave the reader to try this; it gives slightly larger quantiles.

11.3   Cox proportional hazards model

Cox (1972) introduced a less parametric approach known as proportional hazards. There is a baseline hazard function $h_0(t)$ which is modified multiplicatively by covariates (including group indicators), so the hazard function for any individual case is

$$h(t) = h_0(t) \exp \boldsymbol{\beta}^T \mathbf{x}$$

and the interest is mainly in the proportional factors rather than the baseline hazard. Note that the cumulative hazards will also be proportional, so we can examine the hypothesis by plotting survivor curves for sub-groups on log scale. Later we will allow the covariates to depend on time.

The parameter vector $\boldsymbol{\beta}$ is estimated by maximizing a partial likelihood. Suppose one death occurred at time t_j. Then conditional on this event the probability that case i died is

$$\frac{h_0(t) \exp \boldsymbol{\beta} \mathbf{x}_i}{\sum_l I(T_l \geqslant t) h_0(t) \exp \boldsymbol{\beta} \mathbf{x}_l} = \frac{\exp \boldsymbol{\beta} \mathbf{x}_i}{\sum_l I(T_l \geqslant t) \exp \boldsymbol{\beta} \mathbf{x}_l} \qquad (11.10)$$

which does not depend on the baseline hazard. The partial likelihood for $\boldsymbol{\beta}$ is the product of such terms over all observed deaths, and usually contains most of the information about $\boldsymbol{\beta}$ (the remainder being in the observed times of death). However, we need a further condition on the censoring (Fleming & Harrington, 1991, pp. 138–9) that it is independent and *uninformative* for this to be so; the latter means that the likelihood for censored observations in $[t, t + \Delta t)$ does not depend on $\boldsymbol{\beta}$.

The correct treatment of ties cause conceptual difficulties as they are an event of probability zero for continuous distributions. Formally (11.10) may be corrected to include all possible combinations of deaths. As this increases the computational load, it is common to employ the Breslow approximation in which each death is always considered to follow all those it is tied with, but precede censoring events at that time. Let the $\tau_i = I(T_i \geqslant t) \exp \boldsymbol{\beta} \mathbf{x}_i$, and suppose there are d deaths out of m events at time t. Breslow's approximation uses the term

$$\prod_{i=1}^{d} \frac{\tau_i}{\sum_1^m \tau_j}$$

in the partial likelihood at time t. Other options are Efron's approximation

$$\prod_{i=1}^{d} \frac{\tau_i}{\sum_1^m \tau_j - \frac{i}{d} \sum_1^d \tau_j}$$

and the 'exact' partial likelihood

$$\prod_{i=1}^{d} \tau_i \bigg/ \sum \prod_{k=1}^{d} \tau_{j_k}$$

where the sum is over subsets of $1, \ldots, m$ of size d. One of these terms is selected by the method argument of the function coxph, with default efron.

The baseline cumulative hazard $H_0(t)$ is estimated by rescaling the contributions to the number at risk by $\exp \hat{\beta} \mathbf{x}$ in (11.3). Thus in that formula $r(t) = \sum I(T_i \geqslant t) \exp \hat{\beta} \mathbf{x}_i$.

The Cox model is easily extended to allow different baseline hazard functions for different groups, and this is automatically done if they are declared as strata. For our leukaemia example we have:

```
> attach(leuk)
> leuk.cox <- coxph(Surv(time) ~ ag + log(wbc))
> summary(leuk.cox)
    ....
          coef exp(coef) se(coef)     z       p
    ag -1.069     0.343    0.429 -2.49 0.0130
log(wbc)  0.368     1.444    0.136  2.70 0.0069

          exp(coef) exp(-coef) lower .95 upper .95
    ag      0.343      2.913     0.148     0.796
log(wbc)    1.444      0.692     1.106     1.886

Rsquare= 0.377   (max possible= 0.994 )
Likelihood ratio test= 15.6  on 2 df,    p=0.000401
Efficient score test = 16.5  on 2 df,    p=0.000263

> leuk.coxs <- coxph(Surv(time) ~ strata(ag) + log(wbc))
> leuk.coxs
    ....
          coef exp(coef) se(coef)    z       p
log(wbc) 0.391      1.48    0.143 2.74 0.0062
    ....
Likelihood ratio test=7.78  on 1 df, p=0.00529  n= 33

> leuk.coxs1 <- coxph(Surv(time) ~ strata(ag) + log(wbc) +
      ag:log(wbc))
> leuk.coxs1
    ....
              coef exp(coef) se(coef)     z    p
  log(wbc) 0.183      1.20    0.188 0.978 0.33
ag:log(wbc) 0.456      1.58    0.285 1.598 0.11
    ....
> plot(survfit(Surv(time) ~ ag), lty=c(2,3), log=T)
> plot(survfit(leuk.coxs), lty=c(3,2), lwd=2, add=T, log=T)
> leuk.new <- data.frame(wbc=50000)
> plot(survfit(leuk.coxs, newdata=leuk.new), lty=c(5,4),
      log=T, add=T)
> legend(80, 0.8, c("ag absent", "ag present",
    "wbc=50000, abs", "wbc=50000, pres"), lty=c(2,3,4,5) )
```

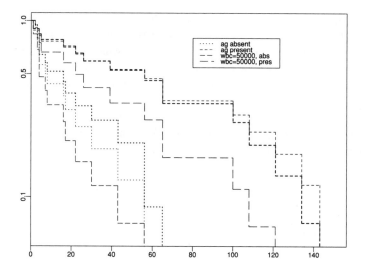

Figure 11.3: Log-survivor curves for the `leuk` dataset. The thick lines are from a Cox model with two strata, the thin lines Kaplan-Meier estimates which ignore the blood counts. Also shown are predicted survivor curves for cases with a blood count of 50,000.

The 'likelihood ratio test' is actually based on (log) partial likelihoods, not the full likelihood, but has similar asymptotic properties.

Note how `survfit` can take the result of a fit of the proportional hazard model. In the first fit the hazards in the two groups differ only by a factor whereas later they are allowed to have separate baseline hazards (which look very close to proportional). There is marginal evidence for a difference in slope within the two strata. Note how straight the log-survivor functions are in Figure 11.3, confirming the good fit of the exponential model for these data. The Kaplan-Meier survivor curves refer to the populations; those from the `coxph` fit refer to a patient in the stratum with an average `log(wbc)`.

The test statistics refer to the whole set of covariates. The likelihood ratio test statistic is the change in deviance on fitting the covariates over just the baseline hazard (by strata); the score test is the expansion at the baseline, and so does not need the parameters to be estimated (although this has been done). Thus the score test is still useful if we do not allow iteration:

```
> leuk.cox <- coxph(Surv(time) ~ ag)
> summary(coxph(Surv(time) ~ log(wbc) +
    offset(predict(leuk.cox, type="lp")), iter.max=0))
    ....
Efficient score test = 7.57  on 1 df,   p=0.00593
```

using the results of a previous fit as a starting point. The R^2 measure quoted by `summary.coxph` is taken from Nagelkirke (1991).

The general proportional hazards model gives estimated (non-intercept) coefficients $\hat{\beta} = (-1.02, 0.36)^T$, compared to the Weibull fit of $(-0.981, 0.298)^T$. The log-logistic had coefficients $(-1.155, 0.609)^T$ which under the approximations of Solomon (1984) would be scaled by $\tau/2$ to give $(-0.789, 0.416)^T$ for a Cox proportional-hazards fit if the log-logistic regression model (an accelerated life model) were the true model.

We next consider the Gehan data. We saw before that the pairs have a negligible effect for the exponential model. Here the effect is a little larger, with $P \approx 8\%$. The Gehan data have a large number of (mainly pairwise) ties.

```
> attach(gehan)
> gehan.cox <- coxph(Surv(time, cens) ~ treat)
> summary(gehan.cox)
    ....
        coef exp(coef) se(coef)    z       p
treat 1.57      4.82     0.412 3.81 0.00014

        exp(coef) exp(-coef) lower .95 upper .95
treat      4.82       0.208      2.15      10.8
    ....
Likelihood ratio test= 16.4  on 1 df,    p=5.26e-05
    ....
> coxph(Surv(time, cens) ~ treat, method="exact")
    ....
        coef exp(coef) se(coef)    z       p
treat 1.63      5.09     0.433 3.76 0.00017

Likelihood ratio test=16.2  on 1 df, p=5.54e-05  n= 42

# The next fit is slow
> coxph(Surv(time, cens) ~ treat+factor(pair), method="exact")
    ....
Likelihood ratio test=45.5  on 21 df, p=0.00148  n= 42
    ....
> 1 - pchisq(45.5-16.2, 20)
[1] 0.082
> detach("gehan")
```

Finally we consider the motors data. The exact fit for the motors dataset is much the slowest, as it has large groups of ties.

```
> attach(motors)
> motor.cox <- coxph(Surv(time, cens) ~ temp)
> motor.cox
    ....
        coef exp(coef) se(coef)    z       p
temp 0.0918      1.1    0.0274 3.36 0.00079
    ....
> coxph(Surv(time, cens) ~ temp, method="breslow")
    ....
```

```
          coef exp(coef) se(coef)    z       p
temp 0.0905       1.09   0.0274 3.3 0.00098
     ....
> coxph(Surv(time, cens) ~ temp, method="exact")
     ....
          coef exp(coef) se(coef)     z       p
temp 0.0947        1.1   0.0274 3.45 0.00056
     ....
> plot( survfit(motor.cox, newdata=data.frame(temp=200),
       conf.type="log-log") )
> summary( survfit(motor.cox, newdata=data.frame(temp=130)) )
time n.risk n.event survival  std.err lower 95% CI upper 95% CI
 408    40      4    1.000 0.000254        0.999            1
 504    36      3    1.000 0.000499        0.999            1
1344    28      2    0.999 0.001910        0.995            1
1440    26      1    0.998 0.002698        0.993            1
1764    20      1    0.996 0.005327        0.986            1
2772    19      1    0.994 0.007922        0.978            1
3444    18      1    0.991 0.010676        0.971            1
3542    17      1    0.988 0.013670        0.962            1
3780    16      1    0.985 0.016980        0.952            1
4860    15      1    0.981 0.020697        0.941            1
5196    14      1    0.977 0.024947        0.929            1
```

The function `survfit` has a special method for `coxph` objects that plots the mean and confidence interval of the survivor curve for an average individual (with average values of the covariates). As we see, this can be over-ridden by giving new data, as shown in Figure 11.4. The non-parametric method is unable to extrapolate to $130°C$ as none of the test examples survived long enough to estimate the baseline hazard beyond the last failure at 5196 hours.

Residuals

The concept of a residual is a difficult one for binary data, especially as here if the event may not be observed because of censoring. A straightforward possibility is to take

$$r_i = \delta_i - \hat{H}(t_i)$$

which is known as the *martingale residual* after a derivation from the mathematical theory given by Fleming & Harrington (1991, §4.5). They show that it is appropriate for checking the functional form of the proportional hazards model, for if

$$h(t) = h_0(t)\phi(x^*) \exp \beta^T \mathbf{x}$$

for an (unknown) function of a covariate x^*, then

$$E[R \mid X^*] \approx [\phi(X^*) - \overline{\phi}] \sum \delta_i / n$$

and this can be estimated by smoothing a plot of the martingale residuals *vs* x^*, for example using `lowess` or the function `scatter.smooth` based on `loess`.

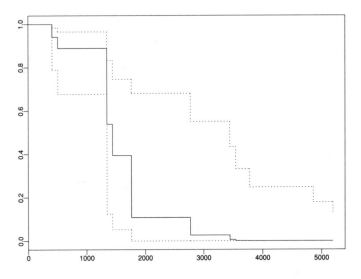

Figure 11.4: The survivor curve for a motor at $200°C$ estimated from a Cox proportional hazards model (solid line) with pointwise 95% confidence intervals (dotted lines).

(The term $\overline{\phi}$ is a complexly weighted mean.) The covariate x^* can be one not included in the model, or one of the terms to check for non-linear effects.

The martingale residuals are the default output of `residuals` on a `coxph` fit.

The martingale residuals can have a very skewed distribution, as their maximum value is 1, but they can be arbitrarily negative. The *deviance residuals* are a transformation

$$\text{sign}\,(r_i)\sqrt{2[-r_i - \delta_i \log(\delta_i - r_i)]}$$

which reduces the skewness, and when squared and summed (approximately) give the deviance. Deviance residuals are best used in plots which will indicate cases not well-fitted by the model.

The *score residuals* are the terms of efficient score for the partial likelihood, this being a sum over cases of

$$L_i = \left[\mathbf{x}_i(t_i) - \overline{\mathbf{x}}(t_i)\right] - \int \left[\mathbf{x}_i(s) - \overline{\mathbf{x}}(s)\right]\hat{h}(s)\,ds$$

where $\overline{\mathbf{x}}(s)$ is defined as the mean weighted by $\exp \boldsymbol{\beta}\mathbf{x}(s)$ over the subjects still in the risk set at time s. For each case the score residual has a term for each parameter, so the result returned is an $n \times p$ matrix. They can be used to examine leverage of individual cases by computing (approximately) the change in $\hat{\boldsymbol{\beta}}$ if the observation was dropped.

The *Schoenfeld residuals* are defined at death times, and are based on sums of the scores up to time t, so define a p-variate residual at each such time, since the process

$$L(t) = \sum_{\text{deaths before } t} \mathbf{x}_i(t) - \overline{\mathbf{x}}(t)$$

jumps at death times. This process may be useful for assessing time trends or lack of proportionality.

Once a type of departure from the base model is discovered or suspected, the proportional hazards formulation is usually flexible enough to allow an extended model to be formulated and the significance of the departure tested within the extended model.

11.4 Further examples

VA lung cancer data

S-PLUS supplies the dataset `cancer.vet` on a Veteran's Administration lung cancer trial used by Kalbfleisch & Prentice (1980), but as it has no on-line help, it is not obvious what it is! It is a matrix of 137 cases with right-censored survival time and the covariates

treatment	standard or test
celltype	one of four cell types
Karnofsky score	of performance on scale 0–100, with high values for relatively well patients
diagnosis	time since diagnosis in months at entry to trial
age	in years
therapy	logical for prior therapy

As there are several covariates, we use the Cox model to establish baseline hazards.

```
> VA.temp <- data.frame(cancer.vet)
> dimnames(VA.temp)[[2]] <- c("treat", "cell", "stime",
    "status", "Karn", "diag.time","age","therapy")
> attach(VA.temp)
> VA <- data.frame(stime, status, treat, age, Karn, diag.time,
    cell=factor(cell), prior=factor(therapy))
> detach("VA.temp")
> attach(VA)
> VA.cox <- coxph(Surv(stime, status) ~ treat + age  + Karn +
    diag.time + cell + prior)
> VA.cox
              coef exp(coef) se(coef)        z       p
   treat  2.95e-01     1.343  0.20755  1.41945 1.6e-01
     age -8.71e-03     0.991  0.00930 -0.93612 3.5e-01
    Karn -3.28e-02     0.968  0.00551 -5.95801 2.6e-09
diag.time  8.18e-05     1.000  0.00914  0.00895 9.9e-01
   cell2  8.62e-01     2.367  0.27528  3.12970 1.7e-03
   cell3  1.20e+00     3.307  0.30092  3.97474 7.0e-05
   cell4  4.01e-01     1.494  0.28269  1.41955 1.6e-01
```

```
    prior  7.16e-02      1.074  0.23231  0.30817 7.6e-01

Likelihood ratio test=62.1  on 8 df, p=1.8e-10  n= 137

> VA.coxs <- coxph(Surv(stime, status) ~ treat + age + Karn +
      diag.time + strata(cell) + prior)
> VA.coxs
               coef exp(coef) se(coef)       z       p
    treat  0.28590     1.331  0.21001  1.361 1.7e-01
      age -0.01182     0.988  0.00985 -1.201 2.3e-01
     Karn -0.03826     0.962  0.00593 -6.450 1.1e-10
diag.time -0.00344     0.997  0.00907 -0.379 7.0e-01
    prior  0.16907     1.184  0.23567  0.717 4.7e-01

Likelihood ratio test=44.3  on 5 df, p=2.04e-08  n= 137

> plot(survfit(VA.coxs), log=T, lty=1:4)
> legend(locator(1), c("squamous", "small", "adeno", "large"),
      lty=1:4)
> plot(survfit(VA.coxs), cloglog=T, lty=1:4)
> cKarn <- factor(cut(Karn, 5))
> VA.cox1 <- coxph(Surv(stime, status) ~ strata(cKarn) + cell)
> plot(survfit(VA.cox1), cloglog=T)
> VA.coxs <- coxph(Surv(stime, status) ~ Karn + strata(cell))
> scatter.smooth(Karn, residuals(VA.coxs))
```

Figures 11.5 and 11.6 show some support for proportional hazards between the cell types, and suggests a Weibull or even exponential distribution.

```
> VA.wei <- survreg(Surv(stime, status) ~ treat + age + Karn +
      diag.time + cell + prior)
> summary(VA.wei)
    ....
Coefficients:
                 Value Std. Error z value        p
(Intercept)   3.490538    0.69117  5.0502 4.41e-07
      treat  -0.228523    0.18684 -1.2231 2.21e-01
        age   0.006099    0.00855  0.7131 4.76e-01
       Karn   0.030068    0.00483  6.2281 4.72e-10
  diag.time  -0.000469    0.00836 -0.0561 9.55e-01
      cell2  -0.826185    0.24631 -3.3542 7.96e-04
      cell3  -1.132725    0.25760 -4.3973 1.10e-05
      cell4  -0.397681    0.25475 -1.5611 1.19e-01
      prior  -0.043898    0.21228 -0.2068 8.36e-01
 Log(scale) -0.074601    0.06631 -1.1250 2.61e-01

Extreme value distribution: Dispersion (scale) = 0.928

> VA.exp <- survreg(Surv(stime, status) ~ Karn + cell,
      dist="exp")
```

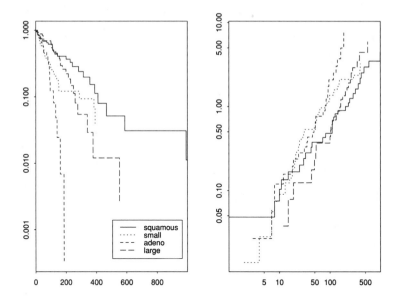

Figure 11.5: Cumulative hazard functions for the cell types in the VA lung cancer trial. The left-hand plot is labelled by survival probability on log scale. The right-hand plot is on log-log scale.

```
> summary(VA.exp)
   ....
Coefficients:
              Value Std. Error z value          p
(Intercept)  3.4222    0.35463    9.65 4.88e-22
       Karn  0.0297    0.00486    6.11 9.95e-10
      cell2 -0.7102    0.24061   -2.95 3.16e-03
      cell3 -1.0933    0.26863   -4.07 4.70e-05
      cell4 -0.3113    0.26635   -1.17 2.43e-01
```

Note that `scale` does not differ significantly from one, so an exponential distribution is an appropriate summary.

Stanford heart transplants

This set of data is analysed by Kalbfleisch & Prentice (1980, §5.5.3). (The data given in Kalbfleisch & Prentice are rounded, but the full data are supplied in our library[3].) It is on survival from early heart transplant operations at Stanford. The new feature is that patients may change treatment during the study, moving from the control group to the treatment group at transplantation, so some of the covariates such as waiting time for a transplant are time-dependent (in the simplest possible way). Patients who received a transplant are treated as two cases, before

---

[3] and with `survival4`, although they may need to be installed from `Examples`.

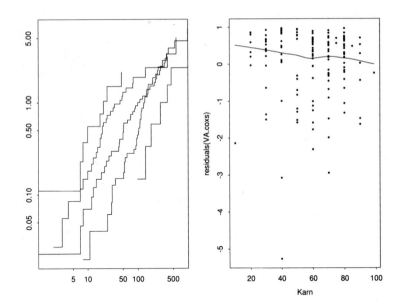

Figure 11.6: Diagnostic plots for the Karnofsky score in the VA lung cancer trial. **Left:** Log-log cumulative hazard plot for 5 groups. **Right:** Martingale residuals vs Karnofsky score, with a smoothed fit.

and after the operation, so cases in the transplant group are in general interval-censored. This is handled by Surv by supplying entry and exit times. For example, patient 4 has the rows

```
start   stop event        age         year   surgery transplant
  0.0   36.0    0  -7.73716632  0.49007529        0          0
 36.0   39.0    1  -7.73716632  0.49007529        0          1
```

which show that he waited 36 days for a transplant and then died after 3 days. The proportional hazards model is fitted from this set of cases, but some summaries need to take account of the splitting of patients.

The covariates are age (in years minus 48), year (after 1 Oct 1967) and an indicator for previous surgery. Rather than use the 6 models considered by Kalbfleisch & Prentice, we do our own model selection.

```
> attach(jasa1)
> coxph(Surv(start, stop, event) ~ transplant*
    (age+surgery+year))
    ....
Likelihood ratio test=18.9  on 7 df, p=0.00868  n= 172
> coxph(Surv(start, stop, event) ~ transplant*(age+year) +
    surgery)
    ....
Likelihood ratio test=18.4  on 6 df, p=0.0053  n= 172
```

```
> stan <- coxph(Surv(start, stop, event) ~ transplant*year +
    age + surgery)
> stan
    ....
                   coef exp(coef) se(coef)      z      p
      transplant -0.6213    0.537   0.5311 -1.17 0.240
            year -0.2526    0.777   0.1049 -2.41 0.016
             age  0.0299    1.030   0.0137  2.18 0.029
         surgery -0.6641    0.515   0.3681 -1.80 0.071
 transplant:year  0.1974    1.218   0.1395  1.42 0.160

Likelihood ratio test=17.1  on 5 df, p=0.00424  n= 172

> stan1 <- coxph(Surv(start, stop, event) ~ strata(transplant)+
    year + year:transplant + age + surgery)
> plot(survfit(stan1), conf.int=T, log=T, lty=c(1,3))
> legend(locator(1), c("before", "after"), lty=c(1,3))

> plot(year[transplant==0], residuals(stan1, collapse=id),
    xlab = "year", ylab="martingale residual")
> lines(lowess(year[transplant==0],
    residuals(stan1, collapse=id)))
> sresid <- resid(stan1, type="score", collapse=id)
> -100 * sresid %*% stan1$var %*% diag(1/stan1$coef)
```

This analysis suggests that survival rates over the study improved *prior* to transplantation, which Kalbfleisch & Prentice suggest could be due to changes in recruitment. The diagnostic plots of Figure 11.7 show nothing amiss. The collapse argument is needed as those patients who received transplants are treated as two cases, and we need the residual per patient.

Now consider predicting the survival of future patients:

```
# Survivor curve for the "average" subject
> summary(survfit(stan))
# Survivor curve for a subject of age 50 on 1/1/71
#    with prior surgery, transplant after 6 months
> stan2 <- data.frame(start=c(0,183), stop=c(183,2*365),
    event=c(1,1), year=c(4,4), age=c(50,50), surgery=c(1,1),
    transplant=c(0,1))
> summary(survfit(stan, stan2, individual=T,
    conf.type="log-log"))
```

time	n.risk	n.event	survival	std.err	lower 95% CI	upper 95% CI
....						
165	43	1	0.1107	0.2030	1.29e-05	0.650
186	41	1	0.1012	0.1926	8.53e-06	0.638
188	40	1	0.0923	0.1823	5.54e-06	0.626
207	39	1	0.0838	0.1719	3.51e-06	0.613
219	38	1	0.0759	0.1617	2.20e-06	0.600
263	37	1	0.0685	0.1517	1.35e-06	0.588
285	35	2	0.0548	0.1221	2.13e-06	0.524

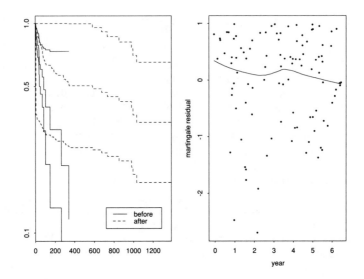

Figure 11.7: Plots for the Stanford heart transplant study. Left: log survivor curves for the two groups. Right: Martingale residuals against calendar time.

308	33	1	0.0486	0.1128	1.24e-06	0.511
334	32	1	0.0431	0.1040	7.08e-07	0.497
340	31	1	0.0380	0.0954	3.97e-07	0.484
343	29	1	0.0331	0.0867	2.10e-07	0.470
584	21	1	0.0268	0.0750	7.30e-08	0.451
675	17	1	0.0205	0.0618	1.80e-08	0.428

The argument `individual=T` is needed to avoid averaging the two cases (which are the same individual). In this case the probability of survival until transplantation is rather low.

Australian AIDS survival

The data on the survival of AIDS patients within Australia are of unusually high quality within that field, and jointly with Dr Patty Solomon we have studied survival up to 1992.[3] There are a large number of difficulties in defining survival from AIDS (acquired immunodeficiency syndrome), in part because as a syndrome its diagnosis is not clear-cut and has almost certainly changed with time. (To avoid any possible confusion, we are studying survival from AIDS and not the HIV infection which is generally accepted as the cause of AIDS.)

The major covariates available were the reported transmission category, and the state or territory within Australia. The AIDS epidemic had started in New South Wales and then spread, so the states will have different profiles of cases

---

[3] We are grateful to the Australian National Centre in HIV Epidemiology and Clinical Research for making these data available to us.

in calendar time. A factor which was expected to be important in survival is the widespread availability of zidovudine (AZT) to AIDS patients from mid-1987 which has enhanced survival, and the use of zidovudine for HIV-infected patients from mid-1990, which it was thought might delay the onset of AIDS without necessarily postponing death further.

The transmission categories were:

hs	male homosexual or bisexual contact
hsid	as **hs** and also intravenous drug user
id	female or heterosexual male intravenous drug user
het	heterosexual contact
haem	haemophilia or coagulation disorder
blood	receipt of blood, blood components or tissue
mother	mother with or at risk of HIV infection
other	other or unknown

The data file gave data on all patients whose AIDS status was diagnosed prior to January 1992, with their status then. Since there is a delay in notification of death, some deaths in late 1991 would not have been reported and we adjusted the endpoint of the study to 1 July 1991. A total of 2843 patients were included, of whom about 1770 had died by the end date. The file contained an ID number, the dates of first diagnosis, birth and death (if applicable), as well as the state and the coded transmission category. The following S code was used to translate the file to a suitable format for a time-dependent-covariate analysis, and to combine the states ACT and NSW (as Australian Capital Territory is a small enclave within New South Wales). The code uses loops and took a few minutes, but this needed to be done only once and was still fast compared to the statistical analyses.

As there are a number of patients who are diagnosed at (strictly, after) death, there are a number of zero survivals. The software used to treat these incorrectly, so all deaths were shifted by 0.9 days to occur after other events the same day. Only the file Aids2 is included in our library. To maintain confidentiality the dates have been jittered, and the smallest states combined.

```
Aids <- read.table("aids.dat")
library(date)  # Therneau's library to manipulate dates
attach(Aids)
diag <- mdy.date(diag.m, diag.d, diag.y, F)
dob <- mdy.date(bir.m, bir.d, bir.y, F)
death <- mdy.date(die.m, die.d, die.y)
# the dates were randomly adjusted here
latedeath <- death >= mdy.date(7,1,91)
death[latedeath] <- mdy.date(7,1,91)
status[latedeath] <- "A"
death[is.na(death)] <- mdy.date(7,1,91)
age <- floor((diag-dob)/365.25)
state <- factor(state)
```

```
levels(state) <- c("NSW","NSW", "Other", "QLD", "Other",
    "Other", "VIC", "Other")
T.categ <- Aids$T
levels(T.categ) <- c("hs", "hsid", "id", "het", "haem",
    "blood", "mother", "other")
Aids1 <- data.frame(state, sex, diag=as.integer(diag),
   death=as.integer(death), status, T.categ,
   age, row.names=1:length(age))
detach("Aids")
rm(death, diag, status, age, dob, state, T.categ, latedeath)
Aids2 <- Aids1[Aids1$diag < as.integer(mdy.date(7, 1, 91)),]
table(Aids2$state)
  NSW Other QLD VIC
 1780   249 226 588

table(Aids2$T.categ)
  hs hsid id het haem blood mother other
2465   72 48 41   46    94      7    70

as.integer(c(mdy.date(7,1,87), (mdy.date(7,1,90)),
    mdy.date(12,31,92)))
[1] 10043 11139 12053


time.depend.covar <- function(data) {
    id <- row.names(data)
    n <- length(id)
    ii <- 0
    stop <- start <- late <- T.categ <- state <- age <-
        status <- sex <- numeric(n)
    idno <- character(n)
    events <- c(0, 10043, 11139, 12053)
    for (i in 1:n) {
        diag <- data$diag[i]
        death <- data$death[i]
        for (j in 1:(length(events)-1)) {
        crit1 <- events[j]
        crit2 <- events[j+1]
        if( diag < crit2 && death >= crit1 ) {
            ii <- ii+1
            start[ii] <- max(0, crit1-diag)
            stop[ii] <- min(crit2, death+0.9)-diag
            if(death > crit2) status[ii] <- 0
            else status[ii] <- codes(data$status[i])-1 # 0 or 1
            state[ii] <- data$state[i]
            T.categ[ii] <- data$T.categ[i]
            idno[ii] <- id[i]
            late[ii] <- j - 1
            age[ii] <- data$age[i]
            sex[ii] <- data$sex[i]
        }}
```

```
     if(i%%100 ==0) print(i)
     }
     levels(state) <- c("NSW", "Other", "QLD", "VIC")
     levels(T.categ) <- c("hs", "hsid", "id", "het", "haem",
         "blood", "mother", "other")
     levels(sex) <- c("F", "M")
     data.frame(idno, zid=factor(late), start, stop, status,
         state, T.categ, age, sex)
   }
   Aids3 <- time.depend.covar(Aids2)
```

Our analysis was based on a proportional hazards model which allowed a proportional change in hazard from 1 July 1987 to 30 June 1990 and another from 1 July 1990; the results show a halving of hazard from 1 July 1987 but a nonsignificant change in 1990. This dataset is large, and the analyses take a long time (1.5 hours in total on a Sun SparcStation IPC) and a lot of memory (up to 10Mb).

```
> attach(Aids3)
> aids.cox <- coxph(Surv(start, stop, status)
      ~ zid + state + T.categ + sex + age)
> summary(aids.cox)
```

	coef	exp(coef)	se(coef)	z	p
zid1	-0.69087	0.501	0.06578	-10.5034	0.00e+00
zid2	-0.78274	0.457	0.07550	-10.3675	0.00e+00
stateOther	-0.07246	0.930	0.08964	-0.8083	4.19e-01
stateQLD	0.18315	1.201	0.08752	2.0927	3.64e-02
stateVIC	0.00464	1.005	0.06134	0.0756	9.40e-01
T.categhsid	-0.09937	0.905	0.15208	-0.6534	5.13e-01
T.categid	-0.37979	0.684	0.24613	-1.5431	1.23e-01
T.categhet	-0.66592	0.514	0.26457	-2.5170	1.18e-02
T.categhaem	0.38113	1.464	0.18827	2.0243	4.29e-02
T.categblood	0.16856	1.184	0.13763	1.2248	2.21e-01
T.categmother	0.44448	1.560	0.58901	0.7546	4.50e-01
T.categother	0.13156	1.141	0.16380	0.8032	4.22e-01
sex	0.02421	1.025	0.17557	0.1379	8.90e-01
age	0.01374	1.014	0.00249	5.5060	3.67e-08

	exp(coef)	exp(-coef)	lower .95	upper .95
zid1	0.501	1.995	0.441	0.570
zid2	0.457	2.187	0.394	0.530
stateother	0.930	1.075	0.780	1.109
stateQLD	1.201	0.833	1.012	1.426
stateVIC	1.005	0.995	0.891	1.133
T.categhsid	0.905	1.104	0.672	1.220
T.categid	0.684	1.462	0.422	1.108
T.categhet	0.514	1.946	0.306	0.863

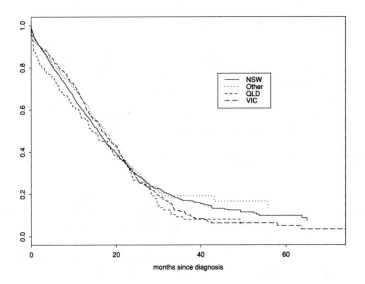

Figure 11.8: Survival of AIDS patients in Australia by state by state.

T.categhaem	1.464	0.683	1.012	2.117
T.categblood	1.184	0.845	0.904	1.550
T.categmother	1.560	0.641	0.492	4.948
T.categother	1.141	0.877	0.827	1.572
sex	1.025	0.976	0.726	1.445
age	1.014	0.986	1.009	1.019

```
Rsquare= 0.045   (max possible= 0.998 )
Likelihood ratio test= 185  on 14 df,   p=0
Efficient score test = 211  on 14 df,   p=0
```

The effect of sex is nonsignificant, and so dropped in further analyses. There is no detected difference in survival during 1990.

Note that Queensland has a significantly elevated hazard relative to New South Wales (which has over 60% of the cases), and that the intravenous drug users have a longer survival, whereas those infected via blood or blood products have a shorter survival, relative to the first category who form 87% of the cases. We can use stratified Cox models to examine these effects (Figures 11.8 and 11.9).

```
> aids1.cox <- coxph(Surv(start, stop, status)
  ~ zid + strata(state) + T.categ + age)
> aids1.surv <- survfit(aids1.cox)
> aids1.surv
              n events mean se(mean) median 0.95CI 0.95CI
 state=NSW 1780   1116  639     17.6    481    450    509
state=Other  249    142  658     42.2    525    436    618
 state=QLD  226    149  519     33.5    439    357    568
 state=VIC  588    355  610     26.3    508    424    618
```

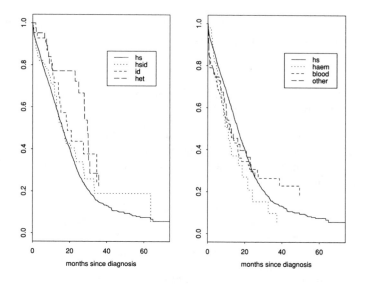

Figure 11.9: Survival of AIDS patients in Australia by transmission category.

```
> plot(aids1.surv, mark.time=F, lty=1:4, xscale=365.25/12,
    xlab="months since diagnosis")
> legend(locator(1), levels(state), lty=1:4)

> aids2.cox <- coxph(Surv(start, stop, status)
    ~ zid + state + strata(T.categ) + age)
> aids2.surv <- survfit(aids2.cox)
> aids2.surv
```

	n	events	mean	se(mean)	median	0.95CI	0.95CI
T.categ=hs	2465	1533	633	15.6	492	473.9	515
T.categ=hsid	72	45	723	86.7	493	396.9	716
T.categ=id	48	19	653	54.3	568	427.9	NA
T.categ=het	40	17	775	57.3	897	753.9	NA
T.categ=haem	46	29	431	53.9	337	209.9	657
T.categ=blood	94	76	583	86.1	358	151.9	666
T.categ=mother	7	3	395	92.6	655	0.9	NA
T.categ=other	70	40	421	40.7	369	24.9	NA

```
> par(mfrow=c(1,2))
> plot(aids2.surv, mark.time=F, lty=1:8, xscale=365.25/12,
    xlab="months since diagnosis", strata=1:4)
> legend(locator(1), levels(T.categ)[1:4], lty=1:4)

> plot(aids2.surv, mark.time=F, lty=1:8, xscale=365.25/12,
    xlab="months since diagnosis", strata=c(1,5,6,8))
> legend(locator(1), levels(T.categ)[c(1,5,6,8)], lty=1:4)
```

The strata argument of plot.survfit (another of our modifications) allows
us to select strata for the plot.

We now consider the possible non-linearity of log-hazard with `age`. First we consider the martingale residual plot (Figure 11.10).

```
cases <- diff(c(0,idno)) != 0
aids.res <- residuals(aids.cox, collapse=idno)
scatter.smooth(age[cases], aids.res, xlab = "age",
    ylab="martingale residual")
```

This shows a slight rise in residual with age over 60, but no obvious effect. The next step is to replace a linear term in age by a step function, with breaks chosen from prior experience. The contrast is chosen explicitly to take as base level the 31–40 age group.

```
age2 <- cut(age, c(-1, 15, 30, 40, 50, 60, 100))
c.age <- factor(age2)
levels(c.age) <- c("0-15", "16-30", "31-40",
    "41-50", "51-60", "61+")
table(c.age)
 0-15 16-30 31-40 41-50 51-60 61+
   39  1022  1583   987   269  85
tmp <- diag(6)
dimnames(tmp) <- list(levels(c.age), levels(c.age))
contrasts(c.age) <- tmp[, -3]

summary(coxph(Surv(start, stop, status) ~ zid + state
    + T.categ + c.age))
    . . . .
                    coef exp(coef) se(coef)         z        p
    . . . .
  c.age0-15  2.51e-01    1.285    0.2888  8.67e-01 3.86e-01
  c.age16-30 -1.02e-01   0.903    0.0625 -1.63e+00 1.04e-01
  c.age41-50  8.00e-02   1.083    0.0615  1.30e+00 1.93e-01
  c.age51-60  3.72e-01   1.450    0.0972  3.82e+00 1.33e-04
   c.age61+   7.12e-01   2.037    0.1566  4.54e+00 5.50e-06
    . . . .
 Likelihood ratio test= 192  on 17 df,    p=0
 Efficient score test = 223  on 17 df,    p=0
detach("Aids3")
```

Beyond this we could fit a smooth function of age via splines, but to save computational time we defer this to the parametric analysis.

We can also consider a parametric analysis. From the survivor curves the obvious model is the Weibull. Since this is both a proportional hazards model and an accelerated life model, we can include the effect of the introduction of zidovudine by assuming a doubling of survival after July 1987. With 'time' computed on this basis we find

```
attach(Aids2)
make.aidsp <- function(){
    cutoff <- 10043
```

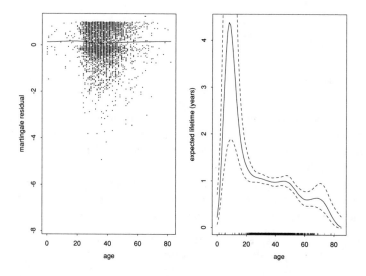

Figure 11.10: Diagnostic plots for age-dependence of AIDS survival in Australia. Left: Martingale residual plot from proportional hazards analysis including a linear effect for age. Right: Predicted survival *vs* age of a NSW hs patient (solid line), with pointwise 95% confidence intervals (dashed lines) and a rug of all observed ages.

```
    btime <- pmin(cutoff, death) - pmin(cutoff, diag)
    atime <- pmax(cutoff, death) - pmax(cutoff, diag)
    survtime <- btime + 0.5*atime
    status <- codes(status) - 1
    data.frame(survtime, status, state, T.categ, age, sex)
}
Aidsp <- make.aidsp()
detach("Aids2")
attach(Aidsp)
aids.wei <- survreg(Surv(survtime+0.9, status) ~ state
    + T.categ + sex + age)
summary(aids.wei, correlation=F)

    ....
Coefficients:
                Value Std. Error  z value         p
  (Intercept)  6.41825    0.2098  30.5970 1.34e-205
   stateOther  0.09387    0.0931   1.0079  3.13e-01
     stateQLD -0.18213    0.0913  -1.9956  4.60e-02
     stateVIC -0.00750    0.0637  -0.1177  9.06e-01
  T.categhsid  0.09363    0.1582   0.5918  5.54e-01
    T.categid  0.40132    0.2552   1.5727  1.16e-01
   T.categhet  0.67689    0.2744   2.4667  1.36e-02
  T.categhaem -0.34090    0.1956  -1.7429  8.14e-02
 T.categblood -0.17336    0.1429  -1.2131  2.25e-01
T.categmother -0.40186    0.6123  -0.6563  5.12e-01
 T.categother -0.11279    0.1696  -0.6649  5.06e-01
```

```
          sex -0.00426      0.1827  -0.0233  9.81e-01
          age -0.01374      0.0026  -5.2862  1.25e-07
    Log(scale)  0.03969      0.0193   2.0571  3.97e-02
```

```
Extreme value distribution: Dispersion (scale) = 1.0405
    . . . .
```

Note that we still have to avoid zero survival. This shows good agreement with the parameters for the Cox model. The parameter α (the reciprocal of the scale) is close to one. For practical purposes the exponential is a good fit, and the parameters are little changed.

We also considered parametric non-linear functions of age by using a spline function. For useful confidence intervals we include the constant term in the predictions, which are for a NSW hs patient (see Figure 11.10).

```
> summary(survreg(Surv(survtime+0.9, status) ~  state
    + T.categ + age), correlation=F)
    . . . .
    Null Deviance: 3974 on 2842 degrees of freedom
Residual Deviance: 3915 on 2830 degrees of freedom   (LL= -3789 )
> t1 <- bs(0:85, knots = c(15,30,40,50,60))
> aids.bs <- survreg(Surv(survtime+0.9,status) ~  state
    + T.categ + t1[age+1,])
> summary(aids.bs, correlation=F)
    . . . .
    Null Deviance: 3998 on 2842 degrees of freedom
Residual Deviance: 3908 on 2823 degrees of freedom   (LL= -3773 )

> cv <- aids.bs$coeff[c(1,12:19)]
> cv.var <- aids.bs$var[c(1,12:19), c(1,12:19)]
> t2 <- cbind(1, t1)
> t3 <- sqrt(diag(t2 %*% cv.var %*% t(t2) ))
> plot(0:85, exp(t2%*%cv)/365.25, type="l", xlab="age",
    ylab = "expected lifetime (years)")
> lines(0:85, exp(t2%*%cv + 1.96 * t3)/365.25, lty=3)
> lines(0:85, exp(t2%*%cv - 1.96 * t3)/365.25, lty=3)
> rug(age+runif(length(age), -0.5, 0.5), 0.015)
```

11.5   Expected survival rates

In medical applications we may want to compare survival rates to those of a standard population, perhaps to standardize the experience of the population under study. As the survival experience of the general population changes with calendar time, this must be taken into account. Therneau's library date (obtainable from statlib) can be used in connection with expected survival to put dates in the correct format (Julian date with class "date").

For a cohort, expected survival is often added to a plot of survivor curves. The function survexp is usually used with a formula generated by ratetable.

The optional argument `times` specifies a vector at which to evaluate survival, by default for all follow times. For example, we could add expected survival for 65-year old US white males to the left plot of Figure 11.7 by

```
> library(date)
> expect <- survexp(~ ratetable(sex=1, year=mdy.date(7,1,1991),
      age=65*365.25), times = seq(0, 1400, 30),
      ratetable=survexp.uswhite)
> lines(expect$time, expect$surv, lty=4)
```

but as the patients are seriously ill, the comparison is not very useful.

Entry and date times can be specified as vectors, when the average survival for the cohort is returned. For individual expected survival, we can use the same form with `type="individual"`, perhaps evaluated at death time.

11.6   Superseded functions

The chapter introduction explained that `survival4` is preferred. For those who *must* use the survival functions in S-PLUS 3.x, we sketch the changes needed.

There are no parametric fitting functions and only the Breslow approximation for fitting proportional hazards models, so the results will differ slightly from those given earlier. The predecessor of `survfit` is `surv.fit`, and that of `coxph` is `coxreg`. We use `model.matrix` to construct design matrices for Cox regressions. (The function `agreg` will compute Cox regressions with time-dependent covariates, but we do not consider its use; the pattern should be clear from `coxreg`.)

```
attach(leuk)
plot(surv.fit(time, rep(1, length(time)), ag), lty=c(2,3))
legend(80, 0.8, c("ag absent", "ag present"), lty=c(2,3))
detach()

attach(gehan)
plot(surv.fit(time, cens, treat))
gehan.surv <- surv.fit(time, cens, treat, conf.type="log-log")
print(gehan.surv, digits=3)
plot(gehan.surv, conf.int=T, lty=c(3,2), log=T,
     xlab="time of remission (weeks)", ylab="survival")
legend(25, 0.1 , c("control","6-MP"), lty=c(2, 3), lwd=2)
surv.diff(time, cens, treat)
detach()

attach(leuk)
coxreg(time, rep(1, length(time)),
    model.matrix( ~ ag + log(wbc))[, -1])
leuk.coxs <- coxreg(time, rep(1, length(time)), log(wbc), ag)
leuk.coxs
coxreg(time, rep(1, length(time)),
```

```
      model.matrix(~ ag * log(wbc))[,3:4], ag)
# three separate plots here
plot(surv.fit(time, rep(1, length(time)), ag), lty=c(2,3), log=T)
plot(surv.fit(time, rep(1, length(time)), ag, coxreg.list=
   leuk.coxs, x=log(wbc)), lty=c(3,2), log=T, ylim=c(0.07,1))
leuk.new <- data.frame(wbc=50000)
plot(surv.fit(time, rep(1, length(time)), ag,
   coxreg.list=leuk.coxs, x=log(wbc), predict.at=log(50000)),
   lty=c(3,2), log=T, ylim=c(0.07,1))
detach()


attach(gehan)
# only Breslow fits are available
coxreg(time, cens, model.matrix(~ treat)[, -1])
coxreg(time, cens, model.matrix(~ treat+factor(pair))[, -1])
1 - pchisq(42.7 - 15.2, 20)
detach()


attach(motors)
motor.cox <- coxreg(time, cens, temp)
motor.cox
plot(surv.fit(time, cens, coxreg.list=motor.cox, x=temp,
   predict.at=200, conf.type="log-log"))
surv.fit(time, cens, coxreg.list=motor.cox, x=temp,
   predict.at=130)
detach()


attach(VA)
coxreg(stime, status, model.matrix( ~ treat + age +
    Karn + diag.time + cell + prior)[,-1])
VA.x <-  model.matrix( ~ treat+age+Karn+diag.time+prior)[,-1]
VA.coxs <- coxreg(stime, status, VA.x, cell)
VA.coxs
plot(surv.fit(stime, status, cell, coxreg.list=VA.coxs,
   x=VA.x), log=T, lty=1:4)
legend(locator(1), c("squamous", "small", "adeno", "large"),
    lty=1:4)
cKarn <- cut(Karn, 5)
VA.x <-  model.matrix( ~ cell)[,-1]
VA.cox1 <- coxreg(stime, status, VA.x, cKarn)
plot(surv.fit(stime, status, cKarn, coxreg.list=VA.cox1,
   x=VA.x), log=T)
VA.coxs <- coxreg(stime, status, model.matrix( ~ Karn)[,-1],
   cell, resid="martingale")
scatter.smooth(Karn, VA.coxs$resid)
detach()
```

Chapter 12

Multivariate Analysis

Multivariate analysis is concerned with datasets which have more than one response variable for each observational or experimental unit. The datasets can be summarized by data matrices X with n rows and p columns, the rows representing the observations or cases, and the columns the variables. The matrix can be viewed either way, depending whether the main interest is in the relationships between the cases or between the variables. Note that for consistency we represent the variables of a case by the *row* vector $\mathbf{x}$.

The main division in multivariate methods is between those methods which assume a given structure dividing the cases into groups, and those which seek to discover structure from the evidence of the data matrix alone. In pattern-recognition terminology the distinction is between *supervised* and *unsupervised* methods. One of our examples is the (in)famous iris data collected by Anderson (1935) and given and analysed by Fisher (1936). This has 150 cases, which are stated to be 50 of each of the three species *Iris setosa, I. virginica* and *I. versicolor*. Each case has four observations on the length and width of its petals and sepals. *A priori* this is a supervised problem, and the obvious questions are to use measurements on a future case to classify it, and perhaps to ask how the variables vary between the species. (In fact, Fisher (1936) used these data to test a genetic hypothesis which placed *I. versicolor* as a hybrid two-thirds of the way from *I. setosa* to *I. virginica*.) However, the classification of species is uncertain, and similar data have been used to identify species by grouping the cases. (Indeed, Wilson (1982) and McLachlan (1992, §6.9) consider whether the iris data can be split into sub-species.) We end the chapter with a similar example on splitting a species of crab.

Krzanowski (1988) and Mardia, Kent & Bibby (1979) are two general references on multivariate analysis. Our treatment is closer to Krzanowski's.

12.1   Graphical methods

The simplest way to examine multivariate data is via a `pairs` plot such as

```
ir <- rbind(iris[,,1], iris[,,2], iris[,,3])
ir.species <- c(rep("s",50), rep("c",50), rep("v",50))
pairs(ir)
```

but it is not easy to identify the groups on the pairs plot. We can make enhanced plots by

```
pairs(ir, panel=function(x,y,...) text(x,y, ir.species, ...))
pairs(ir, panel=function(x,y,...) text(x+runif(150,-0.05,0.05),
    y + runif(150, -0.05, 0.05), ".",
    col=codes(factor(ir.species)), ...))
```

which shows a fairly clear separation between the groups in some of the scatter plots. (Note that we use `text` as `points` does not allow direct specification of the colour of each point. As we can not reproduce colour, we do not use it further in this chapter, but for users with a colour screen it can be very useful, especially at low resolutions where marks are less clear-cut.)

A similar effect can be obtained using the S-PLUS function `brush`,

```
brush(ir)
```

by marking the cases with different symbols (Figure 12.1). Note that coincident points are not marked.

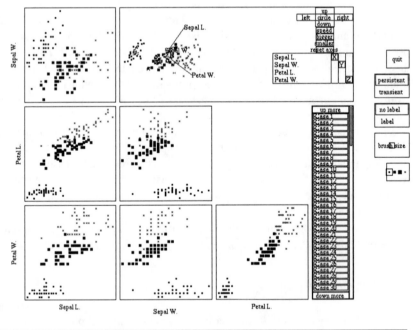

Figure 12.1: `brush` plot of the `iris` data.

The function `brush` also allows combinations of three variables to be considered, and `spin` shows the three-dimensional dynamic plot alone. (The argument `spin=F` to `brush` shows just the dynamic pairs plot and does not include the three-variable plot.) Rather more can be seen from such plots by spinning them than can be seen from the static versions possible in a book. The plot can be rotated in any direction at various speeds and rescaled interactively. (See Section 3.2 for further details.)

Principal component analysis

Linear methods are the heart of classical multivariate analysis, and depend on seeking linear combinations of the variables with desirable properties. For the unsupervised case the main method is *principal component* analysis, which seeks linear combinations of the columns of X with maximal (or minimal) variance. Because the variance can be scaled by rescaling the combination, we constrain the combinations to have unit length.

Let Σ denote the covariance matrix of the data X, which is defined by

$$(n - p)\Sigma = (X - n^{-1}1^T X1)^T (X - n^{-1}1^T X1) = (X^T X - n\overline{\mathbf{x}}\overline{\mathbf{x}}^T)$$

where $\overline{\mathbf{x}} = 1^T X/n$ is the row vector of means of the variables. Then the sample variance of a linear combination $\mathbf{x}\mathbf{a}$ of a row vector $\mathbf{x}$ is $\mathbf{a}^T\Sigma\mathbf{a}$ and this is to be maximized (or minimized) subject to $\|\mathbf{a}\|^2 = \mathbf{a}^T\mathbf{a} = 1$. Since Σ is a non-negative definite matrix, it has an eigendecomposition

$$\Sigma = C^T \Lambda C$$

where Λ is a diagonal matrix of (non-negative) eigenvalues in decreasing order. Let $\mathbf{b} = C\mathbf{a}$, which has the same length as $\mathbf{a}$ (since C is orthogonal). The problem is then equivalent to maximizing $\mathbf{b}^T\Lambda\mathbf{b} = \sum \lambda_i b_i^2$ subject to $\sum b_i^2 = 1$. Clearly the variance is maximized by taking $\mathbf{b}$ to be the first unit vector, or equivalently taking $\mathbf{a}$ to be the column eigenvector corresponding to the largest eigenvalue of Σ. Taking subsequent eigenvectors gives combinations with as large as possible variance which are uncorrelated with those which have been taken earlier. The i th principal component is then the i th linear combination picked by this procedure. (It is only determined up to a change of sign.)

Another point of view is to seek new variables y_j which are rotations of the old variables to explain best the variation in the dataset. Clearly these new variables should be taken to be the principal components, in order. Suppose we use the first k principal components. Then the subspace they span contains the 'best' k-dimensional view of the data. It both has maximal covariance matrix (both in trace and determinant) and best approximates the original points in the sense of minimizing the sum of squared distance from the points to their projections. The first few principal components are often useful to reveal structure in the data.

We can measure how good the k-dimensional view is by

$$\sum_{i=1}^{k} \lambda_i \Big/ \sum_{i=1}^{p} \lambda_i = \sum_{i=1}^{k} \lambda_i \Big/ \text{trace}(\Sigma)$$

The so-called *scree diagram* plots this (or λ_k) against k to look for a good choice of k.

The principal components corresponding to the smallest eigenvalues are the most nearly constant combinations of the variables, and can also be of interest.

Note that the principal components depend on the scaling of the original variables, and this will be undesirable except perhaps if (as in the `iris` data) they are in comparable units. (Even in this case, correlations would often be used.) Otherwise it is conventional to take the principal components of the *correlation* matrix, implicitly rescaling all the variables to have unit sample variance. (The S command `scale` can be used to centre and/or scale a data matrix.)

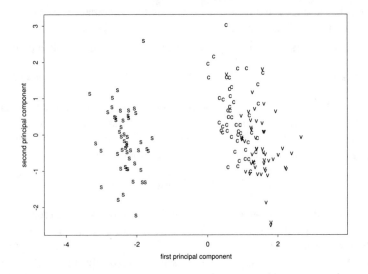

Figure 12.2: First two principal components for the log-transformed `iris` data.

The S command `prcomp` does not work with the covariance matrix Σ but directly with X. To do so it uses the singular value decomposition $X = U\Lambda V^T$ (see page 49). Note that the eigenvalues and eigenvectors of $X^T X$ are Λ^2 and V respectively, and so the principal components can be obtained by applying the SVD to $X - 1\overline{x}$, and this is precisely how `prcomp` works. To obtain the coefficients corresponding to the correlation matrix, use

```
prcomp(scale(X, center=T, scale=T))
```

The function `prcomp` returns in component `rotation` the matrix with columns corresponding to vectors a_i, and in component `sdev` the vector of sample standard deviations of the principal components, that is $\lambda_i/\sqrt{n-1}$. Note that it only returns principal component vectors corresponding to the non-zero eigenvalues, and so cannot be used to find combinations which are exactly zero. Indeed, the function is defined as

```
prcomp <- function(x, retx = TRUE)
{
    s <- svd(scale(x, scale = F), nu = 0)
    rank <- sum(s$d > 0)
    if(rank < ncol(x)) s$v <- s$v[, 1:rank]
    s$d <- s$d/sqrt(max(1, nrow(x) - 1))
    if(retx) list(sdev = s$d, rotation = s$v, x = x %*% s$v)
    else list(sdev = s$d, rotation = s$v)
}
```

By default `prcomp` also returns the X matrix rotated to the principal components. Figure 12.2 shows the first two principal components for the `iris` data based on the covariance matrix, revealing the group structure if it had not already been known. Note that we chose to preserve the scaling of the principal components. A warning: principal component analysis will reveal the gross features of the data, which may already be known, and is often best applied to residuals after the known structure has been removed.

```
> ir.pr <- prcomp(scale(log(ir)))
> eqscplot(ir.pr$x[,1:2], type="n",
    xlab="first principal component",
    ylab = "second principal component")
> text(ir.pr$x[,1:2], ir.species)
> ir.pr$sdev
[1] 1.71246 0.95238 0.36470 0.16568
> ir.pr$rot
          [,1]       [,2]       [,3]       [,4]
[1,]   0.50382 -0.454999 -0.70885  0.191476
[2,]  -0.30237 -0.889144  0.33116 -0.091254
[3,]   0.57679 -0.033788  0.21928 -0.786187
[4,]   0.56750 -0.035456  0.58290  0.580447
```

S-PLUS 3.2 has another function, `princomp`, for principal components, and a `screeplot` function. As `princomp` returns an object of class `"princomp"`, it will be printed and plotted specially. The argument `cor` controls whether the covariance or correlation matrix is used (via re-scaling the variables).

```
> ir.prc <- princomp(log(ir), cor=T)
> options(digits=5)
> ir.prc
Standard deviations:
 Comp. 1 Comp. 2 Comp. 3 Comp. 4
  1.7125 0.95238  0.3647 0.16568
    ....
> summary(ir.prc)
Importance of components:
                         Comp. 1  Comp. 2  Comp. 3    Comp. 4
       Standard deviation 1.71246  0.95238 0.364703 0.1656840
 Proportion of Variance 0.73313  0.22676 0.033252 0.0068628
 Cumulative Proportion 0.73313  0.95989 0.993137 1.0000000
```

```
> plot(ir.prc)
> loadings(ir.prc)
          Comp. 1 Comp. 2 Comp. 3 Comp. 4
Sepal L.    0.504   0.455   0.709   0.191
Sepal W.   -0.302   0.889  -0.331
Petal L.    0.577          -0.219  -0.786
Petal W.    0.567          -0.583   0.580
> eqscplot(ir.prc$scores[,1:2], type="n",
    xlab="first principal component",
    ylab = "second principal component")
> text(ir.prc$scores[,1:2], ir.species)
```

In the terminology of this function, the *loadings* are columns giving the linear combinations a for each principal component, and the *scores* are the data on the principal components. The plot shows the screeplot, a barplot of the variances of the principal components labelled by $\sum_{i=1}^{j} \lambda_i / \text{trace}(\Sigma)$. The result of loadings is rather deceptive, as small entries are suppressed in printing but will be insignificant only if the correlation matrix is used, and that is *not* the default. The prediction method rotates to the principal components. Note that prcomp and princomp give different standard deviations for unscaled data as the former computes variances with divisor $n - 1$, the latter with divisor n.

There are two books devoted solely to principal components, Jackson (1991) and Jolliffe (1986), which we think over-states its value as a technique. Other projection techniques such as projection pursuit (e.g. Huber, 1985; Friedman, 1987; Jones & Sibson, 1987) choose rotations based on more revealing criteria than variances. (These are not implemented in S or S-PLUS, but are available in the package XGobi available for Unix machines running X11 from statlib (see Appendix D) with an S interface.)

Distance methods

A class of methods are based on representing the cases in a low-dimensional Euclidean space so that their proximity reflects the similarity of their variables. To do so we have to produce a measure of similarity. The S function dist uses one of four distance measures between the points in the p-dimensional space of variables. The default is Euclidean distance; others include manhattan (the L_1 distance, summing absolute differences) and maximum. Other distance measures, including some appropriate to categorical variables, are discussed in Section 12.2. Distances are often called *dissimilarities*.

The most obvious of the distance methods is *multi-dimensional scaling*, which seeks a configuration in $\mathbb{R}^d$ such that distances between the points best match those of the distance matrix. Only the classical or metric form of multi-dimensional scaling is implemented in S, which is also known as *principal coordinate analysis*. For the iris data we can use

```
ir.scal <- cmdscale(dist(ir), k=2, eig=T)
eqscplot(ir.scal$points, type="n")
text(ir.scal$points, ir.species, cex=0.6)
```

where care is taken to ensure correct scaling of the axes (see the top left plot of Figure 12.3). Note that a configuration can be determined only up to translation, rotation and reflection, since Euclidean distance is invariant under the group of rigid motions and reflections. An idea of how good the fit is can be obtained by calculating a measure of 'stress':

```
> distp <- dist(ir)
> dist2 <- dist(ir.scal$points)
> sum((distp - dist2)^2)/sum(distp^2)
[1] 0.001747
```

which shows the fit is good. Using classical multi-dimensional scaling with a Euclidean distance as here is precisely equivalent to plotting the first k principal components (without re-scaling to correlations).

A non-metric form of multi-dimensional scaling is Sammon's (1969) non-linear mapping, which given a distance d on n points constructs a k-dimensional configuration with distances $\widetilde{d}$ to minimize a weighted 'stress'

$$E(d, \widetilde{d}) = \sum_{i \neq j} \frac{(d_{ij} - \widetilde{d}_{ij})^2}{d_{ij}}$$

by an iterative algorithm implemented in our function sammon.[1] We have to drop duplicate observations to make sense of $E(d, \widetilde{d})$:

```
ir.sam <- sammon(dist(ir[-143,]))
eqscplot(ir.sam$points, type="n")
text(ir.sam$points, label=ir.species[-143], cex=0.6)
```

Minimum-spanning trees can also be used to represent multivariate data in the plane (Friedman & Rafsky, 1981). A minimum spanning tree (MST) is a series of edges which join all points to form a connected graph of minimum total length. (Such trees are usually unique.) The tree is then represented in the plane so that all edges have the right length, and, locally, distances are roughly preserved. It is a less accurate representation than multi-dimensional scaling but much faster (20 times in this example). The function mstree includes the Friedman–Rafsky algorithm:

```
> ir.mst <- mstree(ir)
> eqscplot(ir.mst$y, ir.mst$x, type="n")
> text(ir.mst$y, ir.mst$x, label=ir.species, cex=0.6)
> dist2 <- dist(cbind(ir.mst$x, ir.mst$y))
> sum((distp - dist2)^2)/sum(distp^2)
[1] 0.029256
```

where we interchanged the axes only to make comparison easier in Figure 12.3.

---

[1] This uses C.

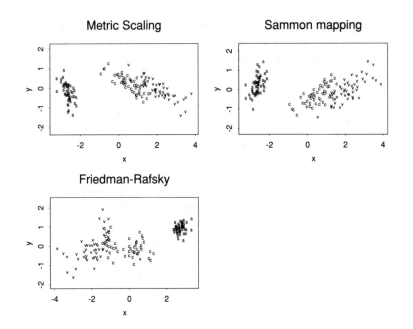

Figure 12.3: Distance-based representations of the `iris` data. The top left plot is by multidimensional scaling, the top right by Sammon's non-linear mapping and the bottom left via minimum-spanning trees. Note that each is defined up to shifts, rotations and reflections.

Representing variables

With a small number of cases, we may want to visualize the dataset. The S dataset `cereal.attitude` gives percentage agreement to 11 statements about each of 8 cereal brands. The statements were

come back to	reasonably priced
tastes nice	a lot of food value
popular with all the family	stays crispy in milk
very easy to digest	helps to keep you fit
nourishing	fun for children to eat
natural flavour	

We may want to see the relationship between the opinions for each brand. The S functions `faces` and `stars` provide us with two ways to do so. Chernoff's (1973) `faces` represents up to 15 variables by features of cartoon faces, because humans are very good at identifying facial features. An example is shown in Figure 12.4. Such plots are subject to manipulation by the order and sign of the variables, and so can deceive!

The function `stars` plots represent each variable by the length of a vector and connect the vectors, as in Figure 12.5. The variables are represented counterclockwise from the x axis.

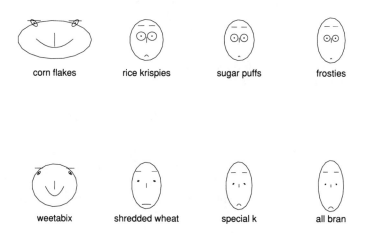

Figure 12.4: A faces plot of the dataset `cereal.attitude`, obtained by
`faces(t(cereal.attitude), labels=dimnames(cereal.attitude)[[2]],`
`ncol=4)` and some spelling corrections. Data from Chakrapani & Ehrenberg (1981).

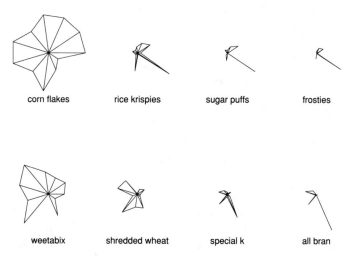

Figure 12.5: A stars plot of the dataset `cereal.attitude`, obtained by
`stars(t(cereal.attitude), labels=dimnames(cereal.attitude)[[2]],`
`ncol=4)` and some spelling corrections.

Biplots

The biplot (due to Gabriel) is another method to represent both the cases and
variables. We suppose that X has been centred to remove column means. The
biplot represents X by two sets of vectors of dimensions n and p producing
a rank-2 approximation to X. The best (in the sense of least-squares) such
approximation is given by replacing Λ in the singular value decomposition of X

by D, a diagonal matrix setting $\lambda_3, \ldots$ to zero, so

$$X \approx \widetilde{X} = [\mathbf{u}_1 \, \mathbf{u}_2] \begin{bmatrix} \lambda_1 & 0 \\ 0 & \lambda_2 \end{bmatrix} \begin{bmatrix} \mathbf{v}_1^T \\ \mathbf{v}_2^T \end{bmatrix} = GH^T$$

where the diagonal scaling factors can be absorbed into G and H in a number of ways. For example, we could take

$$G = [\mathbf{u}_1 \, \mathbf{u}_2] \begin{bmatrix} \lambda_1 & 0 \\ 0 & \lambda_2 \end{bmatrix}^{1-\lambda}, \qquad H = [\mathbf{v}_1 \, \mathbf{v}_2] \begin{bmatrix} \lambda_1 & 0 \\ 0 & \lambda_2 \end{bmatrix}^{\lambda}$$

for $\lambda = 0, 0.5$ or 1. The biplot then consists of plotting the $n + p$ 2-dimensional vectors which form the rows of G and H. The interpretation is based on inner products between vectors from the two sets, which give the elements of $\widetilde{X}$.

Figure 12.6 shows the biplot with $\lambda = 0.5$ for the cereal attitudes data, obtained by

```
X <- scale(t(cereal.attitude), center=T, scale=F)
X.svd <- svd(X)
G <- X.svd$u[,1:2] %*% diag(sqrt(X.svd$d[1:2]))
H <- X.svd$v[,1:2] %*% diag(sqrt(X.svd$d[1:2]))
eqscplot(rbind(G,H), type="n", axes=F, xlab="", ylab="")
points(0, 0, pch=3, mkh=0.5)
points(G, pch=3)
points(H, pch=0)
identify(G, dimnames(X)$brand, cex=0.8)
identify(H, dimnames(X)$statement, cex=0.8)
```

where the transpose is needed for this dataset only.

This puts the vectors on a common scale. Another commonly advocated scaling is to take $\lambda = 1$ and re-scale,

```
G <- X.svd$u[,1:2] * sqrt(dim(X)[1]-1)
H <- X.svd$v[,1:2] %*% diag(X.svd$d[1:2])/sqrt(dim(X)[1]-1)
```

so that the inner products between the rows of H represent (to the extent that $\widetilde{X}$ represents X) the covariances and the Euclidean distance between the rows of G represents the Mahalanobis distance (page 317) between the observations (Jolliffe, 1986, pp. 77–8). In our example this would have the G columns on a much smaller scale than the H columns, making comparison difficult. However, as all the variables are on the same scale (percentages) we can rescale them uniformly so the largest has unit variance:

```
G <- X.svd$u[,1:2] * sqrt(dim(X)[1]-1)
H <- X.svd$v[,1:2] %*% diag(X.svd$d[1:2])/sqrt(dim(X)[1]-1)
H <- H/sqrt(max(diag(var(H))))
eqscplot(rbind(G,H), type="n", axes=F, xlab="", ylab="")
points(0, 0, pch=3, mkh=0.5)
points(G, pch=3)
points(H, pch=0)
identify(G, dimnames(X)$brand, cex=0.8)
identify(H, dimnames(X)$statement, cex=0.8)
```

The plot is then very similar to Figure 12.6. Indeed, the scaling used will only matter when λ_1 and λ_2 are very different, whereas we have `c(1.20, 0.73)`.

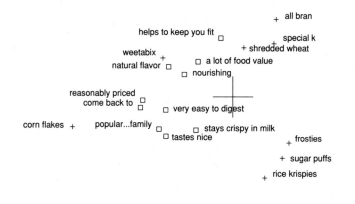

Figure 12.6: Biplot of the `cereal.attitude` data.

In S-PLUS 3.2, biplots can be obtained by

```
biplot(princomp(t(cereal.attitude)), scale=1, expand=1)
```

where `scale` is λ and `expand` scales the columns relative to the rows. (Both default to 1.)

12.2   Cluster analysis

Cluster analysis is concerned with discovering group structure amongst the cases of our n by p matrix. Two general references are Gordon (1981) and Hartigan (1975). Almost all methods are based on a measure of the similarity or dissimilarity between cases. A *dissimilarity coefficient* d is symmetric ($d(A, B) = d(B, A)$), non-negative, and $d(A, A)$ is zero. A similarity coefficient has the scale reversed. Dissimilarities may be a metric:

$$d(A, C) \leqslant d(A, B) + d(B, C)$$

or an *ultrametric*:

$$d(A, B) \leqslant \max\big(d(A, C), d(B, C)\big)$$

but need not be either. We have already seen several dissimilarities calculated by `dist`.

Jardine & Sibson (1971) discuss several families of similarity and dissimilarity measures. For categorical variables most dissimilarities are measures of agreement. The *simple matching coefficient* is the proportion of categorical variables on which the cases differ. The *Jaccard coefficient* applies to categorical variables with a preferred level. It is the proportion of such variables with one of the cases at the preferred level in which the cases differ. The `binary` method of `dist` is of this family, being the Jaccard coefficient if all non-zero levels are preferred. Applied to logical variables on two cases it gives the proportion of variables in which only one is true amongst those which are true on at least one case. There are many variants of these coefficients; Kaufman & Rousseeuw (1990, §2.5) provide a readable summary and recommendations.

Ultrametric dissimilarities have the appealing property that they can be represented by a *dendrogram* such as those shown in Figure 12.7, in which the dissimilarity between two cases can be read off from the height at which they join a single group. Hierarchical clustering methods can be thought of as approximating a dissimilarity by an ultrametric dissimilarity. Jardine & Sibson argue that one method, single-link clustering, uniquely has all the desirable properties of a clustering method. This measures distances between clusters by the dissimilarity of the closest pair, and agglomerates by adding the shortest possible link (i.e. joining the two closest clusters). Other authors disagree, and Kaufman & Rousseeuw (1990, §5.2) give a different set of desirable properties leading uniquely to their preferred method, which views the dissimilarity between clusters as the average of the dissimilarities between members of those clusters. Another popular method is complete-linkage, which views the dissimilarity between clusters as the maximum of the dissimilarities between members.

The S function `hclust` implements these three metrics, selected by its `method` argument which takes values `compact` (the default, for complete-linkage), `average` and `connected` (for single-linkage).

The S dataset `swiss.x` gives five measures of socio-economic data on Swiss provinces about 1888, given by Mosteller & Tukey (1977, pp. 549–551). The data are percentages, so Euclidean distance is a reasonable choice. We use single-link clustering:

```
h <- hclust(dist(swiss.x), method="connected")
plclust(h)
cutree(h, 3)
plclust( clorder(h, cutree(h, 3) ))
```

The first plot suggests three main clusters, and the remaining code re-orders the dendrogram to display those clusters more clearly. Note that there are two main groups, with the point 45 well separated from them.

K-means

The K-means clustering algorithm (MacQueen, 1967; Hartigan, 1975; Hartigan & Wong, 1979) chooses a prespecified number of cluster centres to minimize the within-class sum of squares from those centres. As such it is most appropriate to continuous variables, suitably scaled. The algorithm needs a starting point, so we choose the means of the clusters identified by group-average clustering.

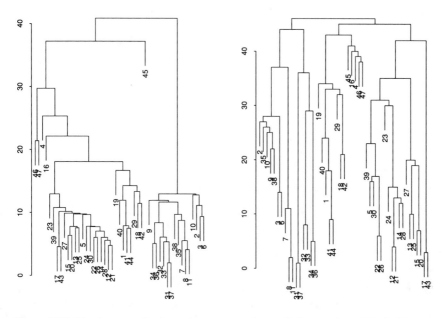

Figure 12.7: Dendrograms for the socio-economic data on Swiss provinces. The left figure is computed by single-link clustering, the right figure by "model-based" clustering.

The clusters *are* altered (cluster 3 contained just point 45), and are shown in principal-component space in Figure 12.8. (The standard deviations show that a two-dimensional representation is reasonable.)

```
h <- hclust(dist(swiss.x), method="average")
initial <- tapply(swiss.x, list(rep(cutree(h, 3),
   ncol(swiss.x)), col(swiss.x)), mean)
km <- kmeans(swiss.x, initial)
swiss.p <- prcomp(swiss.x)
swiss.p$sdev
[1] 43.3668 21.4309  7.6700  3.7278  2.7506
swiss.px <- swiss.p$x[,1:2]
eqscplot(swiss.px, type="n", xlab="first principal component",
   ylab="second principal component")
text(swiss.px, km$cluster)
points(km$centers %*% swiss.p$rot[,1:2], pch=3, mkh=0.5)
```

Model-based clustering

S-PLUS has functions for "model-based" clustering (Banfield and Raftery, 1993) implemented by the functions mclust, mclass and mreloc.[2] For Figure 12.7 we used

---

[2] These are also available from statlib.

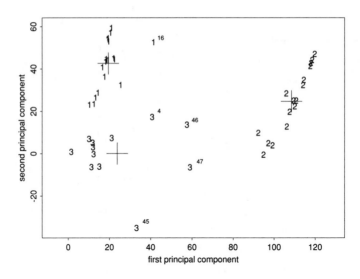

Figure 12.8: The Swiss provinces data plotted on its first two principal components. The labels are the groups assigned by K-means; the crosses denote the group means. Five points are labelled with smaller symbols.

```
h <- mclust(swiss.x, method = "S*")$tree
plclust( clorder(h, cutree(h, 3) ))
```

Note that this works with a data matrix and not with a dissimilarity matrix.

The idea of "model-based" clustering is that the data are independent samples from a series of group populations, but the group labels have been lost. If we knew that the vector γ gave the group labels, and each group had class-conditional pdf $f_i(\mathbf{x}; \theta)$, then the likelihood would be

$$\prod_{i=1}^{n} f_{\gamma_i}(\mathbf{x}_i; \theta) \tag{12.1}$$

Since the labels are unknown, these are regarded as parameters in (12.1), and the likelihood maximized over (θ, γ).

Choosing the class-conditional pdf's to be multivariate normal with different means but a common covariance matrix $\Sigma = \sigma^2 I$ leads to the criterion of minimizing the sum of squares to the cluster centre, that is K-means. The other options available are based on multivariate normals with other constraints on the covariance matrices. For example, the default method S* allows clusters of different sizes and orientations but the same pre-specified 'shape' (the ratio of axes of the ellipsoid), and S is similar but constrains to equal size.

The code returns an "approximate weight of evidence" for the number of clusters, which is this case suggests two or three clusters. A further option is "noise" to allow some of the points to come from a homogeneous Poisson process rather than from one of the cluster groups. (It may help to think of this as a $(k + 1)$ st cluster with a very diffuse distribution.)

Note that the hierarchical clustering given by `mclust` only optimizes the fit criterion in a very crude way. Once the number of clusters is chosen, `mreloc` can be used to optimize further, although in our example no change occurs. If we allow 'noise' two points are re-allocated:

```
h <- mclust(swiss.x, method = "S*", noise=T)
hclass <- mclass(h, 3)
hclass$class
 [1] 1 2 2 4 1 2 2 2 2 2 2 1 1 1 1 1 1 1 1 1 1 1 1 1 1 1 1 1
[29] 1 1 2 2 2 2 2 2 2 2 1 1 1 1 1 1 4 4 4
mreloc(hclass, swiss.x, method = "S*", noise=T)
 [1] 1 2 2 4 1 2 2 2 2 2 2 1 1 1 1 4 1 1 1 1 1 1 1 1 1 1 1 1
[29] 1 1 2 2 2 2 2 2 2 2 1 1 1 4 1 1 4 4 4
```

Note that this is a random algorithm so the results are not repeatable, and that the class number is the lowest-numbered object in the cluster.

12.3 Discriminant analysis

Now suppose that we have a set of g classes, and for each case we know the class (assumed correctly). We can then use the class information to help reveal the structure. Let W denote the within-class covariance matrix, that is the covariance matrix of the variables centred on the class mean, and B denote the between-classes covariance matrix, that is of the predictions by the class means. Let M be the $g \times p$ matrix of class means, and G be the $n \times g$ matrix of class indicator variables (so $g_{ij} = 1$ if and only if case i is assigned to class j). Then the predictions are GM. Let $\bar{x}$ be the means of the variables over the whole sample. Then the sample covariance matrices are

$$W = \frac{(X - GM)^T(X - GM)}{n - g}, \qquad B = \frac{(GM - 1\bar{x})^T(GM - 1\bar{x})}{g - 1} \quad (12.2)$$

Note that B has rank at most $\min(p, g - 1)$.

Fisher (1936) introduced a linear discriminant analysis seeking a linear combination $\mathbf{xa}$ of the variables which has a maximal ratio of the separation of the class means to the within-class variance, that is maximizing the ratio $\mathbf{a}^T B \mathbf{a} / \mathbf{a}^T W \mathbf{a}$. To compute this, choose a scaling $\mathbf{x}S$ of the variables so that they have the identity as their within-group correlation matrix. (One such scaling is take the principal components with respect to W, and re-scale each to unit variance.) On the re-scaled variables the problem is to maximize $\mathbf{a}^T B \mathbf{a}$ subject to $\|\mathbf{a}\| = 1$, and as we saw before, this is solved by taking $\mathbf{a}$ to be the eigenvector of B corresponding to the largest eigenvalue. (Note that re-scaling changes B to $S^T B S$.) The linear combination $\mathbf{a}$ is unique up to a change of sign (unless there are multiple eigenvalues, an event of probability zero). The exact multiple of $\mathbf{a}$ returned by a

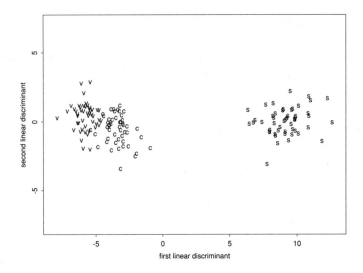

Figure 12.9: The log `iris` data on the first two discriminant axes.

program will depend on its definition of the within-class variance matrix. We use the conventional divisor of $n - g$, but divisors of n and $n - 1$ have been used.

As for principal components, we can take further linear components corresponding to the next largest eigenvalues. There will be at most $r = \min(p, g - 1)$ positive eigenvalues. Note that the eigenvalues are the proportion of the between-classes variance explained by the linear combinations, which may help us to choose how many to use. The corresponding transformed variables are called the *linear discriminants* or *canonical variates*. It is often useful to plot the data on the first few linear discriminants (Figure 12.9). Since the within-group covariances should be the identity, we chose an equal-scaled plot.

```
> a <- lda(log(ir), ir.species)
> a$svd^2/sum(a$svd^2)
[1] 0.9965 0.0035
> a.x <- predict(a, log(ir), dimen=2)$x
> eqscplot(a.x, type="n", xlab="first linear discriminant",
      ylab="second linear discriminant")
> text(a.x, ir.species)
```

This shows that 99.65% of the between-group variance is on the first discriminant axis.

Discrimination for normal populations

An alternative approach to discrimination is *via* probability models. Let π_c denote the prior probabilities of the classes, and $p(\mathbf{x} \,|\, c)$ the densities of distributions of the observations for each class. Then the posterior distribution of the classes after observing $\mathbf{x}$ is

$$p(c \,|\, \mathbf{x}) = \frac{\pi_c p(\mathbf{x} \,|\, c)}{p(\mathbf{x})} \propto \pi_c p(\mathbf{x} \,|\, c)$$

and it is fairly simple to show that the allocation rule which makes the smallest expected number of errors chooses the class with maximal $p(c \,|\, \mathbf{x})$; this is known as the *Bayes rule*.

Now suppose the distribution for class c is multivariate normal with mean $\boldsymbol{\mu}_c$ and covariance Σ_c. Then the Bayes rule minimizes

$$
\begin{aligned}
Q_c &= -2 \log p(\mathbf{x} \,|\, c) - 2 \log \pi_c \\
&= (\mathbf{x} - \boldsymbol{\mu}_c)\Sigma_c^{-1}(\mathbf{x} - \boldsymbol{\mu}_c)^T + \log |\Sigma_c| - 2 \log \pi_c
\end{aligned}
\tag{12.3}
$$

The first term of (12.3) is the *Mahalanobis distance* to the class centre, and can be calculated by the S function `mahalanobis`. The difference between the Q_c for two classes is a quadratic function of $\mathbf{x}$, so the method is known as *quadratic discrimination* and the boundaries of the decision regions are quadratic surfaces in $\mathbf{x}$ space. This is implemented by our function `qda`.

Further suppose that the classes have a common covariance matrix Σ. Differences in the Q_c are then *linear* functions of $\mathbf{x}$, and we can maximize $-Q_c/2$ or

$$L_c = \mathbf{x}\Sigma^{-1}\boldsymbol{\mu}_c^T - \boldsymbol{\mu}_c\Sigma^{-1}\boldsymbol{\mu}_c^T/2 + \log \pi_c \tag{12.4}$$

To use (12.3) or (12.4) we have to estimate $\boldsymbol{\mu}_c$ and Σ_c or Σ. The 'obvious' estimates are used, the sample mean and covariance matrix within each class, and W for Σ.

How does this relate to Fisher's linear discrimination? The latter gives new variables, the linear discriminants, with unit within-class sample variance, and the differences between the group means lie entirely in the first r variables. Thus on these variables the Mahalanobis distance (with respect to $\widehat{\Sigma} = W$) is just

$$\|\mathbf{x} - \boldsymbol{\mu}_c\|^2$$

and only the first r components of the vector depend on c. Similarly, on these variables

$$L_c = \mathbf{x}\boldsymbol{\mu}_c^T - \|\boldsymbol{\mu}_c\|^2/2 + \log \pi_c$$

and we can work in r dimensions. If there are just two classes, there is a single linear discriminant, and

$$L_2 - L_1 = \mathbf{x}(\boldsymbol{\mu}_2 - \boldsymbol{\mu}_1)^T + \text{const}$$

This is an affine function of the linear discriminant, which has coefficient $(\boldsymbol{\mu}_2 - \boldsymbol{\mu}_1)^T$ rescaled to unit length.

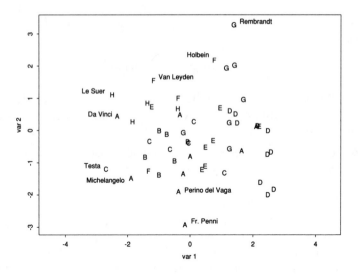

Figure 12.10: The first two linear discriminants for the data of de Piles on painters.

Discrimination between schools of painters

The eighteenth century art critic, de Piles, rated 54 painters on a score of 0–20 on each of composition, drawing, colour and expression. The painters belonged to 8 schools, with class sizes varying from 4 to 10. The data come from Weekes (1986) and are given in our library as data frame `painters`. (They are also considered by Davenport & Studdert-Kennesy, 1972 and Jolliffe, 1986.) We can plot them on the first two canonical variates as shown in Figure 12.10. As the confusion matrices (see page 322) show, we do lose by using just two dimensions. The component `svd` returned by `lda` suggests that all four dimensions are appropriate. The S code used was:

```
attach(painters)
confusion <- function(true, predict)
{
    jt <- table(true, predict)
    jn <- dimnames(jt)[[2]]
    jt1 <- jt[jn, ] # allow for missing groups in predictions
    structure(jt, error = (1 - sum(diag(jt1))/length(true)))
}
a <- lda(painters[,1:4], School)
a$svd
[1] 3.1838 1.9269 1.6497 0.7661
confusion(School, predict(a, painters[,1:4])$class)
  A B C D E F G H
A 5 0 1 2 0 0 0 2
B 4 1 1 0 0 0 0 0
C 0 2 2 0 2 0 0 0
D 0 0 0 9 0 0 1 0
```

```
E 0 0 0 1 4 0 1 1
F 1 0 0 0 0 2 1 0
G 1 0 0 1 1 0 4 0
H 0 0 1 0 0 0 0 3
attr(, "error"):
[1] 0.444444
confusion(School, predict(a, painters[,1:4], dimen=2)$class)
  A B D E F G H
A 6 0 2 0 0 0 2
B 6 0 0 0 0 0 0
C 3 1 1 1 0 0 0
D 0 0 8 0 0 2 0
E 2 0 1 2 0 1 1
F 1 0 0 0 1 1 1
G 1 0 1 1 0 4 0
H 0 0 0 0 1 0 3
attr(, "error"):
[1] 0.555556
a.x <- predict(a, painters[,1:4], dimen=2)$x
eqscplot(a.x, type="n", xlab="var 1", ylab="var 2")
text(a.x, as.character(School))
identify(a.x, row.names(painters))
```

The error rates appear poor, but there are many groups and few data, and this level
of performance is common in problems of this type.

Canonical correlations

Suppose the variables in our data matrix can be split into two sets, and we wish to
relate the sets. For definiteness let X be the matrix for the first set of variables,
and Y that for the second set. Canonical correlation analysis aims to find linear
combinations of each set with maximal correlation. Since we are dealing with
correlations, we do not need to impose a fixed scaling on the linear combinations.

Consider first the population version of the problem. The combined matrix
$[X\,Y]$ has covariance matrix

$$\Sigma = \begin{bmatrix} \Sigma_{xx} & \Sigma_{xy} \\ \Sigma_{yx} & \Sigma_{yy} \end{bmatrix}$$

and the linear combinations $\mathbf{xa}$ and $\mathbf{yb}$ (remember our cases are *row vectors*)
have correlation

$$\text{corr}\,(\mathbf{xa}, \mathbf{yb}) = \frac{\mathbf{a}^T \Sigma_{xy} \mathbf{b}}{\sqrt{(\mathbf{a}^T \Sigma_{xx} \mathbf{a}\,\mathbf{b}^T \Sigma_{yy} \mathbf{b})}}$$

Now rescale the variables to $\mathbf{x}S$ and $\mathbf{y}T$ so that they have identity covariance
matrix within each set (for example by taking the principal components, which
are uncorrelated, and rescaling each to unit variance). On the re-scaled variables

the problem is to maximize $\mathbf{a}^T \Sigma_{xy} \mathbf{b}$ for $\|\mathbf{a}\| = \|\mathbf{b}\| = 1$. Let $U \Lambda V^T$ be the singular-value decomposition of Σ_{xy}. Then

$$\mathbf{a}^T \Sigma_{xy} \mathbf{b} = (U^T \mathbf{a})^T \Lambda (V^T \mathbf{b})$$

which is maximized by taking $\mathbf{a}$ and $\mathbf{b}$ to be the columns of U and V corresponding to the largest singular value. Following a now familiar pattern we can define further pairs corresponding to the smaller singular values. The correlations which are achieved are equal to the singular values. Note that the problem is unchanged by a change of scale on either $\mathbf{a}$ or $\mathbf{b}$, and by a change of sign of both.

The sample version is obtained by replacing the population covariance matrix by its sample estimate. (Here the choice of divisor is irrelevant.) The S function `cancor` follows precisely the algorithm sketched here, and returns a vector of correlations and matrices of the linear combinations (to be of unit centred length). There are optional parameters for the centring of the cross-covariance matrix.

The dataset `swiss.x` has 5 columns of percentages. We divide these into social and educational sets:

```
> cancor(swiss.x[,c(1,4,5)]/100, swiss.x[,c(2,3)]/100)
$cor:
[1] 0.7648296 0.5434899

$xcoef:
           [,1]       [,2]        [,3]
[1,] -0.49049 -0.51314 -0.096209
[2,] -0.14257  0.36880 -0.006300
[3,] -0.63637 -1.37960  4.970178

$ycoef:
           [,1]     [,2]
[1,] 1.766816 -1.8833
[2,] 0.094468  2.1404

$xcenter:
 Agriculture Catholic Infant Mortality
      0.5066  0.41145          0.19943

$ycenter:
 Examination Education
     0.16489   0.10979
```

which shows that the first canonical variate is principally on the army draft examination, and the second contrasts the two educational measurements. For neither is the `Catholic` variable particularly important.

There is a sense in which canonical correlations generalize linear discriminant analysis, for if we perform a canonical correlation analysis on X and G, the matrix of group indicators, we obtain the linear discriminants.

Correspondence analysis

Correspondence analysis is applied to two-way tables of counts, and can be seen as a special case of canonical correlation analysis. Suppose we have an $r \times c$ table N of counts. For example, consider Fisher's (1940) example on colours of eyes and hair of people in Caithness, Scotland:

	fair	red	medium	dark	black
blue	326	38	241	110	3
light	688	116	584	188	4
medium	343	84	909	412	25
dark	98	48	403	681	85

Correspondence analysis seeks 'scores' f and g for the rows and columns which are maximally correlated. Clearly the maximum is one, attained by constant scores, so we seek the largest non-trivial solution. Let X and Y be matrices of the group indicators of the rows and columns respectively. If we ignore means in the variances and correlations, we need to rescale X by $D_r = \sqrt{\mathrm{diag}(n_i./n)}$, and Y by $D_c = \sqrt{\mathrm{diag}(n_{.j}/n)}$, and then take the singular value decomposition $D_r^{-1/2}(X^T Y/n)D_c^{-1/2} = U\Lambda V^T$, so the canonical combinations are the columns of $D_r^{-1/2}U$ and $D_c^{-1/2}V$. Since $X^T Y = N$, we can work directly on the table:

```
corresp <- function(table)
{
    if(!is.matrix(table)) stop("Not a table")
    if(sum(table%%1) > 1e-4)
        warning("non-integer entries in table")
    Dr <- apply(table, 1, sum)/sum(table)
    Dc <- apply(table, 2, sum)/sum(table)
    if(any(Dr==0) || any(Dc==0))
        stop("empty row or column in table")
    Dr <- 1/sqrt(Dr)
    Dc <- 1/sqrt(Dc)
    X <- diag(Dr) %*% (table/sum(table)) %*% diag(Dc)
    dimnames(X) <- dimnames(table)
    X.svd <- svd(X)
    list(cor = X.svd$d[2], rscore = X.svd$u[,2]*Dr,
        cscore = X.svd$v[,2]*Dc)
}
corresp(read.table("Fisher.dat"))
$cor:
[1] 0.447
$rscore:
    blue  light medium dark
 -0.895 -0.986  0.072 1.58
$cscore:
   fair    red medium dark black
 -1.22 -0.522 -0.095 1.32  2.47
```

Note that we take the second set of combinations, as the first will be constant.

Classification

Classification differs from discrimination in being entirely concerned with allocating future cases to one of g classes. We have already seen how this may be done for normal populations. In the terminology of pattern recognition the cases in X with their classifications are known as the *training set*, and future cases form the *test set*. Our primary measure of success is the error (or misclassification) rate, as this is the quantity that the Bayes rule minimizes. Note that we would obtain seriously biased estimates by re-classifying the training set, but that the error rate on a test set randomly chosen from the whole population will be an unbiased estimator.

It may be helpful to know the type of errors made. A *confusion matrix* gives the number of cases with true class i classified as j; this was illustrated for `painters` above.

There are a number of non-parametric classifiers based on non-parametric estimates of the class densities or of the log posterior. None are implemented in S-PLUS, but BDR's library `classif` available from `statlib` implements the k-nearest neighbour classifier and related methods (Devijver & Kittler, 1982) and learning vector quantization (Kohonen, 1990).

We give an example of classification in the next section.

12.4   An example: *Leptograpsus variegatus* crabs

Mahon (Campbell & Mahon, 1974) recorded data on 200 specimens of *Leptograpsus variegatus* crabs on the shore in Western Australia. This occurs in two colour forms, blue and orange, and he collected 50 of each form of each sex and made five physical measurements. These were the carapace (shell) length CL and width CW, the size of the frontal lobe FL and rear width RW, and the body depth BD. Part of the authors' thesis was to establish that the two colour forms were clearly differentiated morphologically, to support classification as two separate species.

We will consider two (fairly realistic) questions:

1. Is there evidence from these morphological data alone of a division into two forms?

2. Can we construct a rule to predict the sex of a future crab of unknown colour form (species). How accurate do we expect the rule to be?

On the second question, the body depth was measured somewhat differently for the females, so should be excluded from the analysis.

The data are physical measurements, so a sound initial strategy is to work on log scale. This has been done throughout. The data are very highly correlated, and pairs and brush plots are none too revealing (try them for yourself).

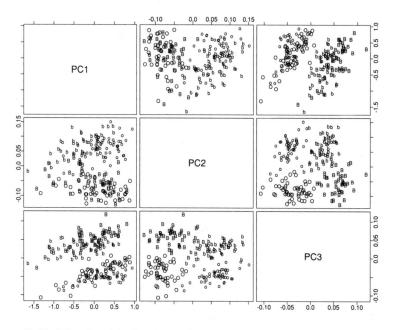

Figure 12.11: Pairs plot of the first three principal components of the `crabs` data. Males are coded as capitals, females as lower case, colours as the initial letter of blue or orange.

```
lcrabs <- log(crabs[,4:8])
crabs.grp <- c("B", "b", "O", "o")[rep(1:4, rep(50,4))]
lcrabs.pr <- prcomp(scale(lcrabs, T, F))
lcrabs.prx <- lcrabs.pr$x
dimnames(lcrabs.prx) <- list(NULL, paste("PC", 1:5, sep=""))
pairs(lcrabs.prx[,1:3], panel=function(x,y,...)
   text(x, y, crabs.grp, cex=0.7))
round(prcomp(scale(lcrabs, T, F))$s, 3)
[1] 0.518 0.075 0.048 0.025 0.009
round(prcomp(scale(lcrabs, T, F))$rot,3)
        [,1]    [,2]    [,3]    [,4]    [,5]
[1,]  0.452  -0.157  -0.438  -0.752   0.114
[2,]  0.387   0.911  -0.098   0.089  -0.062
[3,]  0.453  -0.204   0.371  -0.020  -0.784
[4,]  0.440  -0.072   0.672  -0.021   0.591
[5,]  0.497  -0.315  -0.458   0.652   0.136
```

We started by looking at the principal components. (As the data on log scale *are* very comparable, we did not rescale the variables to unit variance.) The first principal component had by far the largest standard deviation (0.52), with coefficients

```
   FL     RW     CL     CW     BD
0.452  0.387  0.453  0.440  0.497
```

and so is a 'size' effect. A plot of the second and third principal components shows an almost total separation into forms (Figure 12.11) on the third PC, the second PC distinguishing sex. As the coefficients of the third PC are

```
    FL      RW      CL      CW      BD
 -0.438  -0.098   0.371   0.672  -0.458
```

it is contrasting overall size with FL and BD.

To proceed further, for example to do a cluster analysis, we have to remove the dominant effect of size. We used the carapace area as a good measure of size, and divided all measurements by the square root of area. It is also necessary to account for the sex differences, which we can do by analysing each sex separately, or by subtracting the mean for each sex, which we did:

```
cr.scale <- 0.5 * log(crabs$CL * crabs$CW)
slcrabs <- lcrabs - cr.scale
cr.means <- matrix(0, 2, 5)
cr.means[1,] <- apply(slcrabs[crabs$sex=="F",], 2, mean)
cr.means[2,] <- apply(slcrabs[crabs$sex=="M",], 2, mean)
dslcrabs <- slcrabs - cr.means[codes(crabs$sex),]
lcrabs.sam <- sammon(dist(dslcrabs))
eqscplot(lcrabs.sam$points, type="n", xlab="", ylab="")
text(lcrabs.sam$points, crabs.grp)
```

As the Sammon mapping shows (Figure 12.12), Euclidean distance with this set of variables will be sensible. For this set of data, complete-link clustering makes three errors, and K-means and model-based clustering (with the default criterion S*) one or zero errors, depending on the starting point. Note that model-based hierarchical clustering is disastrous; it often does not work well in finding a small number of groups.

```
crabs.h <- cutree(hclust(dist(dslcrabs)),2)
table(crabs$sp, crabs.h)
      1   2
B  100   0
O    3  97
cr.means[1,] <- apply(dslcrabs[crabs.h==1,], 2, mean)
cr.means[2,] <- apply(dslcrabs[crabs.h==2,], 2, mean)
crabs.km <- kmeans(dslcrabs, cr.means)
table(crabs$sp, crabs.km$cluster)
     1    2
B   99    1
O    0  100
eqscplot(lcrabs.sam$points, type="n", xlab="", ylab="")
text(lcrabs.sam$points, crabs.km$cluster)
table(crabs$sp, mreloc(crabs.h, dslcrabs))
      1   2
B  100   0
O    0  100
crabs.mh <- mclass(mclust(dslcrabs), 2)$class
```

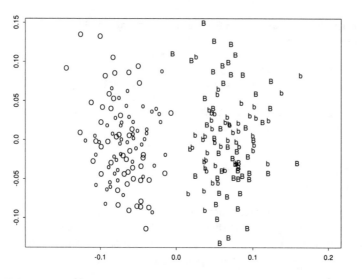

Figure 12.12: Sammon mapping of `crabs` data adjusted for size and sex. Males are coded as capitals, females as lower case, colours as the initial letter of blue or orange.

```
table(crabs$sp, crabs.mh)
     1 2
B   98 2
O  100 0
```

Discriminant analysis for sex

We noted that BD is measured differently for males and females, so it seemed prudent to omit it from the analysis. To start with, we ignore the differences between the forms. Linear discriminant analysis, for what are highly non-normal populations, finds a variable which is essentially $CL^3RW^{-2}CW^{-1}$, a dimensionally neutral quantity. Six errors are made, all for the blue form:

```
dcrabs <- lcrabs[,1:4]
dcrabs.lda <- lda(dcrabs, crabs$sex)
drop(dcrabs.lda$scaling)
[1]  -2.8896 -25.5176  36.3169 -11.8280
dcrabs.pred <- predict(dcrabs.lda, dcrabs)
table(crabs$sex, dcrabs.pred$class)
    F  M
F  97  3
M   3 97
1 - mean(pmax(dcrabs.pred$post[,"F"],
              dcrabs.pred$post[,"M"]))
[1] 0.0265

# 10-fold cross-validation
```

```
rand <- sample(10, dim(crabs)[1], replace = T)
prsum <- 0
cnt <- 0
for (i in unique(rand)) {
    t.lda <- lda(dcrabs[rand != i,], crabs$sex[rand != i])
    t.pr <- predict(t.lda, dcrabs[rand == i,])
    cnt <- cnt + sum(crabs$sex[rand == i] != t.pr$class)
    prsum <- prsum + sum(
                pmax(t.pr$post[,"F"], t.pr$post[,"M"]))
}
c(cnt/200, 1-prsum/200)
[1] 0.035 0.0275
```

The second measure of error rate given is formed by averaging one minus the posterior probability assigned to the selected class. This is a smoother quantity than the error rate count, unbiased if the model is correct but less variable (McLachlan, 1992, §10.5.2). Consider a random pair $(\mathbf{X}, C)$ from the whole population. Then

$$P(\text{correct} \mid \mathbf{X} = \mathbf{x}) = P\big(C = \arg\max_c p(c \mid \mathbf{x}) \mid \mathbf{X} = \mathbf{x}\big) = \max_c p(c \mid \mathbf{x})$$

and so

$$P(\text{correct}) = E\big[\max_c p(c \mid \mathbf{X})\big]$$

Both $I\big[C = \arg\max_c p(c \mid \mathbf{x})\big]$ and $\max_c p(c \mid \mathbf{x})$ are conditionally unbiased estimators of $P(\text{correct} \mid \mathbf{X} = \mathbf{x})$, but the second averages over $P(C \mid \mathbf{X} = \mathbf{x})$ and so has a smaller variance. Another advantage of the smoothed form is that it does not depend on the correctness of the supplied classifications.

It does make sense to take the colour forms into account, especially as the within-group distributions look close to joint normality (look at the data on the linear discriminants). The first two linear discriminants dominate the between-group variation. Figure 12.13 shows the data on those variables.

```
dcrabs.lda4 <- lda(dcrabs, crabs.grp)
dcrabs.lda4$svd
[1] 22.3714 16.4939  2.6048
dcrabs.pr4 <- predict(dcrabs.lda4, dcrabs, dimen=2)
dcrabs.pr2 <- dcrabs.pr4$post %*% c(1,1,0,0)
table(crabs$sex, dcrabs.pr2 > 0.5)
   FALSE TRUE
F     96    4
M      3   97
1 - mean(pmax(1-dcrabs.pr2, dcrabs.pr2))
[1] 0.0269

# repeat 10-fold CV with the same splits
prsum <- 0
cnt <- 0
for (i in unique(rand)) {
    t.lda4 <- lda(dcrabs[rand != i,], crabs.grp[rand != i])
```

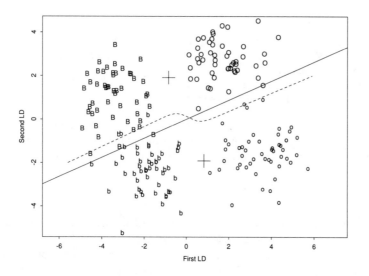

Figure 12.13: Linear discriminants for the `crabs` data. Males are coded as capitals, females as lower case, colours as the initial letter of blue or orange. The crosses are the group means for a linear discriminant for sex (solid line) and the the dashed line is the decision boundary for sex based on four groups.

```
    cl4 <- predict(t.lda4, dcrabs[rand == i,], dimen=2)
    cl2 <- cl4$post %*% c(1,1,0,0)
    cnt <- cnt + sum(codes(crabs$sex[rand == i]) == (cl2 > 0.5))
    prsum <- prsum + sum(pmax(1 - cl2, cl2))
}
c(cnt/200, 1-prsum/200)
[1] 0.02 0.0282
```

We can not represent all the decision surfaces exactly on a plot. However, using the first two linear discriminants as the data will provide a very good approximation; see Figure 12.13.

```
cr.t <- dcrabs.pr4$x[,1:2]
eqscplot(cr.t, type="n", xlab="First LD", ylab="Second LD")
text(cr.t, crabs.grp)
perp <- function(x, y) {
    m <- (x+y)/2
    s <- - (x[1] - y[1])/(x[2] - y[2])
    abline(c(m[2] - s*m[1], s))
}
cr.m <- lda(cr.t, crabs$sex)$means
points(cr.m, pch=3, mkh=0.3)
perp(cr.m[1,], cr.m[2,])

cr.lda <- lda(cr.t, crabs.grp)
x <- seq(-6, 6, 0.25)
```

```
y <- seq(-2, 2, 0.25)
Xcon <- matrix(c(rep(x,length(y)),
                rep(y, rep(length(x),length(y)))),,2)
cr.pr <- predict(cr.lda, Xcon)$post %*% c(1,1,0,0)
contour(x, y, matrix(cr.pr, length(x), length(y)),
   levels=0.5, labex=0, add=T, lty=3)
```

The reader is invited to try quadratic discrimination on this problem. It performs very marginally better than linear discrimination, not surprisingly since the covariances of the groups appear so similar, as can be seen from the result of

```
var(dcrabs[crabs.grp=="0", ])
   . . . .
```

This example is also considered in Section 10.4. The approach taken there is to model the logistic transform of $p(\text{male} \mid \mathbf{x})$ directly, either as a linear function or via a neural network. The cross-validated results are rather better for that approach (1.6% and 1.8%), which make fewer assumptions about the populations.

Chapter 13

Tree-based Methods

The use of tree-based models will be relatively unfamiliar to statisticians, although researchers in other fields have found trees to be an attractive way to express knowledge and aid decision-making. Keys such as Figure 13.1 are common in botany and in medical decision-making, and provide a way to encapsulate and structure the knowledge of experts to be used by less-experienced users. Notice how this tree uses both categorical variables and splits on continuous variables.

The automatic construction of decision trees dates from work in the social sciences by Morgan & Sonquist (1963) and Morgan & Messenger (1973). In statistics Breiman *et al.* (1984) had a seminal influence both in bringing the work to the attention of statisticians and in proposing new algorithms for constructing trees. At around the same time decision tree induction was beginning to be used in the field of *machine learning*, notably by Quinlan (1979, 1983, 1986, 1993). Whereas there is now an extensive literature in machine learning, further statistical contributions are still sparse. The introduction within S of tree-based models described by Clark & Pregibon (1992) has made the methods much more freely available. Their methods are very much in the spirit of exploratory data analysis, with many functions to investigate trees. (On the other hand, it is not possible to enter trees such as Figure 13.1 without cracking the internal structure. It is a tree, and readers are encouraged to draw it.)

Constructing trees may be seen as a type of variable selection. Questions of interaction between variables are handled automatically, and to a large extent so is monotonic transformation of both the x and y variables. These issues are reduced to which variables to divide on, and how to achieve the split.

Figure 13.1 is a *classification* tree since its endpoint is a factor giving the species. Although this the most common use, it is also possible to have *regression* trees in which each terminal node gives a predicted value, as shown in Figure 13.2 for our dataset cpus.

Much of the machine learning literature is concerned with logical variables and correct decisions. The end point of a tree is a (labelled) partition of the space $\mathcal{X}$ of possible observations. In logical problems it is assumed that there *is* a partition of the space $\mathcal{X}$ which will correctly classify all observations, and the task is to find a tree to describe it succinctly. A famous example of Donald Michie (e.g. Michie, 1989) is whether the space shuttle pilot should use the autolander

1.	Leaves subterete to slightly flattened, plant with bulb	2.
	Leaves flat, plant with rhizome	4.
2.	Perianth-tube > 10 mm	**I. × hollandica**
	Perianth-tube < 10 mm	3.
3.	Leaves evergreen	**I. xiphium**
	Leaves dying in winter	**I. latifolia**
4.	Outer tepals bearded	**I. germanica**
	Outer tepals not bearded	5.
5.	Tepals predominately yellow	6.
	Tepals blue, purple, mauve or violet	8.
6.	Leaves evergreen	**I. foetidissima**
	Leaves dying in winter	7.
7.	Inner tepals white	**I. orientalis**
	Tepals yellow all over	**I. pseudocorus**
8.	Leaves evergreen	**I. foetidissima**
	Leaves dying in winter	9.
9.	Stems hollow, perianth-tube 4–7mm	**I. sibirica**
	Stems solid, perianth-tube 7–20mm	10.
10.	Upper part of ovary sterile	11.
	Ovary without sterile apical part	12.
11.	Capsule beak 5–8mm, 1 rib	**I. enstata**
	Capsule beak 8–16mm, 2 ridges	**I. spuria**
12.	Outer tepals glabrous, many seeds	**I. versicolor**
	Outer tepals pubescent, 0–few seeds	**I. × robusta**

Figure 13.1: Key to British species of the genus *Iris*. Simplified from Stace (1991, p. 1140), by omitting parts of his descriptions.

or land manually (Table 13.1). Some enumeration will show that the decision has been specified for 253 out of the 256 possible observations. Some cases have been specified twice. This body of expert opinion needed to be reduced to a simple decision aid, as shown in Figure 13.3. (Table 13.1 appears to result from a decision tree which differs from Figure 13.3 in reversing the order of two pairs of splits.)

Note that the botanical problem is treated as if it were a logical problem, although there will be occasional specimens which do not meet the specification for their species.

13.1 Partitioning methods

The ideas for classification and regression trees are quite similar, but the terminology differs, so we will consider classification first. Classification trees are more familiar and it is a little easier to justify the tree-construction procedure, so we consider them first.

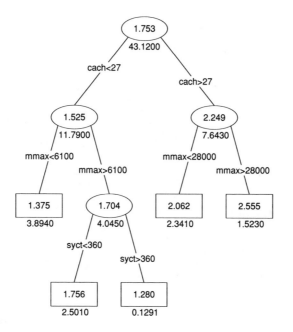

Figure 13.2: A regression tree for the cpu performance data on $\log_{10}$ scale. The value in each node is the prediction for the node, those underneath the nodes indicate the deviance contributions D_i.

Table 13.1: Example decisions for the space shuttle autolander problem.

stability	error	sign	wind	magnitude	visibility	decision
any	any	any	any	any	no	auto
xstab	any	any	any	any	yes	noauto
stab	LX	any	any	any	yes	noauto
stab	XL	any	any	any	yes	noauto
stab	MM	nn	tail	any	yes	noauto
any	any	any	any	Out of range	yes	noauto
stab	SS	any	any	Light	yes	auto
stab	SS	any	any	Medium	yes	auto
stab	SS	any	any	Strong	yes	auto
stab	MM	pp	head	Light	yes	auto
stab	MM	pp	head	Medium	yes	auto
stab	MM	pp	tail	Light	yes	auto
stab	MM	pp	tail	Medium	yes	auto
stab	MM	pp	head	Strong	yes	noauto
stab	MM	pp	tail	Strong	yes	auto

Classification trees

We have already noted that the end-point for a tree is a partition of the space $\mathcal{X}$, and we compare trees by how well that partition corresponds to the correct

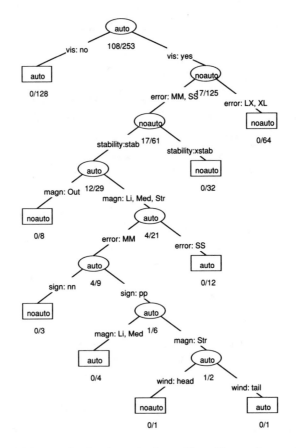

Figure 13.3: Decision tree for shuttle autolander problem. The numbers m/n denote the proportion of training cases reaching that node with the classification in the label.

decision rule for the problem. In logical problems the easiest way to compare partitions is to count the number of errors, or, if we have a prior over the space $\mathcal{X}$, to compute the probability of error.

In statistical problems the distributions of the classes over $\mathcal{X}$ usually overlap, so there is no partition which completely describes the classes. Then for each cell of the partition there will be a probability distribution over the classes, and the Bayes decision rule will choose the class with highest probability. This corresponds to assessing partitions by the overall probability of misclassification. Of course, in practice we do not have the whole probability structure, but a training set of n classified examples which we assume are an independent random sample. Then we can estimate the misclassification rate by the proportion of the training set which is misclassified.

Almost all current tree-construction methods, including those in S, use a one-step lookahead. That is, they choose the next split in an optimal way, without attempting to optimize the performance of the whole tree. (This avoids a combinatorial explosion over future choices, and is akin to a very simple strategy for playing a game such as chess.) However, by choosing the right measure to optimize at each split, we can ease future splits. It does not seem appropriate to use the misclassification rate to choose the splits.

What class of splits should we allow? Both Breiman *et al.*'s CART methodology and the S methods only allow binary splits, which avoids one difficulty in comparing splits, that of normalization by size. For a continuous variable x_j the allowed splits are of the form $x_j < t$ versus $x_j \geq t$. For ordered factors the splits are of the same type. For general factors the levels are divided into two classes. (Note that for L levels there are 2^L possible splits, and if disallow the empty split and ignore the order, there are still $2^{L-1} - 1$. For ordered factors there are only $L - 1$ possible splits.) Some algorithms, including CART but excluding S, allow linear combination of continuous variables to be split, and Boolean combinations to be formed of binary variables.

The justification for the S methodology is to view the tree as providing a probability model (hence the title 'tree-based models' of Clark & Pregibon, 1992). At each node i of a classification tree we have a probability distribution p_{ik} over the classes. The partition is given by the *leaves* of the tree (also known as terminal nodes). Each case in the training set is assigned to a leaf, and so at each leaf we have a random sample n_{ik} from the multinomial distribution specified by p_{ik}.

We now condition on the observed variables $\mathbf{x}_i$ in the training set, and hence we know the numbers n_i of cases assigned to every node of the tree, in particular to the leaves. The conditional likelihood is then proportional to

$$\prod_{\text{cases } j} p_{[j]y_j} = \prod_{\text{leaves } i} \prod_{\text{classes } k} p_{ik}^{n_{ik}}$$

where $[j]$ denotes the leaf assigned to case j. This allow us to define a deviance for the tree as

$$D = \sum_i D_i, \qquad D_i = -2 \sum_k n_{ik} \log p_{ik}$$

as a sum over leaves.

Now consider splitting node s into nodes t and u. This changes the probability model within node s, so the reduction in deviance for the tree is

$$D_s - D_t - D_u = 2 \sum_k \left[n_{tk} \log \frac{p_{tk}}{p_{sk}} + n_{uk} \log \frac{p_{uk}}{p_{sk}} \right]$$

Since we do not know the probabilities, we estimate them from the proportions in the split node, obtaining

$$\hat{p}_{tk} = \frac{n_{tk}}{n_t}, \qquad \hat{p}_{uk} = \frac{n_{uk}}{n_u}, \qquad \hat{p}_{sk} = \frac{n_t \hat{p}_{tk} + n_u \hat{p}_{uk}}{n_s} = \frac{n_{sk}}{n_s}$$

so the reduction in deviance is

$$D_s - D_t - D_u = 2 \sum_k \left[n_{tk} \log \frac{n_{tk} n_s}{n_{sk} n_t} + n_{uk} \log \frac{n_{uk} n_s}{n_{sk} n_u} \right]$$

$$= 2\Big[\sum_k n_{tk} \log n_{tk} + n_{uk} \log n_{uk} - n_{sk} \log n_{sk}$$

$$+ n_s \log n_s - n_t \log n_t - n_u \log n_u\Big]$$

This gives a measure of the value of a split. Note that it is size-biased; there is more value in splitting leaves with large numbers of cases.

The tree construction process takes the maximum reduction in deviance over all allowed splits of all leaves, to choose the next split (Note that for continuous variates the value depends only on the split of the ranks of the observed values, so we may take a finite set of splits.) The tree construction continues until the number of cases reaching each leaf is small (by default $n_i < 10$ in S) or the leaf is homogeneous enough (by default its deviance is less than 1% of the deviance of the root node in S, which is a size-biased measure). Note that as all leaves not meeting the stopping criterion will eventually be split, an alternative view is to consider splitting any leaf and choose the best allowed split (if any) for that leaf, proceeding until no further splits are allowable.

This justification for the value of a split follows Ciampi *et al.* (1987) and Clark & Pregibon, but differs from most of the literature on tree construction. The more common approach is to define a measure of the impurity of the distribution at a node, and choose the split which most reduces the average impurity. Two common measures are the entropy $\sum p_{ik} \log p_{ik}$ and the Gini index

$$\sum_{j \neq k} p_{ij} p_{ik} = 1 - \sum_k p_{ik}^2$$

As the probabilities are unknown, they are estimated from the node proportions. With the entropy measure, the average impurity differs from D by a constant factor, so the tree construction process is the same, except perhaps for the stopping rule. Breiman *et al.* preferred the Gini index.

Regression trees

The prediction for a regression tree is constant over each cell of the partition of $\mathcal{X}$ induced by the leaves of the tree. The deviance is defined as

$$D = \sum_{\text{cases } j} (y_j - \mu_{[j]})^2$$

and so clearly we should estimate the constant μ_i for leaf i by the mean of the values of the training-set cases assigned to that node. Then the deviance is the sum over leaves of D_i, the corrected sum of squares for cases within that node, and the value of a split is the reduction in the residual sum of squares.

The obvious probability model (and that proposed by Clark & Pregibon) is to take a normal $N(\mu_i, \sigma^2)$ distribution within each leaf. Then D is the usual scaled deviance for a Gaussian GLM. However, the distribution at internal nodes of the tree is then a mixture of normal distributions, and so D_i is only appropriate at the leaves. The tree-construction process has to be seen as a hierarchical refinement of probability models, very similar to forwards variable selection in regression. In contrast, for a classification tree, one probability model can be used throughout the tree construction process.

Missing values

One attraction of tree-based methods is the ease with which missing values can be handled. Consider the botanical key of Figure 13.1. We only need to know about a small subset of the 10 observations to classify any case, and part of the art of constructing such trees is to avoid observations which will be difficult or missing in some of the species (or as in capsules, for some of the cases). A general strategy is to 'drop' a case down the tree as far as it will go. If it reaches a leaf we can predict y for it. Otherwise we use the distribution at the node reached to predict y, as shown in Figure 13.2, which has predictions at all nodes.

An alternative strategy (not implemented in S) is used by many botanical keys and can be seen at nodes 9 and 12 of Figure 13.1. A list of characteristics is given, the most important first, and a decision made from those observations which are available. This is codified in the method of *surrogate splits* in which surrogate rules are available at non-terminal nodes to be used if the splitting variable is unobserved.

Tree construction is based on the cases without any missing observations.

S implementation

The S implementation is based on a class `tree` which has a quite complex internal structure. The tree is a regression tree unless the response variable in the model formula is a factor, and the internal structure differs in the two cases. Beware: it is very easy to code the classes numerically and then fit a regression tree by mistake.

The function `tree` constructs trees, and there are `print`, `summary` and `plot` methods for trees. As our first example consider the computer performance data in our data frame `cpus`. Ein-Dor & Feldmesser (1987) studied data on the performance on a benchmark mix of minicomputers and mainframes. The measure was normalized relative to an IBM 370/158–3. There were six machine characteristics, the cycle time (nanoseconds), the cache size (Kb), the main memory size (Kb) and number of channels. (For the latter two there are minimum and maximum possible values; what the actual machine tested had is unspecified.) The original paper set up a linear regression for the square root of performance.

We first consider a tree for the performance and then for log-performance.

```
> cpus.tr <- tree(perf ~ syct+mmin+mmax+cach+chmin+chmax, cpus)
> summary(cpus.tr)
Regression tree:
tree(formula = perf ~ syct+mmin+mmax+cach+chmin+chmax,
    data = cpus)
Variables actually used in tree construction:
```

```
[1] "mmax"  "cach"  "chmax" "mmin"
Number of terminal nodes:  9
Residual mean deviance:  4523 = 904600 / 200
    ....
> print(cpus.tr)
node), split, n, deviance, yval
      * denotes terminal node

  1) root 209 5380000 105.60
    2) mmax<28000 182  585900  60.72
      4) cach<27 141   97850  39.64
        8) mmax<10000 113   36000  32.21 *
        9) mmax>10000 28   30470  69.61 *
      5) cach>27 41  209900 133.20
       10) cach<96.5 34   96490 114.40
         20) mmax<11240 14   12950  72.79 *
         21) mmax>11240 20   42240 143.60 *
       11) cach>96.5 7   43150 224.40 *
    3) mmax>28000 27 1954000 408.30
      6) chmax<59 22  436600 323.20
       12) mmin<12000 15  106500 244.50
         24) cach<56 9   26650 191.60 *
         25) cach>56 6   16700 324.00 *
       13) mmin>12000 7   38480 491.70 *
      7) chmax>59 5  658000 782.60 *
> plot(cpus.tr, type="u");  text(cpus.tr, srt=90)

> cpus.ltr <- tree(log10(perf) ~ syct + mmin + mmax + cach
    + chmin + chmax, cpus)
> summary(cpus.ltr)
Number of terminal nodes:  19
Residual mean deviance:  0.0239 = 4.55 / 190
    ....
> plot(cpus.ltr, type="u");  text(cpus.ltr, srt=90)
```

This output needs some explanation. The model is specified by a model formula with terms separated by +; interactions make no sense for trees. The first tree does not use all of the variables, so the summary informs us. (The second tree does use all 6.) Next we have a compact printout of the tree including the number, sum of squares and mean at the node. The plots are shown in Figure 13.4. The `type` parameter is used to reduce over-crowding of the labels; by default the depths reflect the values of the splits. Plotting uses no labels and allows the tree topology to be studied: split conditions and the values at the leaves are added by `text` (which has a number of other options). The elegant plots such as Figure 13.2 are produced directly in POSTSCRIPT by `post.tree`.[1]

For examples of classification trees we use the `iris` data and the `painters` data discussed in Chapter 12. For `iris` we have

---

[1] In our version of S-PLUS 3.2 the function `make.name` is missing, so argument `file` *must* be given.

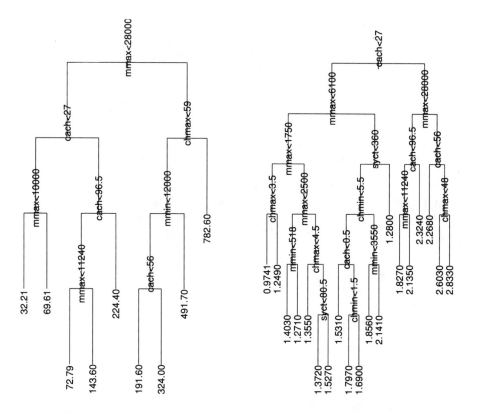

Figure 13.4: Regression trees for the cpu performance data on linear (left) and $\log_{10}$ scale (right).

```
> ir.species <- factor(c(rep("s",50), rep("c", 50),
    rep("v", 50)))
> ird <- data.frame(rbind(iris[,,1], iris[,,2], iris[,,3]))
> ir.tr <- tree(ir.species ~., ird)
> summary(ir.tr)

Classification tree:
tree(formula = ir.species ~ ir)
Variables actually used in tree construction:
[1] "Petal L." "Petal W." "Sepal L."
Number of terminal nodes:  6
Residual mean deviance:   0.125 = 18 / 144
Misclassification error rate: 0.0267 = 4 / 150

> ir.tr
node), split, n, deviance, yval, (yprob)
        * denotes terminal node

 1) root 150 330.0 c ( 0.330 0.33 0.330 )
```

```
   2) Petal.L.<2.45 50    0.0 s ( 0.000 1.00 0.000 ) *
   3) Petal.L.>2.45 100 140.0 c ( 0.500 0.00 0.500 )
      6) Petal.W.<1.75 54   33.0 c ( 0.910 0.00 0.093 )
        12) Petal.L.<4.95 48    9.7 c ( 0.980 0.00 0.021 )
           24) Sepal.L.<5.15 5    5.0 c ( 0.800 0.00 0.200 ) *
           25) Sepal.L.>5.15 43   0.0 c ( 1.000 0.00 0.000 ) *
        13) Petal.L.>4.95 6    7.6 v ( 0.330 0.00 0.670 ) *
      7) Petal.W.>1.75 46    9.6 v ( 0.022 0.00 0.980 )
       14) Petal.L.<4.95 6    5.4 v ( 0.170 0.00 0.830 ) *
       15) Petal.L.>4.95 40   0.0 v ( 0.000 0.00 1.000 ) *
```

The (yprob) give the distribution by class within the node. Note how the second split on Petal length occurs twice, and that splitting node 12 is attempting to classify one case of *I. virginica* without success. Thus viewed as a decision tree we would want to snip off nodes 24, 25, 14, 15. We can do so interactively:

```
> plot(ir.tr)
> text(ir.tr, all=T)
> ir.tr1 <- snip.tree(ir.tr)
node number:  12
   tree deviance =  18.05
   subtree deviance =  22.77
node number:  7
   tree deviance =  22.77
   subtree deviance =  26.99
> ir.tr1
node), split, n, deviance, yval, (yprob)
       * denotes terminal node

  1) root 150 330.0 c ( 0.330 0.33 0.330 )
    2) Petal.L.<2.45 50    0.0 s ( 0.000 1.00 0.000 ) *
    3) Petal.L.>2.45 100 140.0 c ( 0.500 0.00 0.500 )
      6) Petal.W.<1.75 54   33.0 c ( 0.910 0.00 0.093 )
        12) Petal.L.<4.95 48    9.7 c ( 0.980 0.00 0.021 ) *
        13) Petal.L.>4.95 6    7.6 v ( 0.330 0.00 0.670 ) *
      7) Petal.W.>1.75 46    9.6 v ( 0.022 0.00 0.980 ) *
> summary(ir.tr1)

Classification tree:
    . . . .
Number of terminal nodes:  4
Residual mean deviance:   0.185 = 27 / 146
Misclassification error rate: 0.0267 = 4 / 150

par(pty="s")
plot(ird[, 3],ird[, 4], type="n",
   xlab="petal length", ylab="petal width")
text(ird[, 3], ird[, 4], as.character(ir.species))
par(cex=2)
partition.tree(ir.tr1, add=T)
```

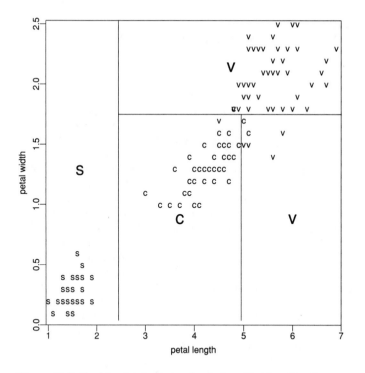

Figure 13.5: Partition for the iris data induced by the snipped tree.

where we clicked twice in succession with mouse button 1 on each of nodes 12 and 7 to remove their subtrees, then with button 2 to quit.

The decision region is now entirely in the petal length – petal width space, so we can show it by `partition.tree`. This example shows the limitations of the one-step-ahead tree construction, for Weiss & Kapouleas (1989) used a rule-induction program to find the set of rules

If Petal length < 3 then *I. setosa*.
If Petal length > 4.9 or Petal width > 1.6 then *I. virginica*.
Otherwise *I. versicolor*.

Compare this with Figure 13.5, which makes one more error.

Now for the `painters` data:

```
> painters.tr <- tree(School ~ . , painters)
> summary(painters.tr)

Classification tree:
tree(formula = School ~ ., data = painters)
Number of terminal nodes:  9
Residual mean deviance:  2.09 = 94.1 / 45
Misclassification error rate: 0.407 = 22 / 54
```

```
> painters.tr
node), split, n, deviance, yval, (yprob)
      * denotes terminal node

  1) root 54 220.0 A ( 0.19 0.11 0.11 0.19 0.13 0.074
                       0.13 0.074 )
    2) Colour<12.5 32 120.0 A ( 0.25 0.19 0.19 0.00 0.12
                               0.094 0.031 0.12 )
      4) Composition<9 6   8.3 A ( 0.50 0.00 0.00 0.00 0.00
                                 0.50 0.00 0.00 ) *
      5) Composition>9 26  88.0 B ( 0.19 0.23 0.23 0.00 0.15
                                 0.00 0.038 0.15 )
       10) Drawing<15.5 18  52.0 B ( 0.00 0.33 0.33 0.00 0.17
                                   0.00 0.056 0.11 )
         20) Colour<9.5 11  23.0 B ( 0.00 0.45 0.36 0.00 0.00
                                   0.00 0.00 0.18 )
           40) Expression<7 5   9.5 C ( 0.00 0.20 0.60 0.00 0.00
                                       0.00 0.00 0.20) *
           41) Expression>7 6  10.0 B ( 0.00 0.67 0.17 0.00 0.00
                                       0.00 0.00 0.17 ) *
         21) Colour>9.5 7  18.0 E ( 0.00 0.14 0.29 0.00 0.43
                                   0.00 0.14 0.00 ) *
       11) Drawing>15.5 8  14.0 A ( 0.62 0.00 0.00 0.00 0.12
                                   0.00 0.00 0.25 ) *
    3) Colour>12.5 22  59.0 D ( 0.091 0.00 0.00 0.45 0.14
                               0.045 0.27 0.00 )
      6) Expression<5.5 10  13.0 D ( 0.10 0.00 0.00 0.80 0.10
                                   0.00 0.00 0.00 )
       12) Composition<7 5   5.0 D ( 0.00 0.00 0.00 0.80 0.20
                                   0.00 0.00 0.00 ) *
       13) Composition>7 5   5.0 D ( 0.20 0.00 0.00 0.80 0.00
                                   0.00 0.00 0.00 ) *
      7) Expression>5.5 12  33.0 G ( 0.083 0.00 0.00 0.17 0.17
                                   0.083 0.50 0.00 )
       14) Composition<12.5 6  16.0 D ( 0.17 0.00 0.00 0.33 0.00
                                       0.17 0.33 0.00 ) *
       15) Composition>12.5 6   7.6 G ( 0.00 0.00 0.00 0.00 0.33
                                       0.00 0.67 0.00 ) *
> plot(painters.tr);  text(painters.tr, all=T)
> painters.tr1 <- snip.tree(painters.tr)
node number:  6
   tree deviance =  94.12
   subtree deviance =  96.89
> summary(painters.tr1)

Classification tree:
snip.tree(tree = painters.tr, nodes = 6)
Number of terminal nodes:  8
Residual mean deviance:  2.11 = 96.9 / 46
Misclassification error rate: 0.407 = 22 / 54
```

```
> tree.screens()
> plot(painters.tr1)
> tile.tree(painters.tr1, painters$School)
> close.screen(all=T)
```

The function `print.tree` does not respect the `width` option, so the output lines have had to be broken by hand. Once more snipping is required.

Fine control

Missing values are handled by the argument `na.action` of `tree`. The default is `na.fail`, to abort if missing values are found; an alternative is `na.omit` which omits all cases with a missing observation. The `weights` and `subset` arguments are available as in all model-fitting functions.

The `predict` method `predict.tree` can be used to predict future cases. This automatically handles missing values by `na.tree.replace`, which adds a new level 'NA' to factors and to continuous variables after first quantizing them, and drops cases down the tree until a NA is encountered or a leaf is reached.

There are three control arguments specified under `tree.control`, but which may be passed directly to `tree`. The `mindev` controls the threshold for splitting a node. The parameters `minsize` and `mincut` control the size thresholds; `minsize` is the threshold for a node size, so nodes of size `minsize` or larger are candidates for a split. Daughter nodes must exceed `mincut` for a split to be allowed. The defaults are `minsize` = 10 and `mincut` = 5.

To make a tree fit exactly, the help page recommends the use of `mindev` = 0 and `minsize` = 2. However, for the shuttle data this will split an already pure node. A small value such as `mindev` = 1e-6 is safer. For example, for the shuttle data:

```
> shuttle.tr <- tree(use ~ ., shuttle, subset=1:253,
                      mindev=1e-6, minsize=2)
> shuttle.tr
node), split, n, deviance, yval, (yprob)
      * denotes terminal node

  1) root 253 350.0  auto ( 0.57 0.43 )
    2) vis: no 128    0.0  auto ( 1.00 0.00 ) *
    3) vis: yes 125  99.0  noauto ( 0.14 0.86 )
      6) error: MM, SS 61  72.0  noauto ( 0.28 0.72 )
       12) stability:stab 29  39.0  auto ( 0.59 0.41 )
         24) magn: Out 8    0.0  noauto ( 0.00 1.00 ) *
         25) magn: Light, Med, Str 21  20.0  auto ( 0.81 0.19 )
           50) error: MM 9  12.0  auto ( 0.56 0.44 )
            100) sign: nn 3    0.0  noauto ( 0.00 1.00 ) *
            101) sign: pp 6   5.4  auto ( 0.83 0.17 )
              202) magn: Light, Med 4  0.0  auto ( 1.00 0.00 ) *
              203) magn: Strong 2   2.8  auto ( 0.50 0.50 )
                406) wind: head 1  0.0  noauto ( 0.00 1.00 ) *
                407) wind: tail 1  0.0  auto ( 1.00 0.00 ) *
```

```
        51) error: SS 12    0.0  auto ( 1.00 0.00 ) *
      13) stability:xstab 32    0.0  noauto ( 0.00 1.00 ) *
      7) error: LX, XL 64    0.0  noauto ( 0.00 1.00 ) *
> post.tree(shuttle.tr)
> shuttle1 <- shuttle[254:256, ]  # 3 missing cases
  stability error sign  wind     magn  vis
1       stab    MM   nn  head   Light  yes
2       stab    MM   nn  head  Medium  yes
3       stab    MM   nn  head  Strong  yes
> predict(shuttle.tr, shuttle1)
   auto  noauto
1     0       1
2     0       1
3     0       1
```

13.2 Cutting trees down to size

With 'noisy' data, that is when the distributions for the classes overlap, it is quite possible to grow a tree which fits the training set well, but which has adapted too well to features of that subset of $\mathcal{X}$. Similarly, regression trees can be too elaborate and over-fit the training data. We need an analogue of variable selection in regression.

The established methodology is tree *pruning*, first introduced by Breiman *et al.* (1984). They considered rooted subtrees of the tree $\mathcal{T}$ grown by the construction algorithm, that is the possible results of snipping off terminal subtrees on $\mathcal{T}$. The pruning process chooses one of the rooted subtrees. Let R_i be a measure evaluated at the leaves, such as the deviance or the number of errors, and let R be the value for the tree, the sum over the leaves of R_i. Let the size of the tree be the number of leaves. Then Breiman *et al.* showed that the set of rooted subtrees of $\mathcal{T}$ which minimize the cost-complexity measure

$$R_\alpha = R + \alpha \, \text{size}$$

is itself nested. That is, as we increase α we can find the optimal trees by a sequence of snip operations on the current tree (just like pruning a real tree). This produces a sequence of trees from the size of $\mathcal{T}$ down to just the root node, but it may prune more than one node at a time.

Pruning is implemented in the function `prune.tree`, which can be asked for the trees for one or more values of α (its argument k) or for a tree of a particular size (its argument `best`). The default measure is the deviance, but for classification trees one can specify `method="misclass"` to use the error count. Unfortunately the version supplied in all current versions of S-PLUS has several errors (discovered after the first printing went to press). Our library MASS contains a replacement function, but *must* be attached by

```
library(MASS, first=T)
```

to obtain the corrected results given in this printing. If C is not available the corrected function runs considerably slower than the original.

Let us prune the cpus tree. Plotting the output of `prune.tree`, the deviance against size, allows us to choose a likely break point (Figure 13.6):

```
> plot(prune.tree(cpus.ltr))
> cpus.ltr1 <- prune.tree(cpus.ltr, best=8)
> plot(cpus.ltr1);    text(cpus.ltr1)
```

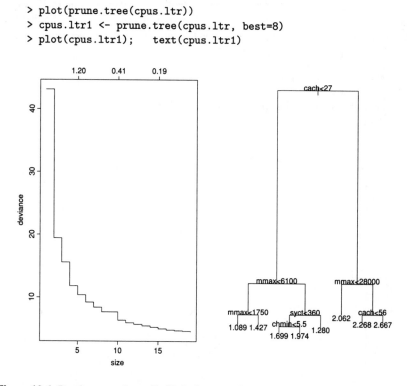

Figure 13.6: Pruning `cpu.ltr`. (Left) deviance *vs* size. (Right) the pruned tree of size 8.

One way to choose the parameter α is to consider pruning as a method of variable selection. Akaike's information criterion (AIC) penalizes minus twice log-likelihood by twice the number of parameters. For classification trees with K classes choosing $\alpha = 2(K - 1)$ will find the rooted subtree with minimum AIC. For regression trees one approximation (Mallows' C_p) to AIC is to replace the minus twice log-likelihood by the residual sum of squares divided by $\hat{\sigma}^2$ (our best estimate of σ^2), so we could take $\alpha = 2\hat{\sigma}^2$. One way to select $\hat{\sigma}^2$ would be from the fit of the full tree model. In our example this suggests $\alpha = 2 \times 0.02394$, which makes no change. Other criteria suggest that AIC and C_p tend to over-fit and choose larger constants in the range 2–6. (Note that counting parameters in this way does not take the selection of the splits into account.)

Cross-validation

We really need a better way to choose the degree of pruning. If a separate validation set is available, we can predict on that set, and compute the deviance versus α for the pruned trees. This will often have a minimum, and we can choose the smallest tree whose deviance is close to the minimum.

If no validation set is available we can make one by splitting the training set. Suppose we split the training set into 10 (roughly) equally sized parts. We can then use 9 to grow the tree and test it on the tenth. This can be done in 10 ways, and we can average the results. This is done by the function `cv.tree`. Note that as ten trees must be grown, the process can be slow, and that the averaging is done for fixed α and not fixed tree size.

```
cpus.cv <- cv.tree(cpus.ltr,, prune.tree)
plot(cpus.cv)
cpus.ltr3 <- prune.tree(cpus.ltr, best=4)
post.tree(cpus.ltr3)
```

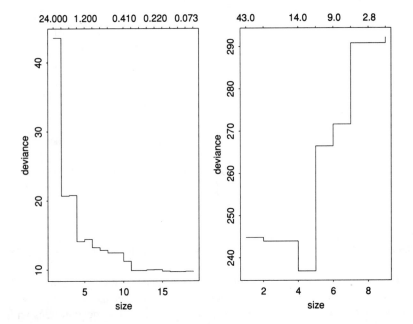

Figure 13.7: Cross-validation plots for pruning (left) `cpu.ltr` and (right) `painters.tr`.

The answers can be far from convincing as Figure 13.7 shows. The algorithm randomly divides the training set, and sometimes a second attempt will give a better answer (but Figure 13.7 shows the clearer of two attempts).

Let us prune the tree grown on the `painters` dataset.

```
painters.cv <- cv.tree(painters.tr,, prune.tree)
plot(painters.cv)
misclass.tree(painters.tr)
[1] 22
misclass.tree(prune.tree(painters.tr, best=4))
[1] 30
```

These re-substitution error counts are optimistically biased; the pruned tree may do better on future cases than the unpruned tree.

There is a difficulty with cross-validating the deviance; if at some leaf a class occurs in the test set but no the training set the deviance will be infinity. (If this occurs at the root node, the deviance will be infinite for all prunings.) To avoid this, zero probabilties are replaced by a very small probability (10^{-3} by default).

Error-rate pruning

Pruning on error-rate is attractive, as it will remove terminal splits which have the same class for each leaf, and allows the use of cross-validation to estimate the error rate. Unfortunately, there are errors in the current implementation. The function `prune.tree` is supposed to return the error rate rather than the deviance for `method="misclass"`, but it only does so if there is no new data. Also, `cv.tree` passes its ... argument only to some of its calls to the pruning function. We can work around these errors by replacing `prune.tree` by the function `p.tree` from our library:

```
cv.tree(painters.tr,, p.tree)
$size:
[1] 9 7 5 3 2 1
$dev:
[1] 38 38 40 45 42 49
    . . . .
p.tree(painters.tr)
$size:
[1] 9 7 5 3 2 1
$dev:
[1] 22 22 26 32 36 44
    . . . .
```

which shows that a pruned tree of size 2 suffices. We have already seen that one node (6) can be removed and in fact node 7 can also be removed, as node 14 has one-third each D and G and could be classified as either. Thus the pruned tree has the same error rate (22/54) on the whole dataset as the full tree. Note that in this example cross-validation gives a 70% higher estimate of the error rate.

13.3   Low birth weights revisited

We return to the example on low birth weights studied in Sections 7.2 and 10.1. As it has a binary response, it is fitted as a classification tree using the binomial log-likelihood. We can continue to use AIC by pruning with $\alpha = 2$. Note that the AIC for this tree, $153.5 + 2 \times 19 = 191.5$, is smaller than for the best logistic regression model of Section 7.2.

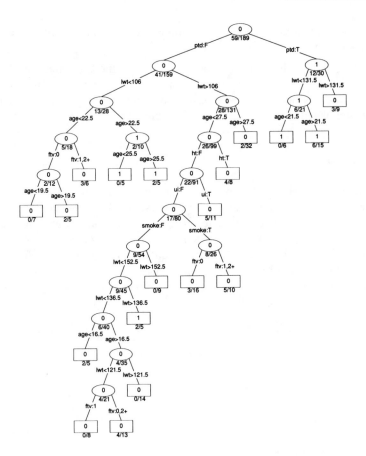

Figure 13.8: AIC-pruned tree for low birth-weight data.

```
> bwt.tr <- tree(low ~ ., bwt)
> summary(bwt.tr)

Classification tree:
tree(formula = factor(low) ~ ., data = bwt)
Number of terminal nodes:  28
Residual mean deviance:  0.907 = 146 / 161
Misclassification error rate: 0.217 = 41 / 189

> bwt.tr1 <- prune.tree(bwt.tr, k=2)
> summary(bwt.tr1)

Classification tree:
Variables actually used in tree construction:
[1] "ptd"   "lwt"   "age"   "ftv"   "ht"   "ui"   "smok‹
Number of terminal nodes:  19
Residual mean deviance:  0.9027 = 153.5 / 170
```

```
Misclassification error rate: 0.2275 = 43 / 189
> p.tree(bwt.tr1)
$size:
[1] 19 11  5  2  1
$dev:
[1] 43 43 44 53 59

> bwt.tr2 <- p.tree(bwt.tr1, best=11)
> bwt.tr3 <- p.tree(bwt.tr1, best=5)
```

The AIC-pruned tree is shown in Figure 13.8, further prunings in Figure 13.9

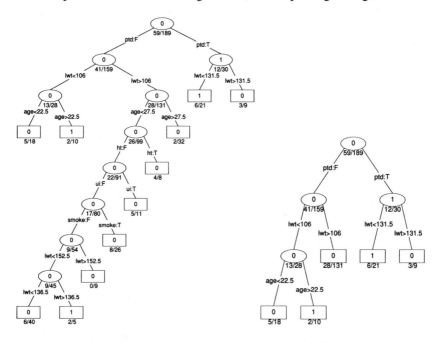

Figure 13.9: Error-rate pruned versions of Figure 13.8. Left with the same rule and right a slightly simplified rule.

We also consider error-rate pruning of the AIC-pruned tree, which will remove splits with identical classifications in both daughters. The tree sequence shows that a further six nodes can be removed at the price of one extra misclassification. Figure 13.9 shows that a compound condition is set up by these six nodes to classify three individuals.

Finally, we consider cross-validated error-rate pruning:

```
> cv.tree(bwt.tr,, p.tree)
$size:
[1] 28 16  8  5  2  1
$dev:
[1] 80 79 78 78 60 64
```

which once again shows a much higher error rate estimate, and suggests the tree of size 2.

For a binary classification such as this one it can be useful to use one of the features of `text.tree`:

```
plot(bwt.tr3)
text(bwt.tr3, label="1")
```

which labels the leaves by the estimated probability of a low birth weight. Unfortunately this option is not provided for `post.tree`.

Chapter 14

Time Series

There are now a large number of books on time series. Our philosophy and notation are close to those of the applied book by Diggle (1990) (from which some of our examples are taken). Brockwell and Davis (1991) and Priestley (1981) provide more theoretical treatments, and Bloomfield (1976) and Priestley are particularly thorough on spectral analysis.

Functions for time series have been included in S for some years, and further time-series support was one of the earliest enhancements of S-PLUS. The interface was changed in S-PLUS 3.2 to use the class mechanism; for information on the earlier interface see Section 14.7.

Regularly-spaced time series are of class rts, and are created by the function rts. (We will not cover the classes cts for series of dates and its for irregularly spaced series.)

Our first running example is lh, a series of 48 observations at 10mins intervals on luteinizing hormone levels for a human female taken from Diggle (1990). This was created by

```
> lh <- rts(scan(n=47))
2.4 2.4 2.4 2.2 2.1 1.5 2.3 2.3 2.5 2.0 1.9 1.7
2.2 1.8 3.2 3.2 2.7 2.2 2.2 1.9 1.9 1.8 2.7 3.0
2.3 2.0 2.0 2.9 2.9 2.7 2.7 2.3 2.6 2.4 1.8 1.7
1.5 1.4 2.1 3.3 3.5 3.5 3.1 2.6 2.1 3.4 3.0 2.9
```

Printing it gives

```
> lh
 1: 2.4 2.4 2.4 2.2 2.1 1.5 2.3 2.3 2.5 2.0 1.9 1.7 2.2 1.8
15: 3.2 3.2 2.7 2.2 2.2 1.9 1.9 1.8 2.7 3.0 2.3 2.0 2.0 2.9
29: 2.9 2.7 2.7 2.3 2.6 2.4 1.8 1.7 1.5 1.4 2.1 3.3 3.5 3.5
43: 3.1 2.6 2.1 3.4 3.0
 start deltat frequency
     1      1         1
```

which shows the attribute vector tspar of the class rts, which is used for plotting and other computations. The components are the start, the label for the first observation, deltat (Δt), the increment between observations, and frequency, the reciprocal of deltat. Note that the final index can be deduced

349

from the attributes and length. Any of `start`, `deltat`, `frequency` and `end` can be specified in the call to `rts`, provided they are specified consistently. In this example the units are known, and we can also specify this:

```
> lh <- rts(scan(n=47), start=1, deltat=1, units="10mins")
  . . . .
> lh
 1: 2.4 2.4 2.4 2.2 2.1 1.5 2.3 2.3 2.5 2.0 1.9 1.7 2.2 1.8
15: 3.2 3.2 2.7 2.2 2.2 1.9 1.9 1.8 2.7 3.0 2.3 2.0 2.0 2.9
29: 2.9 2.7 2.7 2.3 2.6 2.4 1.8 1.7 1.5 1.4 2.1 3.3 3.5 3.5
43: 3.1 2.6 2.1 3.4 3.0
 start deltat frequency
     1      1         1
Time units :  10mins
```

Our second example is a seasonal series. Our dataset `deaths` gives monthly deaths in the UK from a set of common lung diseases for the years 1974 to 1979, from Diggle (1990). This was read into S by

```
> deaths <- rts(scan(n=72), start=1974, frequency=12,
    units="months")
3035 2552 2704 2554 2014 1655 1721 1524 1596 2074 2199 2512
2933 2889 2938 2497 1870 1726 1607 1545 1396 1787 2076 2837
2787 3891 3179 2011 1636 1580 1489 1300 1356 1653 2013 2823
3102 2294 2385 2444 1748 1554 1498 1361 1346 1564 1640 2293
2815 3137 2679 1969 1870 1633 1529 1366 1357 1570 1535 2491
3084 2605 2573 2143 1693 1504 1461 1354 1333 1492 1781 1915
> deaths
        Jan  Feb  Mar  Apr  May  Jun  Jul  Aug  Sep  Oct  Nov
1974: 3035 2552 2704 2554 2014 1655 1721 1524 1596 2074 2199
1975: 2933 2889 2938 2497 1870 1726 1607 1545 1396 1787 2076
1976: 2787 3891 3179 2011 1636 1580 1489 1300 1356 1653 2013
1977: 3102 2294 2385 2444 1748 1554 1498 1361 1346 1564 1640
1978: 2815 3137 2679 1969 1870 1633 1529 1366 1357 1570 1535
1979: 3084 2605 2573 2143 1693 1504 1461 1354 1333 1492 1781

  . . . .
 start   deltat frequency
  1974 0.083333        12
Time units :  months
```

Note how the specification of `units="months"` has triggered a special form of labelling of the print. Quarterly data (with both `frequency=4` and `units="quarters"`) are also treated specially.

There is a series of functions to extract aspects of the time base:

```
> tspar(deaths)
 start   deltat frequency
  1974 0.083333        12
attr(, "units"):
[1] "months"
```

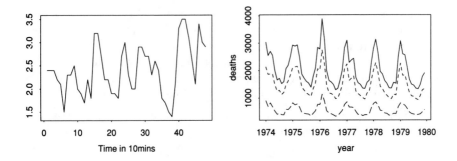

Figure 14.1: Plots by `ts.plot` of `lh` and the three series on deaths by lung diseases. In the right-hand plot the dashed series is for males, the long dashed series for females and the solid line for the total.

```
> start(deaths)
[1] 1974
> end(deaths)
[1] 1979.9
> frequency(deaths)
 frequency
        12
> units(deaths)
[1] "months"
> cycle(deaths)
       Jan Feb Mar Apr May Jun Jul Aug Sep Oct Nov Dec
1974:   1   2   3   4   5   6   7   8   9  10  11  12
1975:   1   2   3   4   5   6   7   8   9  10  11  12
1976:   1   2   3   4   5   6   7   8   9  10  11  12
1977:   1   2   3   4   5   6   7   8   9  10  11  12
1978:   1   2   3   4   5   6   7   8   9  10  11  12
1979:   1   2   3   4   5   6   7   8   9  10  11  12
 start   deltat frequency
  1974 0.083333        12
 Time units :  months
```

Time series can be plotted by `plot`, but the functions `ts.plot`, `ts.lines` and `ts.points` are provided for time-series objects. All can plot several related series together. For example, the `deaths` series is the sum of two series `mdeaths` and `fdeaths` for males and females. Figure 14.1 was created by[1]

```
> par(mfrow = c(2,2))
> ts.plot(lh)
> ts.plot(deaths, mdeaths, fdeaths, lty=c(1,3,4), xlab="year",
      ylab="deaths")
```

---

[1] S-PLUS 3.2 rel 1 for Windows will not plot `lh`; a corrected function is in the `fixes` directory on the disk.

The functions `ts.union` and `ts.intersect` bind together multiple time series. The time axes are aligned and only observations at times which appear in all the series are retained with `ts.intersect`; with `ts.union` the combined series covers the whole range of the components, possibly as `NA` values. The result is a matrix with `tspar` attributes set, or a data frame if argument `dframe=T` is set. We discuss most of the methodology for multiple time series in Section 14.4.

The function `window` extracts a sub-series of a single or multiple time-series, by specifying `start` and/or `end`.

The function `lag` shifts the time axis of a series by k positions, default one. Thus `lag(deaths, k=3)` is the series of deaths one quarter ago. The function `diff` takes the difference between a series and its lagged values, and so returns a series of length $n - k$ with values lost from the beginning (if $k > 0$) or end. It has an argument `differences` which causes the operation to be iterated. For later use, we denote the dth difference of series X_t by $\nabla^d X_t$, and the dth difference at lag s by $\nabla_s^d X_t$.

The function `aggregate` can be used to change the frequency of the time base. For example to obtain quarterly sums or annual means of `deaths`:

```
> aggregate(deaths, 4, sum)
         1    2    3    4
1974: 8291 6223 4841 6785
    ....
> aggregate(deaths, 1, mean)
1974: 2178.3 2175.1 2143.2 1935.8 1995.9 1911.5
```

Each of `lag`, `diff` and `aggregate` can also be applied to multiple time series objects formed by `ts.union` or `ts.intersect`.

14.1 Second-order summaries

The theory for time series is based on the assumption of second-order stationarity after removing any trends (which will include seasonal trends). Thus second moments are particularly important in the practical analysis of time series. We assume that the series X_t runs throughout time, but is observed only for $t = 1, \ldots, n$. We will use the notations X_t and $X(t)$ interchangeably. The series will have a mean μ, often taken to be zero, and the covariance and correlation

$$\gamma_t = \text{cov}\,(X_{t+\tau}, X_\tau), \qquad \rho_t = \text{corr}\,(X_{t+\tau}, X_\tau)$$

do not depend on τ. The covariance is estimated for $t > 0$ from the $n-t$ observed pairs $(X_{1+t}, X_1), \ldots, (X_n, X_{n-t})$. If we just take the standard correlation or covariance of these pairs we will use different estimates of the mean and variance for each of the subseries $X_{1+t}, \ldots, X_n$ and $X_1, \ldots, X_{n-t}$, whereas under our assumption of second-order stationarity these have the same mean and variance. This suggests the estimators

$$c_t = \frac{1}{n} \sum_{s=\min(1,-t)}^{\max(n-t,n)} [X_{s+t} - \overline{X}][X_s - \overline{X}], \qquad r_t = \frac{c_t}{c_0}$$

Note that we use divisor n even though there are $n - |t|$ terms. This is to ensure that the sequence (c_t) is the covariance sequence of some second-order stationary time series.[2] Note that all of γ, ρ, c, r are symmetric functions ($\gamma_{-t} = \gamma_t$ and so on).

The S-PLUS function acf computes and by default plots the sequences (c_t) and (r_t), known as the *autocovariance* and *autocorrelation* functions. The argument type controls which is used, and defaults to the correlation.

Our definitions are easily extended to several time series observed over the same interval. Let

$$\gamma_{ij}(t) = \text{cov}\left(X_i(t+\tau), X_j(\tau)\right)$$

$$c_{ij}(t) = \frac{1}{n} \sum_{s=\min(1,-t)}^{\max(n-t,n)} [X_i(s+t) - \overline{X_i}][X_j(s) - \overline{X_j}]$$

which are not symmetric in t for $i \neq j$. These forms are used by acf for multiple time series:

```
acf(lh)
acf(lh, type="covariance")
acf(deaths)
acf(ts.union(mdeaths, fdeaths))
```

The type may be abbreviated in any unique way, for example cov. The output is shown in Figures 14.2 and 14.3. Note that approximate 95% confidence limits are shown for the autocorrelation plots; these are for an independent series for which $\rho_t = I(t = 0)$. As with a time series *a priori* one is expecting autocorrelation, these limits must be viewed with caution. In particular, if any ρ_t is non-zero, all the limits are invalid.

Note that for a series with a non-unit frequency such as deaths the lags are expressed in the basic time unit, here years. The function acf chooses the number of lags to plot unless this is specified by the argument lag.max. Plotting can be suppressed by setting argument plot=F. The function returns a list which can be plotted subsequently by acf.plot.

The plots of the deaths series show the pattern typical of seasonal series, and the autocorrelations do not damp down for large lags. Note how one of the cross-series is only plotted for negative lags. We have $c_{ji}(t) = c_{ij}(-t)$, so the cross terms are needed for all lags, whereas the terms for a single series are symmetric about 0. The labels are confusing: the plot in row 2 column 1 shows c_{12} for negative lags, a reflection of the plot of c_{21} for positive lags.

---

[2] That is, the covariance sequence is positive-definite.

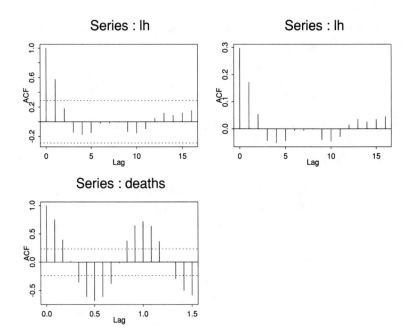

Figure 14.2: `acf` plots for the series `lh` and `deaths`. The top row shows the autocorrelation (left) and autocovariance (right).

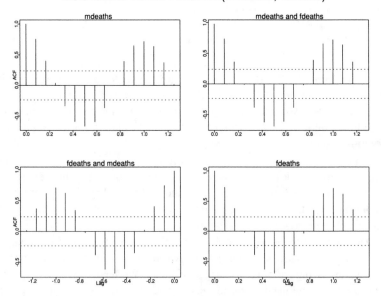

Figure 14.3: Autocorrelation plots for the multiple time series of male and female deaths.

Spectral analysis

The spectral approach to second-order properties is better able to separate short-term and seasonal effects, and also has a sampling theory which is easier to use for non-independent series.

We will only give a brief treatment; extensive accounts are given by Bloomfield (1976) and Priestley (1981). Be warned that accounts differ in their choices of where to put the constants in spectral analysis; we have tried to follow S-PLUS as far as possible.

The covariance sequence of a second-order stationary time series can always be expressed as

$$\gamma_t = \frac{1}{2\pi} \int_{-\pi}^{\pi} e^{i\omega t}\, dF(\omega)$$

for the *spectrum* F, a finite measure on $(-\pi, \pi]$. Under mild conditions which exclude purely periodic components of the series, the measure has a density known as the *spectral density* f, so

$$\gamma_t = \frac{1}{2\pi} \int_{-\pi}^{\pi} e^{i\omega t} f(\omega)\, d\omega = \int_{-1/2}^{1/2} e^{i\omega_f t} f(\omega_f)\, d\omega_f \qquad (14.1)$$

where in the first form the frequency ω is in units of radians/time and in the second form ω_f is in the units of cycles/time, and in both cases time is measured in the units of Δt. If the time series object has a `frequency` greater than one and time is measured in the base units, the spectral density will be divided by `frequency`.

The Fourier integral can be inverted to give

$$f(\omega) = \sum_{-\infty}^{\infty} \gamma_t e^{-i\omega t} = \gamma_0 \left[1 + 2 \sum_{1}^{\infty} \rho_t \cos(\omega t) \right] \qquad (14.2)$$

By the symmetry of γ_t, $f(-\omega) = f(\omega)$, and we need only consider f on $(0, \pi)$. Equations (14.1) and (14.2) are the first place the differing constants appear. Bloomfield and Brockwell & Davis omit the factor $1/2\pi$ in (14.1) which therefore appears in (14.2).

The basic tool in estimating the spectral density is the *periodogram*. For a frequency ω we effectively compute the squared correlation between the series and the sine/cosine waves of frequency ω by

$$I(\omega) = \left| \sum_{t=1}^{n} e^{-i\omega t} X_t \right|^2 / n = \frac{1}{n} \left[\left\{ \sum_{t=1}^{n} X_t \sin(\omega t) \right\}^2 + \left\{ \sum_{t=1}^{n} X_t \cos(\omega t) \right\}^2 \right] \qquad (14.3)$$

Frequency 0 corresponds to the mean, which is normally removed. The frequency π corresponds to a cosine series of alternating ± 1 with no sine series. Bloomfield (but not Brockwell & Davis) has a factor $1/2\pi$ in the definition of the periodogram. S-PLUS appears to divide by the `frequency` to match its view of the spectral density.

The periodogram is related to the autocovariance function by

$$I(\omega) = \sum_{-\infty}^{\infty} c_t e^{-i\omega t} = c_0 \left[1 + 2 \sum_{1}^{\infty} r_t \cos(\omega t) \right]$$

$$c_t = \frac{1}{2\pi} \int_{-\pi}^{\pi} e^{i\omega t} I(\omega) \, d\omega$$

and so conveys the same information. However, each form makes some of that information easier to interpret.

Asymptotic theory shows that $I(\omega) \sim f(\omega)E$ where E has a standard exponential distribution, except for $\omega = 0$ and $\omega = \pi$. Thus if $I(\omega)$ is plotted on log scale, the variation about the spectral density is the same for all $\omega \in (0, \pi)$ and is given by a Gumbel distribution (for that is the distribution of $\log E$). Further, $I(\omega_1)$ and $I(\omega_2)$ will be asymptotically independent at distinct frequencies. Indeed if ω_k is a *Fourier frequency* of the form $\omega_k = 2\pi k/n$, then the periodogram at two Fourier frequencies will be approximately independent for large n. Thus although the periodogram itself does not provide a consistent estimator of the spectral density, if we assume that the latter is smooth, we can average over adjacent independently distributed periodogram ordinates and obtain a much less variable estimate of $f(\omega)$. A kernel smoother is used of the form

$$\hat{f}(\omega) = \frac{1}{h} \int K\left(\frac{\lambda - \omega}{h}\right) I(\omega) \, d\lambda$$

$$\approx \frac{2\pi}{nh} \sum_k K\left(\frac{\omega_k - \omega}{h}\right) I(\omega_k) = \sum_k g_k I(\omega_k)$$

for a probability density K. The parameter h controls the degree of smoothing. To see its effect we approximate the mean and variance of $\hat{f}(\omega)$:

$$\operatorname{var}\left(\hat{f}(\omega)\right) \approx \sum_k g_k^2 f(\omega_k)^2 \approx f(\omega)^2 \sum_k g_k^2 \approx \frac{2\pi}{nh} f(\omega)^2 \int K(x)^2 \, dx$$

$$E\left(\hat{f}(\omega)\right) \approx \sum_k g_k f(\omega_k) \approx f(\omega) + \frac{f''(\omega)}{2} \sum_k g_k (\omega_k - \omega)^2$$

$$\operatorname{bias}\left(\hat{f}(\omega)\right) \approx \frac{f''(\omega)}{2} \frac{2\pi h^2}{n} \int x^2 K(x) \, dx$$

so as h increases the variance decreases but the bias increases. We see that the ratio of the variance to the squared mean is approximately $g^2 = \sum_k g_k^2$. If $\hat{f}(\omega)$ had a distribution proportional to a χ_ν^2, this ratio would be $2/\nu$, so $2/g^2$ is referred to as the equivalent degrees of freedom. Bloomfield and S-PLUS refer to $\sqrt{2 \operatorname{bias}\left(\hat{f}(\omega)\right)/f''(\omega)}$ as the *bandwidth*, which is proportional to h.

To understand these quantities, consider a simple moving average over $2m+1$ Fourier frequencies centred on a Fourier frequency ω. Then the variance is $f(\omega)^2/(2m + 1)$ and the equivalent degrees of freedom is $2(2m + 1)$, as we would expect on averaging $2m + 1$ exponential (or χ_2^2) variates. The bandwidth is approximately

$$\frac{(2m+1)2\pi}{n} \frac{1}{\sqrt{12}}$$

and the first factor is the width of the window in frequency space. (Since S-PLUS works in cycles rather than radians, the bandwidth is about $(2m + 1)/n\sqrt{12}$ frequency on its scale.) The bandwidth is thus a measure of the size of the smoothing window, but rather smaller than the effective width.

The workhorse function for spectral analysis is `spectrum`, which with its default options computes and plots the periodogram on log scale. The function `spectrum` calls `spec.pgram` to do most of the work. (Note: `spectrum` by default removes a linear trend from the series before estimating the spectral density.) For our examples we can use:

```
par(mfrow=c(2,2))
spectrum(lh)
spectrum(deaths)
```

with the result shown in Figure 14.4.

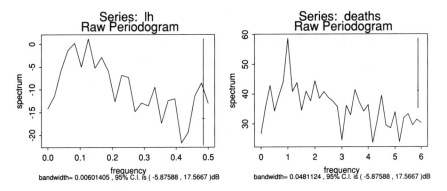

Figure 14.4: Periodogram plots for `lh` and `deaths`.

Note how elaborately labelled the figures are. The plots are on log scale, in units of *decibels*, that is the plot is of $10 \log_{10} I(\omega)$. The function `spec.pgram` returns the bandwidth and (equivalent) degrees of freedom as components `bandwidth` and `df`.

The function `spectrum` also produces smoothed plots, using repeated smoothing with modified Daniell smoothers (Bloomfield, 1976), which are moving averages giving half weight to the end values of the span. Trial-and-error is needed to choose the spans (Figures 14.5 and 14.6):

```
par(mfrow=c(2,2))
spectrum(lh)
spectrum(lh, spans=3)
spectrum(lh, spans=c(3,3))
```

```
spectrum(lh, spans=c(3,5))

spectrum(deaths)
spectrum(deaths, spans=c(3,3))
spectrum(deaths, spans=c(3,5))
spectrum(deaths, spans=c(5,7))
```

The spans should be odd integers, and it helps to produce a smooth plot if they are different and at least two are used. The width of the centre mark on the 95% confidence interval indicator indicates the bandwidth.

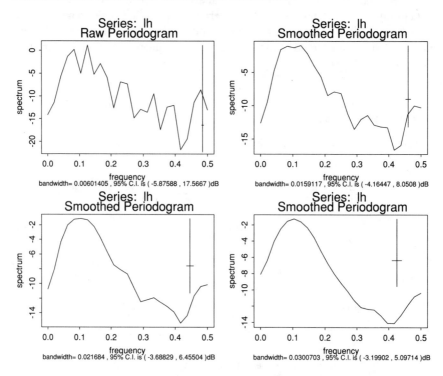

Figure 14.5: Spectral density estimates for `lh`.

The periodogram has other uses. If there are periodic components in the series the distribution theory given above does not apply, but there will be peaks in the plotted periodogram. Smoothing will reduce those peaks, but they can be seen quite clearly by plotting the *cumulative periodogram*

$$U(\omega) = \sum_{0 < \omega_k \leqslant \omega} I(\omega_k) \Big/ \sum_{1}^{\lfloor n/2 \rfloor} I(\omega_k)$$

against ω. The cumulative periodogram is also very useful as a test of whether a particular spectral density is appropriate, as if we replace $I(\omega)$ by $I(\omega)/f(\omega)$, $U(\omega)$ should be a straight line. Further, asymptotically, the maximum deviation

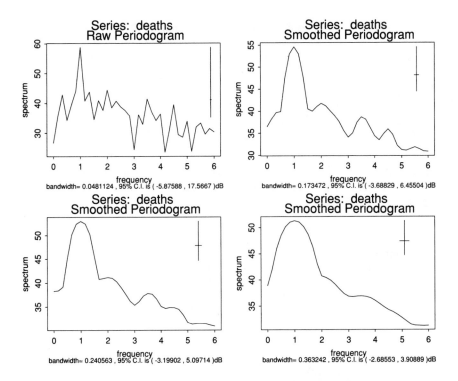

Figure 14.6: Spectral density estimates for `deaths`.

from that straight line has a distribution given by that of the Kolmogorov-Smirnov statistic, with a 95% limit approximately $1.358/[\sqrt{m} + 0.11 + 0.12\sqrt{m}]$ where $m = \lfloor n/2 \rfloor$ is the number of Fourier frequencies included. This is particularly useful for a residual series with f constant, in testing if the series is uncorrelated.

The distribution theory can be made more accurate, and the peaks made sharper, by *tapering* the de-meaned series (Bloomfield, 1976). The magnitude of the first α and last α of the series is tapered down towards zero by a cosine bell, that is X_t is replaced by

$$X_t' = \begin{cases} (1 - \cos \frac{\pi(t-0.5)}{\alpha n})X_t & t \leqslant \alpha n \\ X_t & \alpha n < t < (1 - \alpha)n \\ (1 - \cos \frac{\pi(n-t+0.5)}{\alpha n})X_t & t \geqslant (1 - \alpha)n \end{cases}$$

The proportion α is controlled by the parameter `taper` of `spec.pgram`, and defaults to 10%. It should rarely need to be altered. (The taper function `spec.taper` can be called directly if needed.) Tapering does increase the variance of the periodogram and hence the spectral density estimate, by about 12% for the default taper, but it will decrease the bias near peaks very markedly, if those peaks are not at Fourier frequencies.

We cannot show this directly using `spectrum`, since that plots frequency zero, and as the taper is reduced, the mean of the tapered series and hence $I(0)$ goes to zero. We need to use a taper for the plot used to set the scale:

```
spectrum(deaths)
deaths.spc <- spec.pgram(deaths, taper=0)
lines(deaths.spc$freq, deaths.spc$spec, lty=3)
```

In this example tapering does not help resolve the peaks, but they are at Fourier frequencies. S-PLUS does not supply a function for the cumulative periodogram, but we can write one using the `fft` function to compute a discrete Fourier transform.

```
cpgram <- function(ts, taper=0.1,
    main=paste("Series: ", deparse(substitute(ts)))  )
{
    x <- as.vector(ts)
    x <- x[!is.na(x)]
    x <- spec.taper(scale(x, T, F), p=taper)
    y <- Mod(fft(x))^2/length(x)
    y[1] <- 0
    n <- length(x)
    x <- (0:(n/2))*frequency(ts)/n
    if(length(x)%%2==0) {
        n <- length(x)-1
        y <- y[1:n]
        x <- x[1:n]
    } else y <- y[1:length(x)]
    xm <- frequency(ts)/2
    mp <- length(x)-1
    crit <- 1.358/(sqrt(mp)+0.12+0.11/sqrt(mp))
    oldpty <- par()$pty
    par(pty="s")
    plot(x, cumsum(y)/sum(y), type="s", xlim=c(0, xm),
        ylim=c(0, 1), xaxs="i", yaxs="i", xlab="frequency",
        ylab="", pty="s")
    lines(c(0, xm*(1-crit)), c(crit, 1))
    lines(c(xm*crit, xm), c(0, 1-crit))
    title(main = main)
    invisible(par(pty=oldpty))
}
```

Dropping missing values helps when using the function with residual series, which start with missing values. The results for our examples are shown in Figure 14.7, with 95% confidence bands.

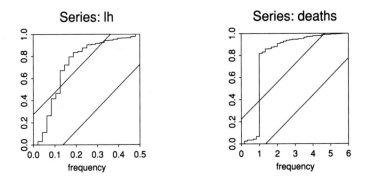

Figure 14.7: Cumulative periodogram plots for `lh` and `deaths`.

14.2   ARIMA models

In the late 1960's Box and Jenkins advocated a methodology for time series based on finite-parameter models for the second-order properties, so this approach is often named after them. Let ϵ_t denote a series of uncorrelated random variables with mean zero and variance σ^2. A moving average process of order q (MA(q)) is defined by

$$X_t = \sum_0^q \beta_j \epsilon_{t-j} \tag{14.4}$$

an autoregressive process of order p (AR(p)) is defined by

$$X_t = \sum_1^q \alpha_i X_{t-i} + \epsilon_t \tag{14.5}$$

and an ARMA(p, q) process is defined by

$$X_t = \sum_1^q \alpha_i X_{t-i} + \sum_0^q \beta_j \epsilon_{t-j} \tag{14.6}$$

(S-PLUS reverses the sign of the MA coefficients for $j > 0$.)

Note that we do not need both σ^2 and β_0 for a MA(q) process, and we will take $\beta_0 = 1$. Some authors put the regression terms of (14.5) and (14.6) on the left-hand side and reverse the sign of α_i. Any of these processes can be given mean μ by adding μ to each observation.

An ARIMA(p, d, q) process (where the I stands for integrated) is a process whose dth difference $\nabla^d X$ is an ARMA(p, q) process.

Equation (14.4) will always define a second-order stationary time series, but (14.5) and (14.6) need not. They need the condition that all the (complex) roots of the polynomial

$$\phi_\alpha(z) = 1 - \alpha_1 z - \cdots - \alpha_p z^p$$

lie outside the unit disc. (The S-PLUS function `polyroot` can be used to check this.) However, there are in general 2^p sets of coefficients in (14.4) which give the same second-order properties, and it is conventional to take the set with roots of

$$\phi_\beta(z) = 1 + \beta_1 z + \cdots + \beta_q z^q$$

outside the unit disc. Let B be the backshift or lag operator defined by $BX_t = X_{t-1}$. Then we conventionally write an ARMA process as

$$\phi_\alpha(B)X = \phi_\beta(B)\epsilon \tag{14.7}$$

The function `arima.sim` simulates an ARIMA process. Simple usage is of the form

```
ts.sim <- arima.sim(list(order=c(1,1,0), ar=0.7), n=200)
```

which generates a series whose first differences follow an AR(1) process.

Model identification

A lot of attention has been paid to *identifying* ARMA models, that is choosing plausible values of p and q by looking at the second-order properties. Much of the literature is reviewed by de Gooijer *et al.* (1985). Nowadays it is computationally feasible to fit all plausible models and choose on the basis of their goodness of fit, but some simple diagnostics are still useful. For an MA(q) process we have

$$\gamma_k = \sigma^2 \sum_{i=0}^{q-|k|} \beta_i \beta_{i+|k|}$$

which is zero for $|k| > q$, and this may be discernible from plots of the ACF. For an AR(p) process the population autocovariances are generally all non-zero, but they satisfy the Yule–Walker equations

$$\rho_k = \sum_1^p \alpha_i \rho_{k-i}, \qquad k > 0 \tag{14.8}$$

This motivates the *partial autocorrelation function*. The partial correlation between X_s and X_{s+t} is the correlation after regression on $X_{s+1}, \ldots, X_{s+t-1}$, and is zero for $t > p$ for an AR(p) process. The PACF can be estimated by solving the Yule–Walker equations (14.8) with $p = t$ and ρ replaced by r, and is given by the `type="partial"` option of `acf`:

```
acf(lh, type="partial")
acf(deaths, type="partial")
```

as shown in Figure 14.8. These are short series, so no definitive pattern emerges, but `lh` might be fitted well by an AR(1) or perhaps an AR(3) process.

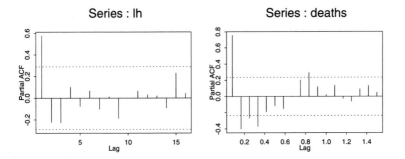

Figure 14.8: Partial autocorrelation plots for the series `lh` and `deaths`.

Model fitting

Selection amongst ARMA processes can be done by Akaike's information criterion (AIC) which penalizes the deviance by twice the number of parameters; the model with the smallest AIC is chosen. (All likelihoods considered assume a Gaussian distribution for the time series.) Fitting can be done by the functions `ar` or `arima.mle`:

```
> lh.ar1 <- ar(lh, F, 1)
> cpgram(lh.ar1$resid, main="AR(1) fit to lh")
> lh.ar <- ar(lh, order.max=9)
> lh.ar$order
[1] 3
> lh.ar$aic
 [1] 18.30668  0.99567  0.53802  0.00000  1.49036  3.21280
 [7]  4.99323  6.46950  8.46258  8.74120
> cpgram(lh.ar$resid, main="AR(3) fit to lh")
> lh1 <- lh - mean(lh)
> lh.arima1 <- arima.mle(lh1, model=list(order=c(1,0,0)),
       n.cond=3)
> arima.diag(lh.arima1)
> lh.arima3 <- arima.mle(lh1, model=list(order=c(3,0,0)),
       n.cond=3)
> arima.diag(lh.arima3)
> lh.arima11 <- arima.mle(lh1, model=list(order=c(1,0,1)),
       n.cond=3)
> arima.diag(lh.arima11)
```

This first fits an AR(1) process and obtains, after removing the mean,

$$X_t = 0.576X_{t-1} + \epsilon_t$$

with $\sigma^2 = 0.208$. It then uses AIC to choose the order amongst AR processes and selects $p = 3$ and fits

$$X_t = 0.653X_{t-1} - 0.064X_{t-2} - 0.227X_{t-3} + \epsilon_t$$

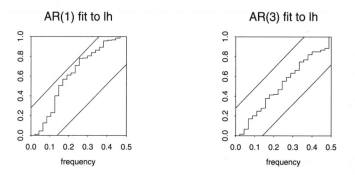

Figure 14.9: Cumulative periodogram plots for residuals of AR models fitted to `lh`.

with $\sigma^2 = 0.196$ and AIC is reduced by 0.996. (For `ar` the component `aic` is the excess over the best fitting model, and it starts from $p = 0$.) The function `ar` by default fits the model by solving the Yule–Walker equations (14.8) with ρ replaced by r. An alternative is to use `method="burg"`.

The function `arima.mle` fits by maximum likelihood and does not include a mean. Confusingly, the `loglik` component returned by `arima.mle` is a measure of the deviance (minus twice the log-likelihood). Equally confusingly, the likelihood considered is not the full likelihood but a likelihood conditional on a set of starting values for the AR and difference terms, so the AIC given cannot be compared between models unless the component `n.cond` is the same.[3] The conditional likelihood conditions on $p + d$ starting values for a non-seasonal series, or `n.cond` if this is larger. The fitted models are

$$X_t = 0.586(0.121)X_{t-1} + \epsilon_t$$

with $\sigma^2 = 0.211$ and AIC $= 59.61$,

$$X_t = 0.658(0.145)X_{t-1} - 0.066(0.174)X_{t-2} - 0.234(0.145)X_{t-3} + \epsilon_t$$

with $\sigma^2 = 0.190$ and AIC $= 59.09$ and

$$X_t = 0.463(0.218)X_{t-1} + \epsilon_t + 0.200(0.241)\epsilon_{t-1}$$

with $\sigma^2 = 0.205$ and AIC $= 60.41$, which shows the MA term is not worthwhile.

The diagnostic plots are shown in Figures 14.9 and 14.10. The cumulative periodograms of the residuals show that the AR(1) process has not removed all the correlation. The bottom panel of the `arima.diag` plots the P-value for the Box-Pierce (1970) *portmanteau test*

$$Q_K = n \sum_{1}^{K} c_k^2 \qquad (14.9)$$

---

[3] This is both unnecessary and unfortunate, as the full likelihood could be computed quite easily; see Brockwell & Davis (1991, §8.7).

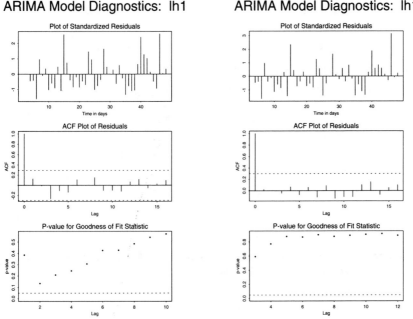

Figure 14.10: Diagnostic plots for ARIMA models fitted to `lh`. (Left) AR(1) and (right) AR(3).

applied to the residuals (and not the Ljung-Box test as a comment in the function suggests). Here the maximum K is set by the parameter `gof.lag` (which defaults to 10) plus $p + q$. Note that although an AR(3) model fits better according to AIC, this is not clear-cut from the diagnostic plots. Although the AIC is smaller, a formal test of the difference in deviances, 4.52, using a χ^2_2 distribution is not significant.

The function `arima.diag` can produce (standardized) residuals from a fit by `arima.mle`, by setting `plot=F`, `acf.resid=F`, `gof.lag=0` and `resid=T` or `std.resid=T`.

The function `arima.mle` can also include differencing and so fit an ARIMA model (the middle integer in `order` is d).

Forecasting

Forecasting is relatively straightforward using the function `arima.forecast`:

```
lh.fore <- arima.forecast(lh1, n=12, model=lh.arima3$model)
lh.fore$mean <- lh.fore$mean + mean(lh)
ts.plot(lh, lh.fore$mean, lh.fore$mean+2*lh.fore$std.err,
    lh.fore$mean-2*lh.fore$std.err)
```

(see Figure 14.11) but the standard errors do not include the effect of estimating the mean and the parameters of the ARIMA model.

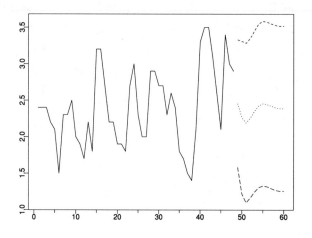

Figure 14.11: Forecasts for 12 periods (of 10 mins) ahead for the series `lh`. The dashed curves are approximate pointwise 95% confidence intervals.

Spectral densities via AR processes

The spectral density of an ARMA process has a simple form: it is given by

$$f(\omega) = \sigma^2 \left| \frac{1 + \sum_s \beta_s e^{-is\omega}}{1 - \sum_t \alpha_t e^{-it\omega}} \right|^2 \tag{14.10}$$

and so we can estimate the spectral density by substituting parameter estimates in (14.10). It is most usual to fit high-order AR models, both because they can be fitted rapidly, and since they can produce peaks in the spectral density estimate by small values of $|1 - \sum \alpha_t e^{-it\omega}|$ (which correspond to nearly non-stationary fitted models since there must be roots of $\phi_\alpha(z)$ near $e^{-i\omega}$).

This procedure is implemented by `spectrum` with `method="ar"`, which calls `spec.ar`. Although popular because it often produces visually pleasing spectral density estimates, it is not recommended (e.g. Thomson, 1990).

Regression terms

The `arima` family of functions can also handle regressions with ARIMA residual processes, that is models of the form

$$X_t = \sum \gamma_i Z_t^{(i)} + \eta_t, \qquad \phi_\alpha(B) \Delta^d \eta_t = \phi_\beta(B) \epsilon \tag{14.11}$$

for one or more external time series $z^{(i)}$. This can be specified for simulation to `arima.sim`, the parameters γ estimated by `arima.mle`, and forecasts computed by `arima.forecast` (provided forecasts of the external series are available). Again, the variability of the parameter estimates $\widehat{\gamma}$ is not taken into account in the computed prediction standard errors.

14.3 Seasonality

For a seasonal series there are two possible approaches. One is to decompose the series, usually into a trend, a seasonal component and a residual, and to apply non-seasonal methods to the residual component. The other is to model all the aspects simultaneously.

Decompositions

Two decomposition algorithms are available. The function `sabl` dates from 1982 and is being superseded by the function `stl`. Both are fairly complex, and the on-line documentation should be consulted for full details and references (principally Cleveland *et al.*, 1990).

The function `stl` can extract a strictly periodic component plus a remainder:

```
deaths.stl <- stl(deaths, "periodic")
ts.plot(deaths, deaths.stl$sea, deaths.stl$rem)
```

as shown in Figure 14.12.

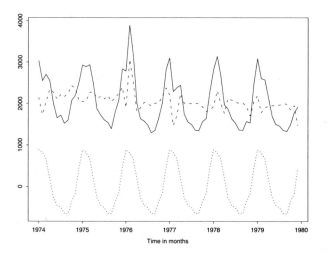

Figure 14.12: `stl` decomposition for the `deaths` series (solid line). The dotted series is the seasonal component, the dashed series the remainder.

The function `monthplot` plots the seasonal component of a series decomposed by `sabl` or `stl`.

We now return to complete the analysis of the `deaths` series by analysing the non-seasonal component. The results are shown in Figure 14.13.

```
> dsd <- deaths.stl$rem
> ts.plot(dsd)
> acf(dsd)
> acf(dsd, type="partial")
```

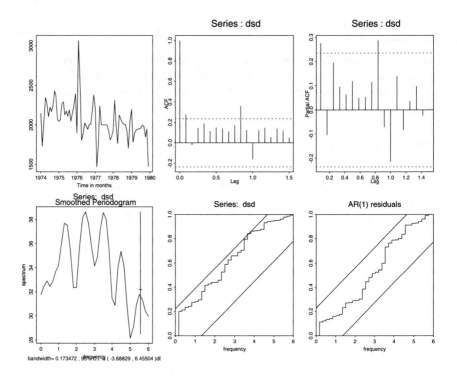

Figure 14.13: Diagnostics for an AR(1) fit to the remainder of an `stl` decomposition of the `deaths` series.

```
> spectrum(dsd, span=c(3,3))
> cpgram(dsd)
> dsd.ar <- ar(dsd)
> dsd.ar$order
[1] 1
> dsd.ar$aic
 [1]   3.64844   0.00000   1.22632   0.40845   1.75574   3.46924
     ....
> dsd.ar$ar
[1,] 0.27469
> cpgram(dsd.ar$resid, main="AR(1) residuals")
> dsd.rar <- ar.gm(dsd)
> dsd.rar$ar
     ....
[1] 0.41493
```

The large jump in the cumulative periodogram at the lowest (non-zero) Fourier frequency is caused by the downward trend in the series. The spectrum has dips at the integers since we have removed the seasonal component and hence all of the frequency and its multiples. (The dip at 1 is obscured by the peak at the Fourier frequency to its left; see the cumulative periodogram.) The plot of the remainder series shows exceptional values for February–March 1976 and 1977. As there are only 6 cycles, the seasonal pattern is difficult to establish at all precisely.

The robust AR-fitting function `ar.gm` helps to overcome the effect of outliers such as February 1976, and produces a considerably higher AR coefficient. The values for February 1976 and 1977 are heavily down-weighted. Unfortunately its output does not mesh well with the other time-series functions.

Seasonal ARIMA models

The function `diff` allows differences at lags greater than one, so for a monthly series the difference at lag 12 is the difference from this time last year. Let s denote the period, often 12. We can then consider ARIMA models for the sub-series sampled s apart, for example for all Januaries. This corresponds to replacing B by B^s in the definition (14.7). Thus an ARIMA $(P, D, Q)_s$ process is a seasonal version of an ARIMA process. However, we may include both seasonal and non-seasonal terms, obtaining a process of the form

$$\Phi_{AR}(B)\Phi_{SAR}(B^s)Y = \Phi_{MA}(B)\Phi_{SMA}(B^s)\epsilon, \qquad Y = (I-B)^d(I-B^s)^D X$$

If we expand this out, we see that it is an ARMA $(p + sP, q + sQ)$ model for Y_t, but parametrized in a special way with large numbers of zero coefficients. It can still be fitted as an ARMA process, and `arima.mle` can handle models specified in this form, by specifying extra terms in the argument `model` with the period set. (Examples follow.)

To identify a suitable model we first look at the seasonally differenced series. Figure 14.14 suggests that this may be over-differencing, but that the non-seasonal term should be an AR(2).

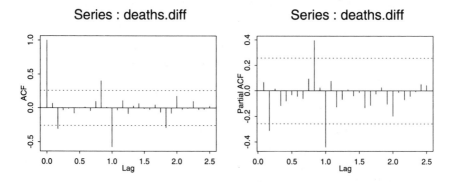

Figure 14.14: Autocorrelation and partial autocorrelation plots for the seasonally differenced `deaths` series. The negative values at lag 12 suggest over-differencing.

```
> deaths.diff <- diff(deaths, 12)
> acf(deaths.diff, 30)
> acf(deaths.diff, 30, type="partial")
```

```
> ar(deaths.diff)
$order:
[1] 12
    ....
$aic:
 [1]  7.8143   9.5471   5.4082   7.3929   8.5839 10.1979 12.1388
 [8] 14.0201 15.7926 17.2504   8.9905 10.9557   0.0000   1.6472
[15]  2.6845   4.4097   6.4047   8.3152
# this suggests the seasonal effect is still present.
> deaths.arima1 <- arima.mle(deaths, model=list(
      list(order=c(2,0,0)), list(order=c(0,1,0), period=12)) )
> deaths.arima1$aic
[1] 847.41
> deaths.arima1$model[[1]]$ar   # the non-seasonal part
[1]   0.12301 -0.30522
> sqrt(diag(deaths.arima1$var.coef))
[1] 0.12504 0.12504
> arima.diag(deaths.arima1, gof.lag=24)
# suggests need a seasonal AR term
> deaths1 <- deaths - mean(deaths)
> deaths.arima2 <- arima.mle(deaths1, model=list(
      list(order=c(2,0,0)),  list(order=c(1,0,0), period=12)) )
> deaths.arima2$aic
[1] 845.38
> deaths.arima2$model[[1]]$ar # non-seasonal part
[1]   0.21601 -0.25356
> deaths.arima2$model[[2]]$ar # seasonal part
[1] 0.82943
> sqrt(diag(deaths.arima2$var.coef))
[1] 0.12702 0.12702 0.07335
> arima.diag(deaths.arima2, gof.lag=24)
> cpgram(arima.diag(deaths.arima2, plot=F, resid=T)$resid)
> deaths.arima3 <- arima.mle(deaths, model=list(
      list(order=c(2,0,0)), list(order=c(1,1,0), period=12)) )
> deaths.arima3$aic   # not comparable to those above
[1] 638.21
> deaths.arima3$model[[1]]$ar
[1]   0.41212 -0.26938
> deaths.arima3$model[[2]]$ar
[1] -0.7269
> sqrt(diag(deaths.arima3$var.coef))
[1] 0.14199 0.14199 0.10125
> arima.diag(deaths.arima3, gof.lag=24)
> arima.mle(deaths1, model=list(list(order=c(2,0,0)),
      list(order=c(1,0,0), period=12)), n.cond=26 )$aic
[1] 664.14
> deaths.arima4 <- arima.mle(deaths1, model=list(
      list(order=c(2,0,0)), list(order=c(2,0,0), period=12)) )
> deaths.arima4$aic
[1] 634.07
```

```
> deaths.arima4$model[[1]]$ar
[1] 0.47821 -0.27873
> deaths.arima4$model[[2]]$ar
[1] 0.17346 0.63474
> sqrt(diag(deaths.arima4$var.coef))
[1] 0.14160 0.14160 0.11393 0.11393
```

The AR-fitting suggests a model of order 12 (of up to 16) which indicates that seasonal effects are still present. The diagnostics from the ARIMA($(2, 0, 0) \times (0, 1, 0)_{12}$) model suggests problems at lag 12. Dropping the differencing in favour of a seasonal AR term gives an AIC which favours the second model. However, the diagnostics suggest that there is still seasonal structure in the residuals, so we include differencing and a seasonal AR term. This gives a seasonal model of the form

$$(I + 0.727B^s)(I - B^s)X = (I - 0.273B^s - 0.727B^{2s})X = \epsilon$$

which suggests a seasonal AR(2) term might be more appropriate.

The best fit found is an ARIMA($(2, 0, 0) \times (2, 0, 0)_{12}$) model, but this does condition on the first 26 observations, including the exceptional value for February 1976 (see Figure 14.1). Fitting the ARIMA($(2, 0, 0) \times (1, 0, 0)_{12}$) model to the same series shows that the seasonal AR term does help the fit considerably.

We consider another example on monthly accidental deaths in the USA 1973–8, from Brockwell & Davis (1991) contained in our dataset `accdeaths`:

```
> dacc <- diff(accdeaths, 12)
> ts.plot(dacc)
> acf(dacc, 30)
> acf(dacc, 30, "partial")
> ddacc <- diff(dacc)
> ts.plot(ddacc)
> acf(ddacc, 30)
> acf(ddacc, 30, "partial")
> ddacc.1 <- arima.mle(ddacc-mean(ddacc),
      model=list(list(order=c(0,0,1)),
      list(order=c(0,0,1), period=12)))
$model[[1]]$ma:
[1] 0.48834
$model[[2]]$ma:
[1] 0.58534
$aic:
[1] 852.72
$loglik:
[1] 848.72
$sigma2:
[1] 94629
> sqrt(diag(ddacc.1$var.coef))
[1] 0.11361 0.10556
> ddacc.2 <- arima.mle(ddacc-mean(ddacc),
```

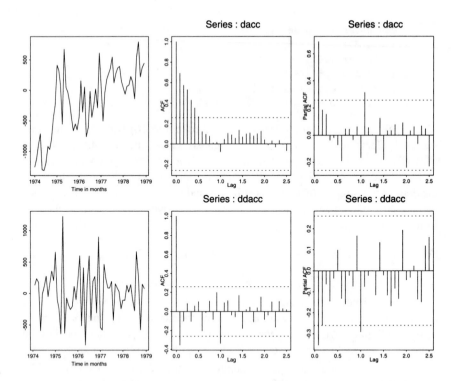

Figure 14.15: Seasonally differenced (top row) and then differenced (bottom row) versions of the accidental deaths series `accdeath` with ACF and PACF plots.

```
        model=list(order=c(0,0,13),
        ma.opt=c(T,F,F,F,F,T,F,F,F,F,F,T,T)),
        max.iter=50, max.fcal=100)
$model$ma:
 [1]   0.60784   0.00000   0.00000   0.00000   0.00000   0.41119
 [7]   0.00000   0.00000   0.00000   0.00000   0.00000   0.67693
[13]  -0.47260
$aic:
 [1] 869.85
$loglik:
 [1] 843.85
$sigma2:
 [1] 70540
> sqrt(diag(ddacc.2$var.coef))
 [1] 0.11473 0.10798 0.10798 0.10798 0.10798 0.10798 0.12052
 [8] 0.10798 0.10798 0.10798 0.10798 0.10798 0.11473
```

The plots (Figure 14.15) suggest the use of $\nabla \nabla_{12} X$, and this has a non-zero mean. The first model fitted is

$$\nabla \nabla_{12} = 28.83 + (1 - 0.488B)(1 - 0.585B^{12})\epsilon$$

and the second model comes from selecting promising non-zero terms in a general MA(13) process, as

$$\nabla\nabla_{12} = 28.83 + (1 - 0.608B - 0.411B^6 - 0.677B^{12} + 0.473B^{13})\epsilon$$

Note that the AIC is wrong; it should be 851.8527 as there are parameters set to zero (although this does not allow for selection). This fit illustrates the ability to constrain coefficients in an ARIMA fit. That standard errors are returned for zero parameters suggests that the standard errors are wrong. Standard likelihood theory suggests deleting rows from the inverse of the information matrix:

```
> dd.VI <- solve(ddacc.2$var.coef)
> sqrt(diag(
    solve(dd.VI[ddacc.2$model$ma.opt,ddacc.2$model$ma.opt])
    ))
[1] 0.096691 0.085779 0.094782 0.095964
```

which shows the power of the S language.

Trading days

Economic and financial series can be affected by the number of trading days in the month or quarter in question. The function `arima.td` computes a multiple time series with seven series giving the number of days in the months and the differences between the number of Saturdays, Sundays, Mondays, Tuesdays, Wednesdays and Thursdays, and the number of Fridays in the month or quarter. This can be used as a regression term in an ARIMA model. It also has other uses. For the `deaths` series we might consider dividing by the number of days in the month to find daily death rates; we already have enough problems with February!

14.4   Multiple time series

The second-order time-domain properties of multiple time series were covered in Section 14.1. The function `ar` will fit AR models to multiple time series, but ARIMA fitting is confined to univariate series. Let $\mathbf{X}_t$ denote a multiple time series, and ϵ_t a correlated sequence of identically distributed random variables. Then a vector AR(p) process is of the form

$$\mathbf{X}_t = \sum_{i}^{p} A_i \mathbf{X}_{t-i} + \epsilon_t$$

for matrices A_i. Further, the components of ϵ_t may be correlated, so we will assume that this has covariance matrix Σ. Again there is a condition on the coefficients, that

$$\det\left[I - \sum_{1}^{p} A_i z^i\right] \neq 0 \text{ for all } |z| \leqslant 1$$

The parameters can be estimated by solving the multiple version of the Yule–Walker equations (Brockwell & Davis, 1991, §11.5), and this is used by `ar.yw`, the function called by `ar`. (The other method, `ar.burg`, also handles multiple series.)

Spectral analysis for multiple time series

The definitions of the spectral density can easily be extended to a pair of series. The cross-covariance is expressed by

$$\gamma_{ij}(t) = \frac{1}{2\pi} \int_{-\pi}^{\pi} e^{i\omega t}\, dF_{ij}(\omega)$$

for a finite complex measure on $(-\pi, \pi]$, which will often have a density f_{ij} so that

$$\gamma_{ij}(t) = \frac{1}{2\pi} \int_{-\pi}^{\pi} e^{i\omega t} f_{ij}(\omega)\, d\omega$$

and

$$f_{ij}(\omega) = \sum_{-\infty}^{\infty} \gamma_{ij}(t) e^{-i\omega t}$$

Note that since $\gamma_{ij}(t)$ is not necessarily symmetric, the sign of the frequency becomes important, and f_{ij} is complex. Conventionally it is written as $c_{ij}(\omega) - i\, q_{ij}(\omega)$ where c is the *co-spectrum* and q is the *quadrature spectrum*. Alternatively we can consider the amplitude $a_{ij}(\omega)$ and phase $\phi_{ij}(\omega)$ of $f_{ij}(\omega)$. Rather than use the amplitude directly, it is usual to work with the *coherence*

$$b_{ij}(\omega) = \frac{a_{ij}(\omega)}{\sqrt{f_{ii}(\omega) f_{jj}(\omega)}}$$

which lies between zero and one.

The *cross-periodogram* is

$$I_{ij}(\omega) = \left[\sum_{s=1}^{n} e^{-i\omega s} X_i(s) \sum_{t=1}^{n} e^{i\omega t} X_j(t) \right] / n$$

and is a complex quantity. It is useless as an estimator of the amplitude spectrum, since if we define

$$J_i(\omega) = \sum_{s=1}^{n} e^{-i\omega s} X_i(s)$$

then

$$|I_{ij}(\omega)| / \sqrt{I_{ii}\omega) I_{jj}(\omega)} = |J_i(\omega) J_j(\omega)^*| / |J_i(\omega)|\, |J_j(\omega)| = 1$$

but smoothed versions can provide sensible estimators of both the coherence and phase.

The function `spec.pgram` will compute the coherence and phase spectra given a multiple time series. The results are shown in Figure 14.16.

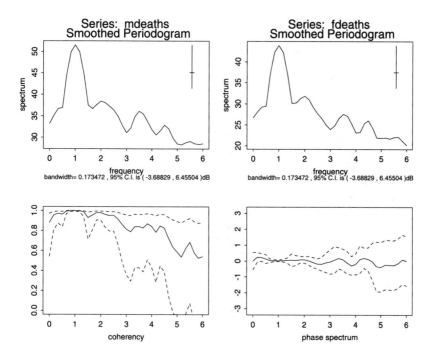

Figure 14.16: Coherence and phase spectra for the two deaths series, with 95% pointwise confidence intervals.

```
spectrum(mdeaths, spans=c(3,3))
spectrum(fdeaths, spans=c(3,3))
mfdeaths.spc <- spec.pgram(ts.union(mdeaths, fdeaths),
    spans=c(3,3))
plot(mfdeaths.spc$freq, mfdeaths.spc$coh, type="l",
    ylim=c(0,1), xlab="coherency", ylab="")
gg <- 2/mfdeaths.spc$df
se <- sqrt(gg/2)
lines(mfdeaths.spc$freq, tanh(atanh(mfdeaths.spc$coh) +
    1.96*se), lty=3)
lines(mfdeaths.spc$freq, tanh(atanh(mfdeaths.spc$coh) -
    1.96*se), lty=3)
plot(mfdeaths.spc$freq, mfdeaths.spc$phase, type="l",
    ylim=c(-pi, pi), xlab="phase spectrum", ylab="")
cl <- asin( pmin( 0.9999, qt(0.95, 2/gg-2)*
    sqrt(gg*(mfdeaths.spc$coh^{-2} - 1)/(2*(1-gg)) ) ) )
lines(mfdeaths.spc$freq, mfdeaths.spc$phase + cl, lty=3)
lines(mfdeaths.spc$freq, mfdeaths.spc$phase - cl, lty=3)
```

These confidence intervals follow Bloomfield (1976, §8.5). At the frequency of 1/year there is a strong signal common to both series, so the coherence is high and both coherence and phase are determined very precisely. At high frequencies there is little information, and the phase cannot be fixed at all precisely.

It is helpful to consider what happens if the series are not aligned:

```
mfdeaths.spc <- spec.pgram(ts.union(mdeaths, lag(fdeaths, 4)),
    spans=c(3,3))
plot(mfdeaths.spc$freq, mfdeaths.spc$coh, type="l",
    ylim=c(0,1), xlab="coherency", ylab="")
gg <- 2/mfdeaths.spc$df
se <- sqrt(gg/2)
lines(mfdeaths.spc$freq, tanh(atanh(mfdeaths.spc$coh) +
    1.96*se), lty=3)
lines(mfdeaths.spc$freq, tanh(atanh(mfdeaths.spc$coh) -
    1.96*se), lty=3)
phase <- (mfdeaths.spc$phase + pi)%%(2*pi) - pi
plot(mfdeaths.spc$freq, phase, type="l",
    ylim=c(-pi, pi), xlab="phase spectrum", ylab="")
cl <- asin( pmin( 0.9999, qt(0.95, 2/gg-2)*
    sqrt(gg*(mfdeaths.spc$coh^{-2} - 1)/(2*(1-gg)) ) ) )
lines(mfdeaths.spc$freq, phase + cl, lty=3)
lines(mfdeaths.spc$freq, phase - cl, lty=3)
```

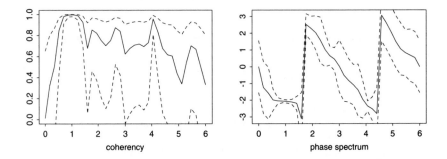

Figure 14.17: Coherence and phase spectra for the re-aligned deaths series, with 95% pointwise confidence intervals.

The results are shown in Figure 14.17. The phase has an added component of slope $2\pi * 4$, since if $X_2(t) = X_1(t - \tau)$,

$$\gamma_{12}(t) = \gamma_{11}(t + \tau), \qquad f_{11}(\omega) = f_{11}(\omega)e^{-i\tau\omega}$$

For more than two series we can consider all the pairwise coherence and phase spectra, which are returned by `spec.pgram`.

14.5 Nottingham temperature data

We now consider a substantial example. The data are mean monthly air temperatures ($^\circ F$) at Nottingham Castle for the months January 1920 – December

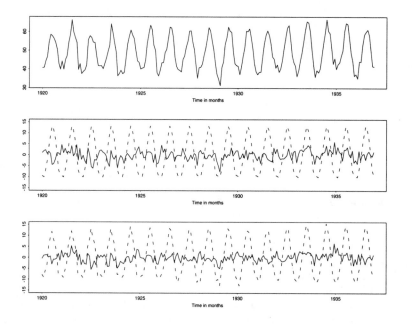

Figure 14.18: Plots of the first 17 years of the `nottem` dataset. Top is the data, middle the `stl` decomposition with a seasonal periodic component and bottom the `stl` decomposition with a 'local' seasonal component.

1939, from *'Meteorology of Nottingham'*, in *City Engineer and Surveyor*. They also occur in Anderson (1976). We will use the years 1920–1936 to forecast the years 1937–1939 and compare with the recorded temperatures. The data are series `nottem` in our library.

```
nott <- window(nottem, end=c(1936,12))
ts.plot(nott)
nott.stl <- stl(nott, "period")
ts.plot(nott.stl$rem-49, nott.stl$sea,
    ylim = c(-15, 15), lty=c(1,3))
nott.stl <- stl(nott, 5)
ts.plot(nott.stl$rem-49, nott.stl$sea,
    ylim = c(-15, 15), lty=c(1,3))
boxplot(split(nott, cycle(nott)), names=month.abb)
```

Figures 14.18 and 14.19 show clearly that February 1929 is an outlier. It *is* correct—it was an exceptionally cold month in England. The `stl` plots show that the seasonal pattern is fairly stable over time. Since the value for February 1929 will distort the fitting process, we altered it to a low value for February of 35. We first model the remainder series:

```
> nott[110] <- 35
> nott.stl <- stl(nott, "period")
> nott1 <- nott.stl$rem - mean(nott.stl$rem)
```

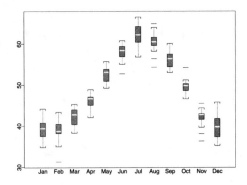

Figure 14.19: Monthly boxplots of the first 17 years of the `nottem` dataset.

```
> acf(nott1)
> acf(nott1,, "partial")
> cpgram(nott1)
> ar(nott1)$aic
 [1] 13.67432  0.00000  0.11133  2.07849  3.40381  5.40125
> plot(0:23, ar(nott1)$aic, xlab="order", ylab="AIC",
      main="AIC for AR(p)")
> nott1.ar1 <- arima.mle(nott1, model=list(order=c(1,0,0)))
> nott1.ar1$model$ar:
[1] 0.27255
> sqrt(nott1.ar1$var.coef)
ar(1) 0.06753
> nott1.fore <- arima.forecast(nott1, n=36,
      model=nott1.ar1$model)
> nott1.fore$mean <- nott1.fore$mean + mean(nott.stl$rem) +
                         nott.stl$sea[1:36]
> ts.plot(window(nottem, 1937), nott1.fore$mean,
      nott1.fore$mean+2*nott1.fore$std.err,
      nott1.fore$mean-2*nott1.fore$std.err, lty=c(3,1,2,2))
> title("via Seasonal Decomposition")
```

(see Figures 14.20 and 14.21) all of which suggest an AR(1) model. (Remember that a seasonal term has been removed, so we expect negative correlation at lag 12.) The confidence intervals in Figure 14.21 for this method ignore the variability of the seasonal terms. We can easily make a rough adjustment. Each seasonal term is approximately the mean of 17 approximately independent observations (since 0.2725516^{12} is negligible). Those observations have variance about 5.05 about the seasonal term, so the seasonal term has standard error about $\sqrt{5.05/17} = 0.55$, compared to the 2.25 for the forecast. The effect of estimating the seasonal terms is in this case negligible. (Note that the forecast errors are correlated with errors in the seasonal terms.)

We now move to the Box-Jenkins methodology of using differencing:

```
> acf(diff(nott,12), 30)
```

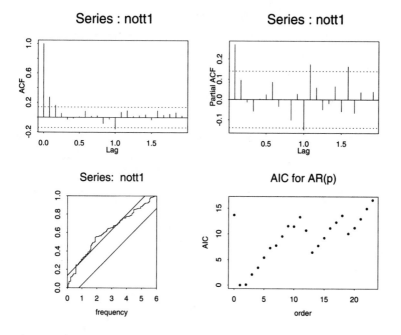

Figure 14.20: Summaries for the remainder series of the `nottem` dataset.

```
> acf(diff(nott,12), 30, "partial")
> cpgram(diff(nott,12))
> nott.arima1 <- arima.mle(nott,
      model=list(list(order=c(1,0,0)), list(order=c(2,1,0),
      period=12)))
> nott.arima1
$model[[1]]$ar:
[1] 0.32425
$model[[2]]$ar:
[1] -0.87576 -0.31311
> sqrt(diag(nott.arima1$var.coef))
[1] 0.073201 0.073491 0.073491
> arima.diag(nott.arima1, gof.lag=24)
> nott.fore <- arima.forecast(nott, n=36,
      model=nott.arima1$model)
> ts.plot(window(nottem, 1937), nott.fore$mean,
      nott.fore$mean+2*nott.fore$std.err,
      nott.fore$mean-2*nott.fore$std.err, lty=c(3,1,2,2))
> title("via Seasonal ARIMA model")
```

(see Figures 14.21 and 14.22) which produces slightly wider confidence intervals, but happens to fit the true values slightly better. The fitted model is

$$(I + 0.875B^s + 0.313B^{2s})(I - B^s)(I - 0.324B)X = \epsilon$$

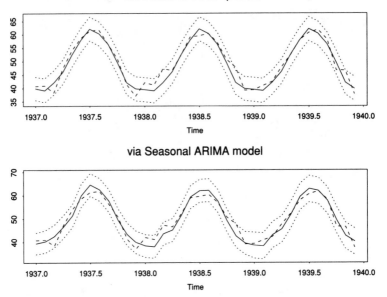

Figure 14.21: Forecasts (solid), true values (dashed) and approximate 95% confidence intervals for the `nottem` series. The upper plot is via a seasonal decomposition, the lower plot via a seasonal ARIMA model.

14.6   Other time-series functions

S-PLUS has a number of time-series functions which are used less frequently and we have not yet discussed. This section is only cursory.

Many of the other functions implement various aspects of filtering, that is converting one times series into another while emphasising some features and de-emphasising others. A linear filter is of the form

$$Y_t = \sum_j a_j X_{t-j}$$

which is implemented by the function `filter`. The coefficients are supplied, and it is assumed that they are non-zero only for $j \geqslant 0$ (`sides=1`) or $-m \leqslant j \leqslant m$ (`sides=2`, the default). A linear filter affects the spectrum by

$$f_Y(\omega) = \left| \sum a_s e^{-is\omega} \right|^2 f_X(\omega)$$

and filters are often described by aspects of the gain function $\left| \sum a_s e^{-is\omega} \right|$. Kernel smoothers such as `ksmooth` are linear filters when applied to regularly-spaced time series.

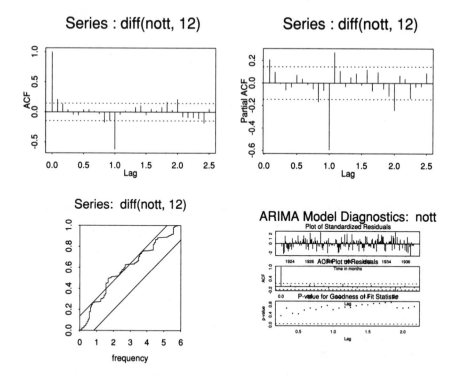

Figure 14.22: Seasonal ARIMA modelling of the `nottem` series. The top row shows the ACF and partial ACF of the yearly differences, which suggest an AR model with seasonal and non-seasonal terms. The bottom row shows the cumulative periodogram of the differenced series and the output of `arima.diag` for the fitted model.

Another way to define a linear filter is recursively (as in exponential smoothing), and this can be done by `filter`, using

$$Y_t = \sum_{s=1}^{\ell} a_s Y_{t-s}$$

in which case ℓ initial values must be specified by the argument `init`.

Converting an ARIMA process to the innovations process ϵ is one sort of recursive filtering, implemented by the function `arima.filt`.

A large number of smoothing operations such as `lowess` can be regarded as filters, but they are non-linear. The functions `acm.filt`, `acm.ave` and `acm.smo` provide filters resistant to outliers.

Complex demodulation is a technique to extract approximately periodic components from a time series. It is discussed in detail by Bloomfield (1976, Chapter 7) and implemented by the S-PLUS function `demod`.

Some time series exhibit correlations which never decay exponentially, as they would for an ARMA process. One way to model this phenomena is fractional differencing (Brockwell & Davis, 1991, §13.2). Suppose we expand ∇^d by a binomial expansion:

$$\nabla^d = \sum_{j=0}^{\infty} \frac{\Gamma(j-d)}{\Gamma(j+1)\Gamma(-d)} B^j$$

and use the right-hand side as the definition for non-integer d. This will only make sense if the series defining $\nabla^d X_t$ is mean-square convergent. A fractional ARIMA process is defined for $d \in (-0.5, 0.5)$ by the assumption that $\nabla^d X_t$ is an ARMA(p, q) process, so

$$\phi(B)\nabla^d X = \theta(B)\epsilon, \qquad \text{so} \qquad \phi(B)X = \theta(B)\nabla^{-d}\epsilon$$

and we can consider it also as an ARMA(p, q) process with fractionally integrated noise. The spectral density is of the form

$$f(\omega) = \sigma^s \left| \frac{\theta(e^{-i\omega})}{\phi(e^{-i\omega})} \right|^2 \times |1 - e^{-i\omega}|^{-2d}$$

and the behaviour as ω^{-2d} at the origin will help identify the presence of fractional differencing.

The functions `arima.fracdiff` and `arima.fracdiff.sim` implement fractionally-differenced ARIMA processes.

14.7 Backwards compatibility

In versions of S-PLUS prior to 3.2, time series were distinguished by an attribute, `tsp`, rather than by a class. As `tsp` had three components, `start`, `end` and `frequency`, such series can easily be coerced to class `rts` by the function `as.rts`. It may be useful, however, to set the `units` and re-create the series via `rts`.

Some readers may only have the earlier versions available to them. The changes required are minimal. Use the function `ts` to create a time series, `tsplot` to plot it, and use `tsmatrix` rather than `ts.intersect`. (There is no direct equivalent of `ts.union`.) For example:

```
> deaths <- ts(scan(n=72), start=1974, frequency=12)
3035 2552 2704 2554 2014 1655 1721 1524 1596 2074 2199 2512
 . . . .
> lh <- ts(scan(n=47))
2.4 2.4 2.4 2.2 2.1 1.5 2.3 2.3 2.5 2.0 1.9 1.7
 . . . .
> tsplot(lh)
> tsplot(deaths, mdeaths, fdeaths, lty=c(1,3,4), xlab="year",
       ylab="deaths")
> acf(tsmatrix(mdeaths, fdeaths))
```

Chapter 15

Spatial Statistics

Spatial statistics is a recent and graphical subject which is ideally suited to implementation in S; S itself includes one spatial interpolation method, `akima`, and `loess` which can be used for two-dimensional smoothing, but the specialist methods of spatial statistics have been added and are given in our library `spatial`. The main references for spatial statistics are Ripley (1981, 1988), Diggle (1983), Upton & Fingleton (1985) and Cressie (1991). Not surprisingly, our notation is closest to Ripley (1981).

Two versions of the trend surface and kriging library are provided, one using C and one purely in S. The S version is slower, and has function names capitalized to avoid confusion. Windows users who have not installed `library(spatial)` and others who are unable to use the dynamically-loaded C code please note that they must use

```
Surf.ls  Surf.gls  Trmat  Prmat  Semat  Correlogram  Variogram
```

rather than the non-capitalized names in the example code. The internal structure of the objects is incompatible between the slow and fast versions.

15.1 Interpolation and kriging

We provide three examples of datasets for spatial interpolation. The dataset `topo` contains 52 measurements of topographic height (in feet) within a square of side 310 feet (labelled in 50 feet units). The datasets `shkap` and `npr1` are measurements on oil fields in the (then) USSR and in the USA. Both contain permeability measurements (a measure of the ease of oil flow in the rock) and `npr1` also has porosity (the volumetric proportion of the rock which is pore space).

Suppose we are given n observations $Z(\mathbf{x}_i)$ and we wish to map the process $Z(\mathbf{x})$ within a region D. (The sample points $\mathbf{x}_i$ are usually, but not always, within D.) Although our treatment will be quite general, our S code assumes D to be a two-dimensional region, which covers the majority of examples. There are however applications to the terrestrial sphere and in three dimensions in mineral and oil applications.

Trend surfaces

One of the earliest methods was that of *trend surfaces*, polynomial regression surfaces of the form

$$f((x,y)) = \sum_{r+s \leqslant p} a_{rs} x^r y^s \qquad (15.1)$$

where the parameter p is the order of the surface. There are $P = (p+1)(p+2)/2$ coefficients. Originally (15.1) was fitted by least-squares, and could for example be fitted using `lm` with `poly` which will give polynomials in one or more variables. There will however be difficulties in prediction, and `predict.gam` must be used to ensure that the correct orthogonal polynomials are generated. This is rather inefficient in applications such as ours in which the number of points at which prediction is needed may far exceed n. Our function `surf.ls` implicitly rescales x and y to $[-1, 1]$, which ensures that the first few polynomials are far from collinear. We show some low-order trend surfaces for the `topo` dataset in Figure 15.1, generated by:

```
library(spatial)
par(pty="s")
topo.ls <- surf.ls(2, topo)
trsurf <- trmat(topo.ls, 0, 6.5, 0, 6.5, 30)
contour(trsurf, levels=seq(600,1000,25))
points(topo)
title("Degree=2")
topo.ls <- surf.ls(3, topo)
    ....
topo.ls <- surf.ls(4, topo)
    ....
topo.ls <- surf.ls(6, topo)
```

Figure 15.1 shows trend surfaces for the `topo` dataset. The highest degree, 6, has 28 coefficients fitted from 52 points. The higher-order surfaces begin to show the difficulties of fitting by polynomials in two or more dimensions, when inevitably extrapolation is needed at the edges.

There are several other ways to show trend surfaces in S. Figure 15.2 shows a greyscale plot from `image` and a perspective plot from `persp`. They were generated by

```
topo.ls <- surf.ls(4, topo)
trsurf <- trmat(topo.ls, 0, 6.5, 0, 6.5, 30)
image(trsurf)
points(topo)
points(perspp(topo$x, topo$y, topo$z, persp(trsurf)))
```

One difficulty with fitting trend surfaces is that in most applications the observations are not regularly spaced, and sometimes they are most dense where the surface is high (for example in mineral prospecting). This makes it important to take the spatial correlation of the errors into consideration. We thus suppose that

$$Z(\mathbf{x}) = \mathbf{f}(\mathbf{x})^T \boldsymbol{\beta} + \epsilon(\mathbf{x})$$

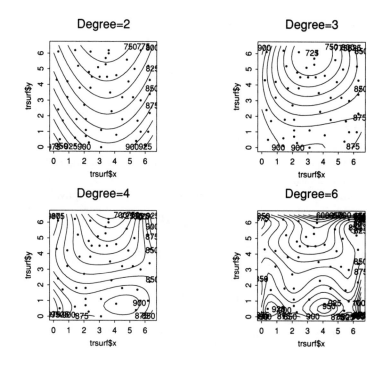

Figure 15.1: Trend surfaces for the `topo` dataset, of degrees 2, 3, 4 and 6.

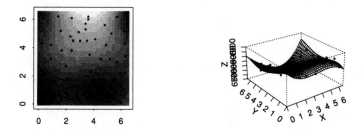

Figure 15.2: The quartic trend surfaces for the `topo` dataset.

for a parametrized trend term such as (15.1) and a zero-mean spatial stochastic process $\epsilon(\mathbf{x})$ of errors. We assume that $\epsilon(\mathbf{x})$ possesses second moments, and has covariance matrix

$$C(\mathbf{x}, \mathbf{y}) = \text{cov}\left(\epsilon(\mathbf{x}), \epsilon(\mathbf{y})\right)$$

(this assumption will be relaxed slightly later). Then the natural way to estimate $\boldsymbol{\beta}$ is by *generalized least squares*, that is to minimize

$$[Z(\mathbf{x}_i) - \mathbf{f}(\mathbf{x}_i)^T\boldsymbol{\beta}]^T[C(\mathbf{x}_i, \mathbf{x}_j)]^{-1}[Z(\mathbf{x}_i) - \mathbf{f}(\mathbf{x}_i)^T\boldsymbol{\beta}]$$

We need some simplified notation. Let $\mathbf{Z} = F\beta + \epsilon$ where

$$F = \begin{bmatrix} \mathbf{f}(x_1)^T \\ \vdots \\ \mathbf{f}(x_n)^T \end{bmatrix}, \qquad \mathbf{Z} = \begin{bmatrix} Z(\mathbf{x}_1) \\ \vdots \\ Z(\mathbf{x}_n) \end{bmatrix}, \qquad \epsilon = \begin{bmatrix} \epsilon(\mathbf{x}_1) \\ \vdots \\ \epsilon(\mathbf{x}_n) \end{bmatrix}$$

and let $K = [C(\mathbf{x}_i, \mathbf{x}_j)]$. We assume that K is of full rank. Then the problem is to minimize

$$[\mathbf{Z} - F\beta]^T K^{-1}[\mathbf{Z} - F\beta] \tag{15.2}$$

The Choleski decomposition (Golub & Van Loan, 1989; Nash, 1990) finds a lower-triangular matrix L such that $K = LL^T$. (The S function chol is unusual in working with $U = L^T$.) Then minimzing (15.2) is equivalent to

$$\min_{\beta} \|L^{-1}[\mathbf{Z} - F\beta]\|^2$$

which reduces the problem to one of ordinary least squares. To solve this we use the QR decomposition (Golub & Van Loan, 1989) of $L^{-1}F$ as

$$QL^{-1}F = \begin{bmatrix} R \\ 0 \end{bmatrix}$$

for an orthogonal matrix Q and upper-triangular $P \times P$ matrix R. Write

$$QL^{-1}\mathbf{Z} = \begin{bmatrix} \mathbf{Y}_1 \\ \mathbf{Y}_2 \end{bmatrix}$$

as the upper P and lower $n - P$ rows. Then $\hat{\beta}$ solves

$$R\hat{\beta} = \mathbf{Y}_1$$

which is easy to compute as R is triangular.

Trend surfaces for the topo data fitted by generalized least squares are shown later (Figure 15.4), where we discuss the choice of the covariance function C.

Local trend surfaces

We have commented on the difficulties of using polynomials as global surfaces. There are two ways to make their effect local. The first is to fit a polynomial surface for each predicted point, using only the nearby data points. The function loess is of this class, and provides a wide range of options. By default it fits a quadratic surface by weighted least-squares, the weights ensuring that 'local' data points are most influential. We will only give details for the span parameter α less than one. Let $q = \lfloor \alpha n \rfloor$, and let δ denote the Euclidean distance to the qth nearest point to $\mathbf{x}$. Then the weights are

$$w_i = \left[1 - \left(\frac{d(\mathbf{x}, \mathbf{x}_i)}{\delta}\right)^3\right]_+^3$$

for the observation at $\mathbf{x}_i$. ([ ]$_+$ denotes the positive part.) Full details of loess are given by Cleveland *et al.* (1992). For our example we have:

```
topo.loess <- loess(z ~ x * y, topo, degree=2, span = 0.25)
topo.mar <- list(x = seq(0, 6.5, 0.1), y=seq(0, 6.5, 0.1))
topo.lop <- predict(topo.loess, expand.grid(topo.mar))
contour(topo.mar$x,topo.mar$y,topo.lop,
   levels = seq(700,1000,25))
points(topo)
topo.loess <- loess(z ~ x * y, topo, degree=1, span = 0.25)
   ....
```

Although loess allows a wide range of smoothing via its parameter span, it is designed for exploratory work and has no way to choose the smoothness except to 'look good'.

The Dirichlet tessellation of a set of points is the set of *tiles*, each of which is associated with a data point, and is the set of points nearer to that data point than any other. There is an associated triangulation, the Delaunay triangulation, in which data points are connected by an edge of the triangulation if and only if their Dirichlet tiles share an edge. (Algorithms and examples are given in Ripley (1981,

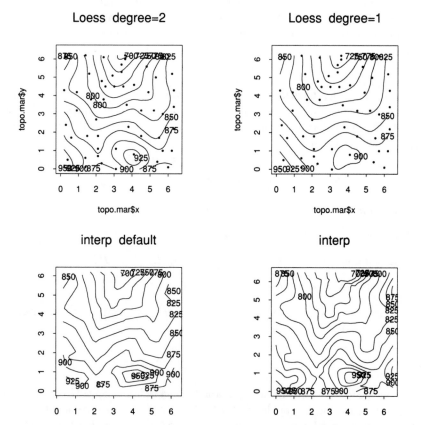

Figure 15.3: loess and interp surfaces for the topo dataset.

§4.3).) Akima's (1978) fitting method fits a fifth-order trend surface within each triangle of the Delaunay triangulation; details are given in Ripley (1981, §4.3). The S implementation is the function `interp`; Akima's example is in datasets `akima.x`, `akima.y` and `akima.z`. The method is forced to interpolate the data, and has no flexibility at all to choose the smoothness of the surface. The arguments `ncp` and `extrap` control details of the method: see the on-line help for details. For Figure 15.3 we used

```
contour(interp(topo$x, topo$y, topo$z),xlab="", ylab="",
   levels = seq(700,1000,25))
contour(interp(topo$x, topo$y, topo$z, topo.mar$x, topo.mar$y,
   ncp=4, extrap=T), xlab="", ylab="",
   levels = seq(700,1000,25))
```

Kriging

Kriging is the name of a technique developed by Matheron in the early 1960s for mining applications which has been independently discovered many times. Journel & Huijbregts (1978) give a comprehensive guide to its application in the mining industry. In its full form, *universal kriging*, it amounts to fitting a process of the form

$$Z(\mathbf{x}) = \mathbf{f}(\mathbf{x})^T \beta + \epsilon(\mathbf{x})$$

by generalized least squares, predicting the value at $\mathbf{x}$ of both terms and taking their sum. Thus it differs from trend-surface prediction which predicts $\epsilon(\mathbf{x})$ by zero. In what is most commonly termed *kriging*, the trend surface is of degree zero, that is a constant.

Our derivation of the predictions is given by Ripley (1981, pp. 48–50). Let $k(\mathbf{x}) = [C(\mathbf{x}, \mathbf{x}_i)]$. The computational steps are

1. Form $K = [C(\mathbf{x}_i, \mathbf{y}_i)]$, with Cholesky decomposition L.

2. Form F and $\mathbf{Z}$

3. Minimize $\|L^{-1}\mathbf{Z} - L^{-1}F\beta\|^2$, reducing $L^{-1}F$ to R.

4. Form $\mathbf{W} = \mathbf{Z} - F\hat{\beta}$, and $\mathbf{y}$ such that $L(L^T y) = \mathbf{W}$.

5. Predict $Z(\mathbf{x})$ by $\hat{Z}(\mathbf{x}) = \mathbf{y}^T k(\mathbf{x}) + \mathbf{f}(\mathbf{x})^T \hat{\beta}$, with error variance given by $C(\mathbf{x}, \mathbf{x}) - \|\mathbf{e}\|^2 + \|\mathbf{g}\|^2$ where

$$L\mathbf{e} = k(\mathbf{x}), \qquad R^T \mathbf{g} = \mathbf{f}(\mathbf{x}) - (L^{-1}F)^T \mathbf{e}$$

This recipe involves only linear algebra and so can be implemented in S. However, for the topographic data predicting a 51×51 grid the C version is about 10 times faster and the standard error calculations are 15 times faster. For the `topo` data we have (Figure 15.4):

```
par(mfrow=c(2,2), pty="s")
topo.ls <- surf.ls(2, topo)
trsurf <- trmat(topo.ls, 0, 6.5, 0, 6.5, 30)
contour(trsurf, levels=seq(700, 1000, 25))
points(topo)
title("LS trend surface")
topo.gls <- surf.gls(2, expcov, topo, d=0.7)
trsurf <- trmat(topo.gls, 0, 6.5, 0, 6.5, 30)
contour(trsurf, levels=seq(700, 1000, 25))
points(topo)
title("GLS trend surface")

prsurf <- prmat(topo.gls, 0, 6.5, 0, 6.5, 50)
contour(prsurf, levels=seq(700, 1000, 25))
points(topo)
title("Kriging prediction")
sesurf <- semat(topo.gls, 0, 6.5, 0, 6.5, 30)
contour(sesurf, levels=c(20, 25))
points(topo)
title("Kriging s.e.")
```

Covariance estimation

To use either generalized least squares or kriging we have to know the covariance function C. We will assume that

$$C(\mathbf{x}, \mathbf{y}) = c(d(\mathbf{x}, \mathbf{y})) \qquad (15.3)$$

where $d()$ is Euclidean distance. (An extension known as *geometric anisotropy* can be incorporated by re-scaling the variables, as we did for the Mahalanobis distance in Chapter 12.) We can compute a *correlogram* by dividing the distance into a number of bins and finding the covariance between pairs whose distance falls into that bin, then dividing by the overall variance.

Choosing the covariance is very much an iterative process, as we need the covariance of the residuals, and the fitting of the trend surface by generalized least squares depends on the assumed form of the covariance function. Further, as we have residuals their covariance function is a biased estimator of c. In practice it is important to get the form right for small distances, for which the bias is least.

Although $c(0)$ must be one, there is no reason why $c(0+)$ should not be less than one. This is known in the kriging literature as a *nugget effect* since it could arise from a very short-range component of the process $Z(\mathbf{x})$. A more general explanation is measurement error. In any case, if there is a nugget effect, the predicted surface will have spikes at the data points, and so effectively will not interpolate but smooth.

The kriging literature tends to work with the *variogram* rather than the covariance function. More properly termed the semi-variogram, this is defined by

$$V(\mathbf{x}, \mathbf{y}) = \frac{1}{2} E[Z(\mathbf{x}) - Z(\mathbf{y})]^2$$

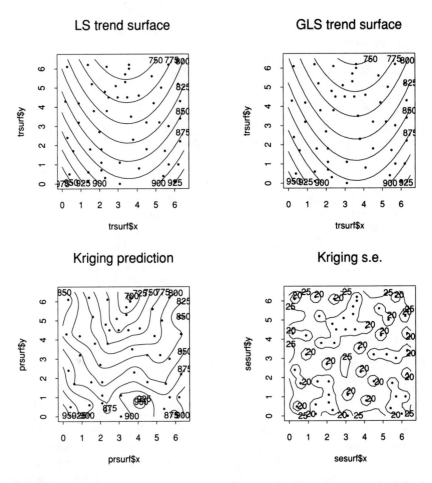

Figure 15.4: Trend surfaces by least squares and generalized least squares, and a kriged surface and standard error of prediction, for the `topo` dataset.

and is related to C by

$$V(\mathbf{x}, \mathbf{y}) = \frac{1}{2}[C(\mathbf{x}, \mathbf{x}) + C(\mathbf{y}, \mathbf{y})] - C(\mathbf{x}, \mathbf{y}) = c(0) - c(d(\mathbf{x}, \mathbf{y}))$$

under our assumption (15.3). However, since different variance estimates will be used in different bins, the empirical versions will not be so exactly related. Much heat and little light emerges from discussions of their comparison.

There are a number of standard forms of covariance functions which are commonly used. A nugget effect can be added to each. The exponential covariance has

$$c(r) = \sigma^2 \exp -r/d$$

the so-called Gaussian covariance is

$$c(r) = \sigma^2 \exp -(r/d)^2$$

and the spherical covariance is in two dimensions

$$c(r) = \sigma^2 \left[1 - \frac{2}{\pi} \left(\frac{r}{d} \sqrt{1 - \frac{r^2}{d^2}} + \sin^{-1} \frac{r}{d} \right) \right]$$

and in three dimensions (but also valid as a covariance function in two)

$$c(r) = \sigma^2 \left[1 - \frac{3d}{2r} + \frac{d^3}{2r^3} \right]$$

for $r \leqslant d$ and zero for $r > d$. Note that this is genuinely local, since points at a greater distance than d from $\mathbf{x}$ are given zero weight at step 5 (although they do affect the trend surface).

We promised to relax the assumption of second-order stationarity slightly. As we only need to predict residuals, we only need a covariance to exist in the space of linear combinations $\sum a_i Z(\mathbf{x}_i)$ which are orthogonal to the trend surface. For degree 0, this corresponds to combinations with sum zero. It is possible that the variogram is finite, without the covariance existing, and there are extensions to more general trend surfaces given by Matheron (1973) and reproduced by Cressie (1991, §5.4). In particular, we can always add a constant to c without affecting the predictions (except perhaps numerically). Thus if the variogram v is specified, we work with covariance function $c = \text{const} - v$ for a suitably large constant. The main advantage is in allowing us to use certain functional forms which do not correspond to covariances, such as

$$v(d) = d^\alpha, 0 \leqslant \alpha < 2 \quad \text{or} \quad d^3 - \alpha d$$

The variogram $d^2 \log d$ corresponds to a thin-plate spline in $\mathbb{R}^2$ (see Wahba, 1990 and the review in Cressie, 1991, §3.4.5).

Our functions `correlogram` and `variogram` allow the empirical correlogram and variogram to be plotted and functions `expcov`, `gaucov` and `sphercov` compute the exponential, Gaussian and spherical covariance functions (the latter in 2 and 3 dimensions) and can be used as arguments to `surf.gls`. For our running example we have

```
topo.kr <- surf.ls(2, topo)
correlogram(topo.kr, 25)
d <- seq(0, 7, 0.1)
lines(d, expcov(d, 0.7))
variogram(topo.kr, 25)
```

see Figure 15.5. We then consider fits by generalized least squares.

```
topo.kr <- surf.gls(2, expcov, topo, d=0.7)
correlogram(topo.kr, 25)
lines(d, expcov(d, 0.7))
lines(d, gaucov(d, 1.0, 0.3), lty=3) # try nugget effect
topo.kr <- surf.gls(2, gaucov, topo, d=1, alph=0.3)
```

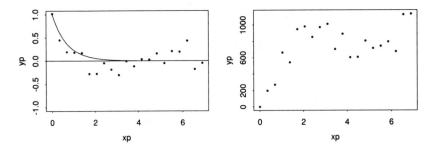

Figure 15.5: Correlogram (left) and variogram (right) for the residuals of `topo` dataset from a least-squares quadratic trend surface.

```
prsurf <- prmat(topo.kr, 0, 6.5, 0, 6.5, 50)
contour(prsurf, levels=seq(700, 1000, 25))
points(topo)
sesurf <- semat(topo.kr, 0, 6.5, 0, 6.5, 25)
contour(sesurf, levels=c(15, 20, 25))
points(topo)

topo.kr <- surf.ls(0, topo)
correlogram(topo.kr, 25)
lines(d, gaucov(d, 2, 0.05))
topo.kr <- surf.gls(0, gaucov, topo, d=2, alph=0.05, nx=10000)
prsurf <- prmat(topo.kr, 0, 6.5, 0, 6.5, 50)
contour(prsurf, levels=seq(700, 1000, 25))
points(topo)
sesurf <- semat(topo.kr, 0, 6.5, 0, 6.5, 25)
contour(sesurf, levels=c(15, 20, 25))
points(topo)
```

We first fit a quadratic surface by least-squares, then try one plausible covariance function (Figure 15.7). Re-fitting by generalized least squares suggests this function and another with a nugget effect, and we predict the surface from both. The first was shown in Figure 15.4, the second in Figure 15.6. We also consider not using a trend surface but a longer-range covariance function, also shown in Figure 15.6. (The small nugget effect is to ensure numerical stability as without it the matrix K is very ill-conditioned; the correlations at short distances are very near one. We increased `nx` for a more accurate look-up table of covariances.)

15.2   Point process analysis

A spatial point pattern is a collection of n points within a region $D \subset \mathbb{R}^2$. The number of points is thought of as random, and the points are considered to be generated by a stationary isotropic point process in $\mathbb{R}^2$. (This means that there is no preferred origin or orientation of the pattern.) For such patterns probably

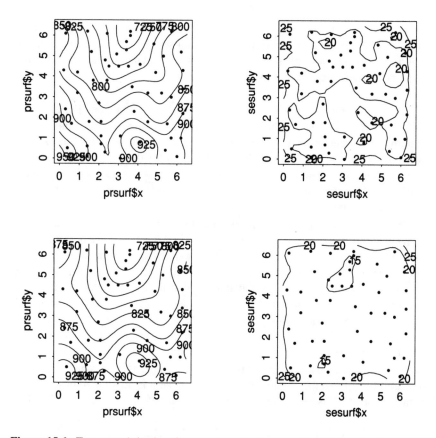

Figure 15.6: Two more kriged surfaces and standard errors of prediction for the `topo` dataset. The top row uses a quadratic trend surface and a nugget effect. The bottom row is without a trend surface.

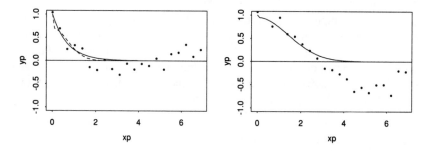

Figure 15.7: Correlograms for the `topo` dataset. (Left) residuals from quadratic trend surface showing exponential covariance (solid) and Gaussian covariance (dashed). (Right) raw data with fitted Gaussian covariance function.

the most useful summaries of the process are the first and second moments of the counts $N(A)$ of the numbers of points within a set $A \subset D$. The first moment can be specified by a single number, the *intensity* λ giving the expected number of points per unit area, obviously estimated by n/a where a denotes the area of D.

The second moment can be specified by Ripley's K function. For example, $\lambda K(t)$ is the expected number of points within distance t of a point of the pattern. The benchmark of complete randomness is the Poisson process, for which $K(t) = \pi t^2$, the area of the search region for the points. Values larger than this indicate clustering on that distance scale, and smaller values indicate regularity. This suggests working with $L(t) = \sqrt{K(t)/\pi}$, which will be linear for a Poisson process.

We only have a single pattern from which to estimate K or L. The definition in the previous paragraph suggests an estimator of $\lambda K(t)$; average over all points of the pattern the number seen within distance t of that point. This would be valid but for the fact that some of the points will be outside D and so invisible. There are a number of edge-corrections available, but that of Ripley (1976) is both simple to compute and rather efficient. This considers a circle centred on the point $\mathbf{x}$ and passing through another point $\mathbf{y}$. If the circle lies entirely within D, the point is counted once. If a proportion $p(\mathbf{x}, \mathbf{y})$ of the circle lies within D, the point is counted as $1/p$ points. (We may want to put a limit on for small p, to reduce the variance at the expense of some bias.) This gives an estimator $\lambda \widehat{K}(t)$ which is unbiased for t up to the circumradius of D (so that it is possible to observe two points $2t$ apart). Since we do not know λ, we estimate it by $\hat{\lambda} = n/a$. Finally

$$\widehat{K}(t) = \frac{a}{n^2} \sum_{\mathbf{x} \in D, d(\mathbf{y},\mathbf{x}) \leqslant t} \frac{1}{p(\mathbf{x}, \mathbf{y})}$$

and obviously we estimate $L(t)$ by $\sqrt{\widehat{K}(t)/\pi}$. We find that on square-root scale the variance of the estimator varies little with t.

Our first example is the Swedish pines data of Ripley (1981, §8.6). This records 72 trees within a 10-metre square. Figure 15.8 shows that $\widehat{L}$ is not straight, and comparison with simulations from a binomial process (a Poisson process conditioned on $N(D) = n$, so n independently uniformly distributed points within D) shows that the lack of straightness is significant. The upper two panels of Figure 15.8 were produced by the following code:

```
library(spatial)
pines <- ppinit("pines.dat")
par(mfrow=c(2,2), pty="s")
plot(pines, xlim=c(0,10), ylim=c(0,10), xlab="", ylab="",
    xaxs="i", yaxs="i")
plot(Kfn(pines,5), type="s", xlab="distance", ylab="L(t)")
lims <- Kenvl(5, 100, Psim(72))
lines(lims$x, lims$l, lty=2)
lines(lims$x, lims$u, lty=2)
```

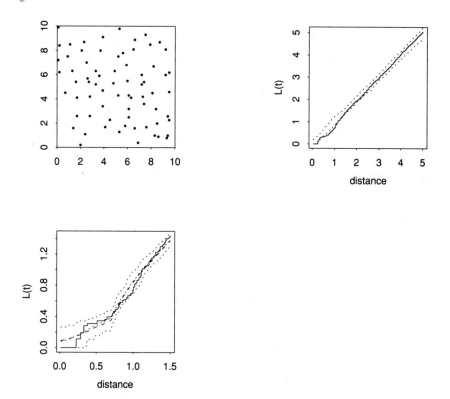

Figure 15.8: The Swedish pines dataset from Ripley (1981), with two plots of $L(t)$. That at the upper right shows the envelope of 100 binomial simulations, that at the lower left the average and the envelope (dotted) of 100 simulations of a Strauss process with $c = 0.2$ and $R = 0.7$. Also shown (dashed) is the average for $c = 0.15$. All units are in metres.

The function `ppinit` reads the data from the file and also the coordinates of a rectangular domain D. The latter can be reset, or set up for simulations, by the function `ppregion`. (It *must* be set for each session.) The function `Kfn` returns an estimate of $L(t)$ and other useful information for plotting, for distances up to its second argument `fs` (for full-scale).

The functions `Kaver` and `Kenvl` return the average and, for `Kenvl`, also the extremes of K-functions (on L scale) for a series of simulations. The function `Psim(n)` simulates the binomial process on n points within the domain D which has already been set.

Alternative processes

We need to consider alternative point processes to the Poisson. One of the most useful for regular point patterns is the so-called Strauss process, which is simulated by `Strauss(n, c, r)`. This has a density of n points proportional to

$$c^{\text{number of } R\text{-close pairs}}$$

and so has $K(t) < \pi t^2$ for $t \leqslant R$ (and up to about $2R$). For $c = 0$ we have a 'hard-core' process which never generates pairs closer than R and so can be envisaged as laying down the centres of non-overlapping discs of diameter $r = R$.

Figure 15.8 also shows the average and envelope of the L-plots for a Strauss process fitted to the pines data by Ripley (1981). There the parameters were chosen by trial-and-error based on a knowledge of how the L-plot changed with (c, R). Ripley (1988) considers the estimation of c for known R by the pseudo-likelihood. This is done by our function pplik and returns an estimate of about $c = 0.15$ ('about' since it uses numerical integration). As Figure 15.8 shows, the difference between $c = 0.2$ and $c = 0.15$ is small. We used the following code:

```
ppregion(pines)
plot(Kfn(pines,1.5), type="s", xlab="distance", ylab="L(t)")
lims <- Kenvl(1.5, 100, Strauss(72, 0.2, 0.7))
lines(lims$x, lims$a, lty=2)
lines(lims$x, lims$l, lty=2)
lines(lims$x, lims$u, lty=2)
pplik(pines, 0.7)
lines(Kaver(1.5, 100, Strauss(72, 0.15, 0.7)), lty=3)
```

which took about 45 seconds.

The theory is given by Ripley (1988, p. 67). For a point $\xi \in D$ let $t(\xi)$ denote the number of points of the pattern within distance t of ξ. Then the pseudo-likelihood estimator solves

$$\frac{\int_D t(\xi) c^t(\xi) \, d\xi}{\int_D c^t(\xi) \, d\xi} = \frac{\#(R\text{-close pairs})}{n} = \frac{n\widehat{K}(R)}{a}$$

and the left-hand side is an increasing function of c. The function pplik uses the S-PLUS function uniroot to find a solution in the range $(0, 1]$.

Other processes for which simulation functions are provided are the binomial process (Psim(n)) and Matérn's sequential spatial inhibition process (SSI(n, r)) which sequentially lays down centres of discs of radius r that do not overlap existing discs.

References

Abbey, S. (1988) Robust measures and the estimator limit. *Geostandards Newsletter* **12**, 241–248.

Aitchison, J. (1986) *The Statistical Analysis of Compositional Data.* London: Chapman and Hall.

Aitkin, M. (1978) The analysis of unbalanced cross classifications (with discussion). *Journal of the Royal Statistical Society A* **141**, 195–223.

Aitkin, M. and Clayton, D. (1980) The fitting of exponential, Weibull, and extreme value distributions to complex censored survival data using GLIM. *Applied Statistics* **29**, 156–163.

Akaike, H. (1974) A new look at statistical model identification. *IEEE Transactions on Automatic Control* **AU–19**, 716–722.

Akima, H. (1978) A method of bivariate interpolation and smooth surface fitting for irregularly distributed data points. *ACM Transactions on Mathematical Software* **4**, 148–159.

Analytical Methods Committee (1987) Recommendations for the conduct and interpretation of co-operative trials. *The Analyst* **112**, 679–686.

Analytical Methods Committee (1989a) Robust statistics — how not to reject outliers. Part 1. Basic concepts. *The Analyst* **114**, 1693–1697.

Analytical Methods Committee (1989b) Robust statistics — how not to reject outliers. Part 2. Inter-laboratory trials. *The Analyst* **114**, 1699–1702.

Andersen, P. K., Borgan, Ø., Gill, R. D. and Keiding, N. (1993) *Statistical Models based on Counting Processes.* New York: Springer-Verlag.

Anderson, E. (1935) The irises of the Gaspe Peninsula. *Bulletin of the American Iris Society*, **59**, 2–5.

Anderson, O. D. (1976) *Time Series Analysis and Forecasting. The Box-Jenkins Approach.* London: Butterworths.

Atkinson, A. C. (1985) *Plots, Transformations and Regression.* Oxford: Oxford University Press.

Atkinson, A. C. (1986) Comment: Aspects of diagnostic regression analysis. *Statistical Science* **1**, 397-402.

Atkinson, A. C. (1988) Transformations unmasked. *Technometrics* **30**, 311–318.

Azzalini, A. and Bowman, A. W. (1990) A look at some data on the Old Faithful geyser. *Applied Statistics* **39**, 357-365.

Banfield, J. D. and Raftery, A. E. (1993) Model-based Gaussian and non-Gaussian clustering. *Biometrics* **49**, 803–821.

Barnett, V. and Lewis, T. (1985) *Outliers in Statistical Data.* Second Edition. Chichester: John Wiley and Sons.

Bassett, G. W. and Koenker, R. (1978) Asymptotic theory of least absolute error regression. *Journal of the American Statistical Association* **73**, 618–622.

Bates, D. M. and Chambers, J. M. (1992) Nonlinear models. Chapter 10 of Chambers and Hastie (1992).

Bates, D. M. and Watts, D. G. (1980) Relative curvature measures of nonlinearity (with discussion). *Journal of the Royal Statistical Society, Series B* **42**, 1–25.

Bates, D. M. and Watts, D. G. (1988) *Nonlinear Regression Analysis and its Applications.* New York: John Wiley and Sons.

Beale, E. M. L. (1960) Confidence intervals in non-linear estimation (with discussion). *Journal of the Royal Statistical Society B* **22**, 41–88.

Becker, R. A. and Chambers, J. M. (1988) Auditing of data analyses. *SIAM Journal of Scientific and Statistical Computing* **9**, 747–760.

Becker, R. A., Chambers, J. M. and Wilks, A. R. (1988) *The NEW S Language.* New York: Chapman and Hall. (Formerly Monterey: Wadsworth and Brooks/Cole.)

Bie, O, Borgan, Ø. and Liestøl, K. (1987) Confidence intervals and confidence bands for the cumulative hazard rate function and their small sample properties. *Scandinavian Journal of Statistics* **14**, 221–233.

Bishop, Y. M. M., Fienberg, S. E. and Holland, P. W. (1975) *Discrete Multivariate Analysis.* Cambridge, MA.: MIT Press.

Bloomfield, P. (1976) *Fourier Analysis of Time Series: An Introduction.* New York: John Wiley and Sons.

Bloomfield, P. and Steiger, W. L. (1983) *Least Absolute Deviations: Theory, Applications, and Algorithms.* Boston: Birkhauser.

Borgan, Ø. and Liestøl, K. (1990) A note on confidence intervals and bands for the survival function based on transformations. *Scandinavian Journal of Statistics* **17**, 35–41.

Box, G. E. P. and Cox, D. R. (1964) The analysis of transformations (with discussion). *Journal of the Royal Statistical Society B* **26**, 211–252.

Box, G. E. P. Hunter, W. G. and Hunter, J. S. (1978) *Statistics for Experimenters.* New York: John Wiley and Sons.

Box, G. E. P. and Pierce, D. A. (1970) Distribution of residual autocorrelations in autoregressive-integrated moving average time series models. *Journal of the American Statistical Association* **65**, 1509–1526.

Breiman, L. and Friedman, J. H. (1985) Estimating optimal transformations for multiple regression and correlations (with discussion). *Journal of the American Statistical Association* **80**, 580–619.

Breiman, L., Friedman, J. H., Olshen, R. A. and Stone, C. J. (1984) *Classification and Regression Trees.* Monterey: Wadsworth and Brooks/Cole.

Brockwell, P. J. and Davis, R. A. (1991) *Time Series: Theory and Methods.* Second Edition. New York: Springer-Verlag.

Brownlee, K. A. (1965) *Statistical Theory and Methodology in Science and Engineering.* Second Edition. New York: John Wiley and Sons.

Campbell, N. A. and Mahon, R. J. (1974) A multivariate study of variation in two species of rock crab of genus *Leptograpsus. Australian Journal of Zoology* **22**, 417–425.

Chakrapani, T. K. and Ehrenberg, A. S. C. (1981) An alternative to factor analysis in marketing research – Part 2: Between group analysis. *Professional Marketing Research Society Journal* **1**, 32–38.

Chambers, J. M. and Hastie, T. J. (eds) (1992) *Statistical Models in S.* New York: Chapman and Hall. (Formerly Monterey: Wadsworth and Brooks/Cole.)

Chernoff, H. (1973) The use of faces to represent points in k-dimensional space graphically. *Journal of the American Statistical Association* **68**, 361–368.

Ciampi, A., Chang, C.-H., Hogg, S. and McKinney, S. (1987) Recursive partitioning: a versatile method for exploratory data analysis in biostatistics. In *Biostatistics* (ed I. B. McNeil and G. J. Umphrey), pp. 23–50. New York: Reidel.

Clark, L. A. and Pregibon, D. (1992) Tree-based models. Chapter 9 of Chambers and Hastie (1992).

Cleveland, R. B., Cleveland, W. S., McRae, J. E. and Terpenning, I. (1990) STL: A seasonal-trend decomposition procedure based on loess (with discussion). *Journal of Official Statistics* **6**, 3–73.

Cleveland, W. S. (1993) *Visualizing Data*. Summit, NJ: Hobart Press.

Cleveland, W. S., Grosse, E. and Shyu, W. M. (1992) Local regression models. Chapter 8 of Chambers and Hastie (1992).

Collett, D. (1991) *Modelling Binary Data*. London: Chapman and Hall.

Cox, D. R. (1972) Regression models and life-tables (with discussion). *Journal of the Royal Statistical Society B* **34**, 187–220.

Cox, D. R. and Oakes, D. (1984) *Analysis of Survival Data*. London: Chapman and Hall.

Cox, D. R. and Snell, E. J. (1989) *The Analysis of Binary Data*. Second Edition. London: Chapman and Hall.

Cressie, N. A. C. (1991) *Statistics for Spatial Data*. New York: John Wiley and Sons.

Cybenko, G. (1989) Approximation by superpositions of a sigmoidal function. *Mathematics of Controls, Signals, and Systems* **2**, 303–314.

Daniel, C. and Wood, F. S. (1971) *Fitting Equations to Data*. New York: John Wiley and Sons.

Darroch, J. N. and Ratcliff, D. (1972) Generalized iterative scaling for log-linear models. *Annals of Mathematical Statistics* **43**, 1470–1480.

Davenport, M. and Studdert-Kennedy, G. (1972) The statistical analysis of aesthetic judgement: an exploration. *Applied Statistics* **21**, 324–333.

Davies, P. L. (1993) Aspects of robust linear regression. *Annals of Statistics* **21**, 1843–1899.

Davison, A. C. and Snell, E. J. (1991) Residuals and diagnostics. Chapter 4 of Hinkley, Reid and Snell (1991).

Deming, W. E. and Stephan, F. F. (1940) On a least-squares adjustment of a sampled frequency table when the expected marginal totals are known. *Annals of Mathematical Statistics* **11**, 427–444.

Devijver, P. A. and Kittler, J. V. (1982) *Pattern Recognition: A Statistical Approach*. Englewood Cliffs, NJ: Prentice/Hall.

Diaconis, P. and Shahshahani, M (1984) On non-linear functions of linear combinations. *SIAM Journal of Scientific and Statistical Computing* **5**, 175–191.

Diggle, P. J. (1983) *Statistical Analysis of Spatial Point Patterns*. London: Academic Press.

Diggle, P. J. (1990) *Time Series: A Biostatistical Introduction*. Oxford: Oxford University Press.

Dixon, W. J. (1960) Simplified estimation for censored normal samples. *Annals of Mathematical Statistics* **31**, 385–391.

Draper, N. R. and Smith, H. (1981) *Applied Regression Analysis*. Second Edition. New York: John Wiley and Sons.

Efron, B. (1982) *The Jackknife, the Bootstrap, and Other Resampling Plans*. Philadelphia: Society for Industrial and Applied Mathematics.

Efron, B. and Tibshirani, R. (1993) *An Introduction to the Bootstrap*. New York: Chapman and Hall.

Ein-Dor, P. and Feldmesser, J. (1987) Attributes of the performance of central processing units: A relative performance prediction model, *Communications of the ACM* **30**, 308–317.

Emerson, J. D. and Hoaglin, D. C. (1983) Analysis of two-way tables by medians. In Hoaglin *et al.* (1983), pp. 165–210.

Emerson, J. D. and Wong, G. Y. (1985) Resistant non-additive fits for two-way tables. In Hoaglin *et al.* (1985), pp. 67–124.

Everitt, B. S. and Hand, D. J. (1981) *Finite Mixture Distributions*. London: Chapman and Hall.

Feigl, P. and Zelen, M. (1965) Estimation of exponential survival probabilities with concomitant information. *Biometrics* **21**, 826–838.

Firth, D. (1991) Generalized linear models. Chapter 3 of Hinkley, Reid and Snell (1991).

Fisher, R. A. (1925) Theory of statistical estimation. *Proceedings of the Cambridge Philosophical Society* **22**, 700–725.

Fisher, R. A. (1936) The use of multiple measurements in taxonomic problems. *Annals of Eugenics (London)* **7**, 179–188.

Fisher, R. A. (1940) The precision of discriminant functions. *Annals of Eugenics (London)* **10**, 422–429.

Fisher, R. A. (1947) The analysis of covariance method for the relation between a part and the whole. *Biometrics* **3**, 65–68.

Fleming, T. R. and Harrington, D. P. (1981) A class of hypothesis tests for one and two sample censored survival data. *Communications in Statistics* **A10(8)**, 763–94.

Fleming, T. R. and Harrington, D. P. (1991) *Counting Processes and Survival Analysis*, New York: John Wiley and Sons.

Freedman, D. and Diaconis, P. (1981) On the histogram as a density estimator: L_2 theory. *Zeitschrift für Wahrscheinlichkeitstheorie und verwandte Gebiete* **57**, 453–476.

Friedman, J. H. (1987) Exploratory projection pursuit. *Journal of the American Statistical Association* **82**, 249–266.

Friedman, J. H. (1991) Multivariate adaptive regression splines (with discussion). *Annals of Statistics* **19**, 1–141.

Friedman, J. H. and Rafsky, L. C. (1981) Graphics for the multivariate two-sample problem (with discussion). *Journal of the American Statistical Association* **76**, 277–295.

Friedman, J. H. and Stuetzle, W. (1981) Projection pursuit regression. *Journal of the American Statistical Association* **76**, 817–823.

Funahashi, K. (1989) On the approximate realization of continuous mappings by neural networks. *Neural Networks* **2**, 183–192.

Gehan, E. A. (1965) A generalized Wilcoxon test for comparing arbitrarily singly-censored samples. *Biometrika* **52**, 203–223.

Golub, G. H. and Van Loan, C. F. (1989) *Matrix Computations*. Second Edition. Baltimore: Johns Hopkins University Press.

Goodall, C. (1983) M-estimators of location: An outline of the theory. In Hoaglin *et al.* (1983), pp. 339–403.

Goodman, L. A. (1978) *Analyzing Qualitative/Categorical Data: Log-Linear Models and Latent-Structure Analysis.* Cambridge, MA: Abt Books.

de Gooijer, J. G., Abraham, B., Gould, A. and Robinson, L. (1985) Methods for determining the order of an autoregressive-moving average process: A survey. *International Statistical Review* **53**, 301–329.

Gordon, A. D. (1981) *Classification: Methods for the Exploratory Analysis of Multivariate Data.* London: Chapman and Hall.

Haberman, S. J. (1978) *Analysis of Qualitative Data. Volume 1: Introductory Topics.* New York: Academic Press.

Haberman, S. J. (1979) *Analysis of Qualitative Data. Volume 2: New Developments.* New York: Academic Press.

Hampel, F. R., Ronchetti, E. M., Rousseeuw, P. J. and Stahel, W. A. (1986) *Robust Statistics. The Approach Based on Influence Functions.* New York: John Wiley and Sons.

Härdle, W. (1991) *Smoothing Techniques with Implementation in* S. New York: Springer-Verlag-Verlag.

Hartigan, J. A. (1975) *Clustering Algorithms.* New York: John Wiley and Sons.

Hartigan, J. A. and Wong, M. A. (1979) A K-means clustering algorithm. *Applied Statistics* **28**, 100–108.

Hastie, T. J. and Tibshirani, R. J. (1990) *Generalized Additive Models.* London: Chapman and Hall.

Heiberger, R. M. (1989) *Computation for the Analysis of Designed Experiments.* New York: John Wiley and Sons.

Heiberger, R. M. and Becker, R. A. (1992) Design of an S function for robust regression using iteratively reweighted least squares. *Journal of Computational and Graphical Statistics* **1**, 181–196.

Hertz, J., Krogh, A. and Palmer, R. G. (1991) *Introduction to the Theory of Neural Computation.* Redwood City, CA: Addison-Wesley.

Hettmansperger, T. P. and Sheather, S. J. (1992) A cautionary note on the method of least median squares. *American Statistician* **46**, 79–83.

Hinkley, D. V., Reid, N. and Snell, E. J. eds (1991) *Statistical Theory and Modelling. In honour of Sir David Cox, FRS.* London: Chapman and Hall.

Hoaglin, D. C., Mosteller, F. and Tukey, J. W., eds (1983) *Understanding Robust and Exploratory Data Analysis.* New York: John Wiley and Sons.

Hoaglin, D. C., Mosteller, F. and Tukey, J. W., eds (1985) *Exploring Data Tables: Trends and Shapes.* New York: John Wiley and Sons.

Hoaglin, D. C., Mosteller, F. and Tukey, J. W., eds (1991) *Fundamentals of the Exploratory Analysis of Variance.* New York: John Wiley and Sons.

Hornik, K., Stinchcombe, M. and White, H. (1989) Multilayer feedforward networks are universal approximators. *Neural Networks* **2**, 359–366.

Hosmer, D. W. and Lemeshow, S. (1989) *Applied Logistic Regression.* New York: John Wiley and Sons.

Huber, P. J. (1981) *Robust Statistics.* New York: John Wiley and Sons.

Huber, P. J. (1985) Projection pursuit (with discussion). *Annals of Statistics* **13**, 435–525.

Iglewicz, B. (1983) Robust scale estimators and confidence intervals for location. In Hoaglin *et al.* (1983), pp. 405–431.

Jackson, J. E. (1991) *A User's Guide to Principal Components.* New York: John Wiley and Sons.

Jardine, N. and Sibson, R. (1971) *Mathematical Taxonomy.* London: John Wiley and Sons.

John, P. W. M. (1971) *Statistical Design and Analysis of Experiments.* New York: Macmillan.

Jolliffe, I. T. (1986) *Principal Component Analysis.* New York: Springer-Verlag.

Jones, M. C. and Sibson, R. (1987) What is projection pursuit? (with discussion). *Journal of the Royal Statistical Society A* **150**, 1–36.

Journel, A. G. and Huijbregts, Ch. J. (1978) *Mining Geostatistics.* London: Academic Press.

Kalbfleisch, J. D. and Prentice, R. L. (1980) *The Statistical Analysis of Failure Time Data.* New York: John Wiley and Sons.

Kaufman, L. and Rousseeuw, P. J. (1990) *Finding Groups in Data. An Introduction to Cluster Analysis.* New York: John Wiley and Sons.

Klein, J. P. (1991) Small sample moments of some estimators of the variance of the Kaplan-Meier and Nelson-Aalen estimators. *Scandinavian Journal of Statistics* **18**, 333–340.

Knuth, D. E. (1968) *The Art of Computer Programming, Volume 1: Fundamental Algorithms.* Reading, MA: Addison-Wesley.

Kohonen, T. (1990) The self-organizing map. *Proceedings IEEE* **78**, 1464–1480.

Krzanowski, W. J. (1988) *Principles of Multivariate Analysis. A User's Perspective.* Oxford: Oxford University Press.

Lawless, J. F. (1982) *Statistical Models and Methods for Lifetime Data.* New York: John Wiley and Sons.

Lawless, J. F. (1987) Negative binomial and mixed poisson regression. *Canadian Journal of Statistics* **15**, 209–225.

MacQueen, J. (1967) Some methods for classification and analysis of multivariate observations. In *Proc. 5th Berkeley Symp. Math. Statist. Probab.* **1**, pp. 281–297. Berkeley, CA: University of California Press.

Mandel, J. (1969) A method of fitting empirical surfaces to physical or chemical data. *Technometrics* **11**, 411–429.

Marazzi, A. (1993) *Algorithms, Routines and S Functions for Robust Statistics.* Pacific Grove, CA: Wadsworth and Brooks/Cole.

Mardia, K. V., Kent, J. T. and Bibby, J. M. (1979) *Multivariate Analysis.* London: Academic Press.

Matheron, G. (1973) The intrinsic random functions and their applications. *Advances in Applied Probability* **5**, 439–468.

McCullagh, P. and Nelder, J. A. (1989) *Generalized Linear Models.* Second Edition. London: Chapman and Hall.

McCulloch, W. S. and Pitts, W. (1943) A logical calculus of ideas immanent in neural activity. *Bulletin of Mathematical Biophysics* **5**, 115–133.

McLachlan, G. J. (1992) *Discriminant Analysis and Statistical Pattern Recognition.* New York: John Wiley and Sons.

McLachlan, G. J. and Basford, K. E. (1988) *Mixture Models: Inference and Applications to Clustering.* New York: Marcel Dekker.

Michie, D. (1989) Problems of computer-aided concept formation. In *Applications of Expert Systems 2*, ed. J. R. Quinlan, Glasgow: Turing Institute Press / Addison-Wesley, pp. 310–333.

Miller, R. G. (1981) *Survival Analysis*, New York: John Wiley and Sons.

Morgan, J. N. and Messenger, R. C. (1973) *THAID: a Sequential Search Program for the Analysis of Nominal Scale Dependent Variables*. Survey Research Center, Institute for Social Research, University of Michigan.

Morgan, J. N. and Sonquist, J. A. (1963) Problems in the analysis of survey data, and a proposal. *Journal of the American Statistical Association* **58**, 415–434.

Mosteller, F. and Tukey, J. W. (1977) *Data Analysis and Regression*. Reading, MA: Addison-Wesley.

Nagelkirke, N. J. D. (1991) A note on a general definition of the coefficient of determination. *Biometrika* **78**, 691–2.

Narula, S. C. and Wellington, J. F. (1982) The minimum sum of absolute errors regression: A state of the art survey. *International Statistical Review* **50**, 317–326.

Nash, J. C. (1990) *Compact Numerical Methods for Computers. Linear Algebra and Function Minimization*. Second Edition. Bristol: Adam Hilger.

Nelson, W. D. and Hahn, G. J. (1972) Linear estimation of a regression relationship from censored data. Part 1 – simple methods and their application (with discussion). *Technometrics* **14**, 247–276.

Peto, R. and Peto, J. (1972) Asymptotically efficient rank invariant test procedures (with discussion). *Journal of the Royal Statistical Society A* **135**, 185–206.

Plackett, R. L. (1974) *The Analysis of Categorical Data*. London: Griffin.

Priestley, M. B. (1981) *Spectral Analysis and Time Series*. London: Academic Press.

Quinlan, J. R. (1979) Discovering rules by induction from large collections of examples. In *Expert Systems in the Microelectronic Age*, ed. D. Michie, Edinburgh: Edinburgh University Press.

Quinlan, J. R. (1983) Learning efficient classification procedures and their application to chess end-games. In *Machine Learning* eds R. S. Michalski, J. G. Carbonell and T. M. Mitchell, pp. 463–482. Palo Alto: Tioga.

Quinlan, J. R. (1986) Induction of decision trees. *Machine Learning* **1**, 81–106.

Quinlan, J. R. (1993) *C4.5: Programs for Machine Learning*. San Mateo, CA: Morgan Kaufmann.

Rao, C. R. (1971a) Estimation of variance and covariance components—MINQUE theory. *Journal of Multivariate Analysis* **1**, 257–275.

Rao, C. R. (1971b) Minimum variance quadratic unbiased estimation of variance components. *Journal of Multivariate Analysis* **1**, 445–456.

Rao, C. R. (1973) *Linear Statistical Inference and its Applications*. Second Edition. New York: John Wiley and Sons.

Rao, C. R. and Kleffe, J. (1988) *Estimation of Variance Components and Applications*. Amsterdam: North-Holland.

Ratkowsky, D. A. (1983) *Nonlinear Regression Modeling: A Unified Practical Approach*. New York: Marcel Dekker.

Ratkowsky, D. A. (1990) *Handbook of Nonlinear Regression Models*. New York: Marcel Dekker.

Ries, P. N. and Smith, H. (1963) The use of chi-square for preference testing in multidimensional problems. *Chemical Engineering Progress* **59**, 39–43.

Ripley, B. D. (1976) The second-order analysis of stationary point processes. *Journal of Applied Probability* **13**, 255–266.

Ripley, B. D. (1981) *Spatial Statistics*. New York: John Wiley and Sons.

Ripley, B. D. (1987) *Stochastic Simulation*. New York: John Wiley and Sons.

Ripley, B. D. (1988) *Statistical Inference for Spatial Processes*. Cambridge: Cambridge University Press.

Ripley, B. D. (1993) Statistical aspects of neural networks. In *Networks and Chaos – Statistical and Probabilistic Aspects* eds O. E. Barndorff-Nielsen, J. L. Jensen and W. S. Kendall, pp. 40–123. London: Chapman and Hall.

Ripley B. D. (1994) Neural networks and flexible regression and discrimination. *Advances in Applied Statistics* 2, pp. 39–57, Abingdon: Carfax.

Roeder, K. (1990) Density estimation with confidence sets exemplified by superclusters and voids in galaxies. *Journal of the American Statistical Association* 85, 617–624.

Ross, G. J. S. (1970) The efficient use of function minimization in non-linear maximum-likelihood estimation. *Applied Statistics* 19, 205–221.

Ross, G. J. S. (1990) *Nonlinear Estimation*. New York: Springer-Verlag.

Rousseeuw, P. J. and Leroy, A. M. (1987) *Robust Regression and Outlier Detection*. New York: John Wiley and Sons.

Rousseeuw, P. J. and van Zomeren, B. C. (1990) Unmasking multivariate outliers and leverage points (with discussion). *Journal of the American Statistical Association* 85, 633–651.

Sammon, J. W. (1969) A non-linear mapping for data structure analysis. *IEEE Trans. Comput* C-18, 401–409.

Scheffé, H. (1959) *The Analysis of Variance*. New York: John Wiley and Sons.

Scott, D. W. (1979) On optimal and data-based histograms. *Biometrika* 66, 605–610.

Scott, D. W. (1992) *Multivariate Density Estimation. Theory, Practice, and Visualization*. New York: John Wiley and Sons.

Seber, G. A. F. and Wild, C. J. (1989) *Nonlinear Regression*. New York: John Wiley and Sons.

Sheather, S. J. and Jones, M. C. (1991) A reliable data-based bandwidth selection method for kernel density estimation. *Journal of the Royal Statistical Society B* 53, 683–690.

Silverman, B. W. (1985) Some aspects of the spline smoothing approach to non-parametric regression curve fitting. *Journal of the Royal Statistical Society B* 47, 1–52.

Silverman, B. W. (1986) *Density Estimation for Statistics and Data Analysis*. London: Chapman and Hall.

Solomon, P. J. (1984) Effect of misspecification of regression models in the analysis of survival data. *Biometrika* 71, 291–298.

Stace, C. (1991) *New Flora of the British Isles*. Cambridge: Cambridge University Press.

Staudte, R. G. and Sheather, S. J. (1990) *Robust Estimation and Testing*. New York: John Wiley and Sons.

Stevens, W. L. (1948) Statistical analysis of a non-orthogonal tri-factorial experiment. *Biometrika* 35, 346–367.

Stone, M. (1974) Cross-validatory choice and assessment of statistical predictions (with discussion). *Journal of the Royal Statistical Society B* 36, 111-147.

Street, J. O., Carroll, R. J. and Ruppert, D. (1988) A note on computing robust regression estimates via iteratively reweighted least squares. *American Statistician* 42, 152–154.

Thisted, R. A. (1988) *Elements of Statistical Computing. Numerical Computation.* New York: Chapman and Hall.

Thomson, D. J. (1990) Time series analysis of Holocene climate data. *Philosophical Transactions of the Royal Society A* **330**, 601–616.

Tibshirani, R. (1988) Estimating transformations for regression via additivity and variance stabilization. *Journal of the American Statistical Association* **83**, 394–405.

Titterington, D. M., Smith, A. F. M and Makov, U. E. (1985) *Statistical Analysis of Finite Mixture Distributions.* Chichester: John Wiley and Sons.

Tsiatis, A. A. (1981) A large sample study of Cox's regression model. *Annals of Statistics* **9**, 93–108.

Tukey, J. W. (1960) A survey of sampling from contaminated distributions. In *Contributions to Probability and Statistics* eds I. Olkin, S. Ghurye, W. Hoeffding, W. Madow and H. Mann, pp. 448–485. Stanford: Stanford University Press.

Upton, G. J. G. and Fingleton, B. J. (1985) *Spatial Data Analysis by Example. Volume 1.* Chichester: John Wiley and Sons.

Velleman, P. F. and Hoaglin, D. C. (1981) *Applications, Basics, and Computing of Exploratory Data Analysis.* Boston: Duxbury.

Wahba, G. (1990) *Spline Models for Observational Data.* Philadelphia: SIAM.

Weekes, A. J. (1986) *A Genstat Primer.* London: Edward Arnold.

Weisberg, S. (1985) *Applied Linear Regression.* Second Edition. New York: John Wiley and Sons.

Weiss, S. M. and Kapouleas, I. (1989) An empirical comparison of pattern recognition, neural nets and machine learning classification methods. *Proc. 11th International Joint Conference on Artificial Intelligence, Detroit, 1989*, pp. 781–787.

White, H. (1992) *Artificial Neural Networks: Approximation and Learning Theory.* Oxford: Basil Blackwell.

Whittaker, J. (1990) *Graphical Models in Applied Multivariate Statistics.* Chichester: John Wiley and Sons.

Wichmann, B. A. and Hill, I. D. (1982) Algorithm AS183. An efficient and portable pseudo-random number generator. *Applied Statistics* **31**, 188–190. (Correction **33**, 123.)

Wilkinson, G. N. and Rogers, C. E. (1973) Symbolic description of factorial models for analysis of variance. *Applied Statistics* **22**, 392–399.

Williams, E. J. (1959) *Regression Analysis.* New York: John Wiley and Sons.

Wilson, S. R. (1982) Sound and exploratory data analysis. In *COMPSTAT 1982, Proceedings in Computational Statistics* eds H. Caussinus, P. Ettinger and R. Tamassone, pp. 447-450. Vienna: Physica-Verlag.

Wood, L. A. and Martin, G. M. (1964) Compressibility of natural rubber at pressures below 500 kg/cm^2. *Journal of Research National Bureau of Standards* **68A**, 259–268.

Yates, F. (1935) Complex experiments. *Journal of the Royal Statistical Society (Supplement)* **2**, 181–247.

Yates, F. (1937) *The Design and Analysis of Factorial Experiments.* Technical communication No. 35, Imperial Bureau of Soil Science, Harpenden, England.

Appendix A

Datasets and Software

The diskette supplied with this book contains our datasets and S code. The methods of installation differ for Unix and Windows users:

Unix The disk is a DOS format 1.44Mb diskette. Copy the files VR2.Z and scripts.Z to a suitable place in the Unix file system (and ensure that the suffix is capital Z). Then execute the commands

```
uncompress VR2.Z
sh VR2
```

to unpack the software into three subdirectories. Type

```
./INS
```

which should install all the software. If you do not have a C compiler, ./INS Sonly will install only the software which does not need compiled code (nor dynamic loading), and where possible substitute pure S versions. Those who need to use dyn.load2 (for S-PLUS, only HP 700/800 users at the time of writing), should run ./INS dyn.load2. Finally,

```
./INS tidy
```

removes source files no longer needed. The Install files contain hints and other ways of installation if these should fail. The directory test contains some tests of the compiled code. The software can then be used as a private library (see Appendix D), or if privileged access is available, the directories MASS, nnet and spatial can be linked or moved to the directory SHOME/library, where SHOME can be found by Splus HOME. It that case the file README should be appended to SHOME/library/README.

Windows The subdirectory MASS on the disk needs to be copied to a suitable place such as the main library directory, for example using XCOPY /S or the Windows File Manager. In a standard installation, the right place is

```
c:\spluswin\library\mass
```

Important: if you have S-PLUS 3.1, copy the contents of the subdirectory MASS31 to the same place, accepting all replacements of functions with the same name.

If desired, also copy across subdirectories nnet, spatial and scripts. Once installed the directories can be used as libraries; for further information consult Appendix D.

The disk also contains a file scripts.Z which can be unpacked to files giving the in-text code for each chapter, a directory scripts for Windows users and a directory fixes giving patches to various versions of S-PLUS.

A.1 Directories

There are three directories. Part of the spatial library is also contained in the MASS library for Windows (for compatibility with the first printing and earlier versions of S-PLUS for Windows).

MASS

This contains all the datasets and a number of S functions, as well as number of other datasets which we have used in learning or teaching.

abbey	nickel in syenite rock
accdeaths	US accidental deaths 1973–8
Aids2	Australian AIDS survival data (not in Windows)
animals	body and brain weights from Rousseeuw & Leroy (1987)
birthwt	data on low birth weights for babies from Hosmer & Lemeshow (1989)
cats	cat weights from Fisher (1947)
cement	dataset on heat evolved in setting cements
chem	copper in wholemeal flour
coop	co-operative trial in analytical chemistry
cpus	dataset on performance of cpus
crabs	measurements on *Leptograpsus* crabs
DDT	DDT in kale
deaths	time series on UK lung deaths 1974–9 from Diggle (1990)
mdeaths, fdeaths	as above, for males and females
faithful	duration and waiting time for eruption from Old Faithful
forbes	Forbes' dataset on boiling points, from Atkinson (1985)
galaxies	velocities of 82 galaxies, from Roeder (1990)
gehan	remission times on leukemia patients (censored)
hills	dataset on times of Scottish hill races
immer	bivariate randomized block experiment on barley
jasa1	Stanford heart transplants
leuk	(uncensored) survival times on leukemia patients
lh	dataset on luteinizing hormone from Diggle (1990)
mammals	body weight (kg) and brain weight (g) of mammals, from We berg (1985)

mcycle	motorcycle impact data from Silverman (1985)
melanoma	melanoma survival from Andersen *et al.* (1993)
michelson	Michelson's measurements of the speed of light
minn38	status of Minnesota high-school leavers of 1938
motors	accelerated life testing on motorettes
newcomb	passage times for light
nottem	time-series of temperatures in Nottingham, 1920–1939
oats	split-plot trial of oat varieties
painters	de Piles' scores for painters
phones	Belgian 'phone calls 1950–1973
quine	school absences from Aitkin (1978)
road	dataset on road deaths in the US
rock	dataset on relating permeability to physical measurements
rotifer	numbers of rotifers of two species falling out of suspension for different fluid densities
rubber	experimental data on rubber wear
ships	ship damage incidents, from McCullagh & Nelder (1989)
shrimp	shrimp in shrimp cocktail
shuttle	decision tree problem from Michie (1989)
steam	steam pressure data from Draper & Smith (1981)
stormer	Stormer viscometer calibration data from Williams (1959)
survey	survey of Adelaide University students
trees	black cherry trees heights, diameters and volumes
wtloss	weight loss by an obese patient
write.matrix	write matrix or numeric data frame
eqscplot	equally-scaled plots
ttest	two-sample t test
area	integration routine
fbeta	test for area
print.abbrev	print method for abbreviate class
polynom	polynomials via horner.c (not **Windows**)
nclass.scott	choose number of histogram classes
nclass.FD	choose number of histogram classes
hist.scott	histogram with other bin choices
hist.FD	histogram with other bin choices
frequency.polygon	
nclass.freq	choose number of classes for frequency polygon
ucv, bcv	cross-validation functions for density
width.SJ	bandwidth choice for density
boxcox	Box-Cox transformations for regression
stdres	standardized residuals for linear models
studres	studentized residuals for linear models
vcov	variance-covariance summary for regression-like objects
dose.p	function for LD50-like calculations
anova.negbin	functions for full negative binomial family
glm.convert	ditto
glm.nb	ditto
negative.binomial	ditto
summary.negbin	ditto

`theta.md`	ditto
`theta.mm`	ditto
`rnegbin`	simulation from the full negative binomial distribution
`mad`	re-scaled median absolute deviation (not **Windows**)
`IQR`	inter-quartile range
`huber`	Huber location with MAD scale
`hubers`	Huber proposal 2
`rlm`	robust linear model fitting (by Huber M-estimator)
`hsreg`	workhorse for `rlm`
`print.rlm`	print function
`summary.rlm`	summary function
`deviance.nls`	deviance function for `nls` fits
`rms.curv`	curvature measures for non-linear regressions
`deriv3`	symbolic differentiations
`D, make.call`	extended functions
`plot.survfit`	modified function to plot survivor curves
`lda`	linear discrimination
`predict.lda`	predict function for `lda`
`qda`	quadratic discrimination
`predict.qda`	predict function for `qda`
`sammon`	Sammon's non-linear mapping (C only).
`corresp`	correspondence analysis
`p.tree`	prune tree by mis-classifications
`cpgram`	cumulative periodogram

The files `houses.dat` (Section 2.4) and `horner.c` (Section 4.7) are supplied in this directory.

nnet

Software for feed-forward neural networks with a single hidden layer.

`nnet`	fit neural network using variable metric optimizer
`predict.nnet`	predict future observations
`print.nnet`	print method
`summary.nnet`	summary method
`nnet.Hess`	evaluate Hessian
`which.is.max`	select maximum of vector, breaking ties at random
`class.ind`	generates indicator matrix for factor

This can only be used under **Windows** in version S-PLUS 3.2 or later.

spatial

As well as the following S functions, this directory contains a number of datasets of point patterns, described in the text file `PP.files` (`PP.fil` under **Windows**). Where two functions are given, the capitalized function is the pure S version.

`correlogram, Correlogram`

```
variogram, Variogram
expcov                  covariance functions
gaucov
sphercov
surf.ls, Surf.ls        LS trend surface
surf.gls, Surf.gls      GLS trend surface
trmat, Trmat            evaluate trend surface
prmat, Prmat            kriged surface
semat, Semat            s.e. of prediction

topo                    topographic dataset
npr1                    US Naval Petroleum Reserve 1 dataset
shkap                   USSR petroleum reservoir dataset

ppinit                  initialize from datafile
ppregion                set region
Kfn                     evaluate K function
Kenvl                   form envelopes
Kaver                   form average
Psim                    Poisson/binomial simulation
SSI                     sequential inhibition process simulation
Strauss                 Strauss process simulation
pplik                   pseudo-likelihood estimation for Strauss process
```

This can only be used under **Windows** in version **S-PLUS** 3.2 or later. The kriging datasets and the pure S code are also in library `MASS` in the **Windows** version of the software, for compatibility with earlier versions.

A.2 Sources of machine-readable versions

Both the datasets and S code described in this book are available from a number of electronic file servers. They are distributed as a *shar* archive identical in structure to that on the disk. These are text files which when uncompressed and used as the argument to the Bourne shell function `sh` unpack to form a set of files. (For users of non-Unix systems a number of *uncompress* and *unshar* programs are widely available.)

Servers are available in the USA, England and Australia. Please use the one nearest to you. These will contain the latest version of the software, including corrections and enhancements.

USA The files are available by anonymous ftp, gopher and World Wide Web from `statlib`. Use ftp to `lib.stat.cmu.edu` with user `statlib` and directory `S/Modapplstat` or send email to `statlib@stat.cmu.edu` with body

```
send modapplstat from S
```

for detailed instructions.

England The files are available by anonymous ftp or World Wide Web from the directory

> `markov.stats.ox.ac.uk:/pub/Sbook`

Fetch the `README` for further details.

Australia The files are available by anonymous ftp from the directory

> `attunga.stats.adelaide.edu.au:pub/Sbook`

Fetch the `README` for further details.

A.3   Caveat

These datasets and software are provided in good faith, but none of the authors, publishers or distributors warrant their accuracy nor can be held responsible for the consequences of their use.

The software has been tested on many platforms, but we shall be unable to assist with platform-dependent difficulties.

Appendix B

Common S-PLUS Functions

This Appendix contains the call sequences and short descriptions of about 250 commonly used S-PLUS functions, with cross references to descriptions in the main text. Note that the call sequences have been abbreviated to include only the most commonly used arguments and are possibly subject to change; for full current details see the online help.

`.C(NAME, ...)`

> Page 113. Function to call an external C routine which has been loaded previously, for example by `dyn.load`.

`.First.lib(section, lib.loc)`

> Page 434. If a function named `.First.lib` exists in the library section, library will execute this function when you attach the library section. The arguments are the name of the library directory and the name of the section.

`.Fortran(NAME, ...)`

> Page 113. Function to call an external FORTRAN routine which has been loaded previously, for example by `dyn.load`.

`abline(a, b)`
`abline(coef)`
`abline(reg)`
`abline(h=, v=)`

> Page 66. Adds a line to a plot, often a regression line.

`ace(x, y, wt, monotone=NULL, linear=NULL)`

> Page 258. Performs a form of nonlinear regression which nonparametrically transforms both the y and the x variables to produce an additive model. The transformations are chosen to maximize the correlation between the transformed y and the sum of the transformed x's.

`acf(x, lag.max, type="correlation", plot=T)`

> Page 353. Estimates and plots autocovariance, autocorrelation or partial autocorrelation function.

`aggregate(x, nf=1, fun=sum, ndeltat=1)`

> Page 352. Returns a shorter time series with observations that are the result of `fun` applied to a partition of the original time series.

`all(..., na.rm=F)`

> Page 86. Gives a single logical value that is the 'and' of the logical values that are input.

`anova(object, ...)`
> Page 149. Produces an object of class `"anova"` from a fit of a statistical model, usually an analysis of variance or deviance.

`any(..., na.rm=F)`
> Page 86. Gives a single logical value that is the 'or' of the logical values that are input.

`aov(formula, data)`
> Page 147. Compute the analysis of variance for the specified model.

`apply(X, MARGIN, FUN, ...)`
> Page 47. Returns a vector or array by applying a specified function FUN to sections of an array.

`ar(x, aic=T, order.max, method="yule-walker")`
> Page 363. Fits a model of the form $x_t = \alpha_1 x_{t-1} + \cdots + \alpha_p x_{t-p} + e(t)$.

`ar.gm(x, order=1)`
> Page 369. Computes robust generalized M-estimates of autoregressive (AR) parameters and robust location and scale for time series data.

`args(x)`
> Page 26. Displays the argument names and their defaults for a function.

`arima.diag(z, gof.lag=1)`
> Page 364. Computes diagnostics for an ARIMA model. The diagnostics include the autocorrelation function of the residuals, the standardized residuals, and the portmanteau goodness-of-fit test statistic.

`arima.forecast(x, model, n, xreg, reg.coef)`
> Page 365. Forecasts a univariate time series using an ARIMA model. Also estimates standard errors.

`arima.mle(x, model, n.cond, xreg)`
> Page 363. Fit a univariate ARIMA model by Gaussian maximum (conditional) likelihood, conditioning on at least the first `n.cond` observations.

`arima.sim(model, n=100, xreg, reg.coef)`
> Page 362. Simulates a univariate ARIMA time series. By default the innovations are Gaussian.

`array(data=NA, dim, dimnames=NULL)`
> Page 42. Creates an array. Arrays are matrices and higher-dimensional generalizations of matrices.

`assign(x, value, frame, where, ...)`
> Page 109. Assigns a value to a name. The location of the assignment can either be a specific frame (useful when writing functions) or a data directory. This is not often used directly except to write to frame 0, the session frame.

`attach(what=NULL, pos=2, name)`
> Page 38. Adds a directory, data frame or list to the S search list.

`attr(x, which)`
> Page 20. Returns a specific attribute of an object. This function can also be used to create a new attribute.

`attributes(x)`
> Returns or changes all of the attributes of an object.

`avas(x, y, wt, monotone, linear)`
> Page 258. Computes a form of nonlinear regression which transforms both the dependent and independent variables to produce an additive model with constant residual variance.

`axis(side, at, labels=T, ticks=T)`

Page 72. Adds an axis to the current plot. The side, positioning of tick marks, labels and other options can be specified.

`biplot.default(obs, bivars, var.axes=T)`

Page 311. Produces a plot in which both the observations and the variables are represented in a two dimensional space.

`boxcox(model, lambda, plotit=T, ...)`

Page 171. Computes and plots log-likelihoods for the Box-Cox transformation $y^\lambda - 1$.

`boxplot(...)`

Page 129. Produces side by side boxplots from one or more vectors.

`browser(object, ...)`

Page 99. Allows the user to inspect the contents of an object or function frame.

`brush(x, collab, rowlab, hist=F, spin=T)`

Pages 68 and 302. Creates, for certain graphics devices, a matrix of all two-dimensional scatterplots of multivariate data plus optional histograms and three-dimensional spinning plot, all of which may have points highlighted interactively.

`bs(x, df, knots, degree=3, intercept=FALSE)`

Page 248. Generate a B-spline basis matrix for polynomial splines.

`c(...)`

Pages 19, 22 and 39. Concatenates objects into a vector or a list. The result is a list if one or more of the objects is a list, and a vector otherwise.

`cancor(x, y, xcenter, ycenter)`

Page 320. Performs canonical correlation analysis to find maximally correlated linear relationships between two groups of multivariate data. By default the data is centered using means.

`cat(..., file, sep=" ", fill=F)`

Page 55. Coerces arguments to mode character and then prints to standard output, or to a specified file.

`cbind(...)`

Page 32. Returns a matrix that is pieced together from several vectors and/or matrices.

`close.screen(n, all=F)`

Page 71. Closes one or all the screens of a split screen.

`cmdscale(d, k=2, eig=F, add=F)`

Page 306. Performs classical (metric) multi-dimensional scaling to represent data in a low-dimensional Euclidean space. (This is also known as principal coordinates analysis.)

`codes(object)`

Page 23. Returns a numeric vector which contains the codes used by an ordered factor. If an unordered factor is given, the codes for the sorted levels of the factor are returned.

`coef(object)`
`coefficients(object)`

Page 148. Extract the coefficients from a fitted model.

`contour(x, y, z, nlevels=5, levels=pretty(z, nlevels), add=F)`

Page 69. Makes a contour plot or adds contours.

`contrasts(x)`

Page 156. Returns the matrix which is the `contrasts` attribute of an object.

`coplot(formula, data, given.values, panel=points)`

Page 74. The coplot function produces a Conditioning Plot which shows how a response depends on a predictor given another predictor.

`corresp(table)`

Page 321. Correspondence analysis assigns scores to the rows and columns of a two-way table in such a way that they are maximally correlated amongst non-constant scores.

`cov.mve(x, cor=F, print=T)`

Page 222. Find robust multivariate estimates of location and covariance matrix from a data matrix via the minimum volume ellipsoid estimate. The result uses the mve to reject points before returning estimates based on the cleaned data.

`coxph(formula, data, subset, iter.max=10,`
`      method=c("efron","breslow","exact"),)`

Page 280. Fit a Cox proportional hazards model in survival analysis. Time-dependent variables, time-dependent strata, multiple events per subject, and other extensions are incorporated using the counting process formulation.

`cut(x, breaks, labels)`

Page 248. Creates a category object by dividing continuous data into intervals. Either specific cut points or the number of equal width intervals can be specified. Warning: this creates intervals of the form $(a, b]$, that is the cutpoint is in the interval to its left.

`cutree(tree, k=0, h=0)`

Page 312. An auxiliary function for cluster analysis. It returns a vector of group numbers for the observations that were clustered. Either the number of groups desired or a clustering height may be specified.

`cv.tree(object, rand, FUN=shrink.tree)`

Page 344. Cross-validates the tree sequence obtained by either shrinking or pruning a tree.

`data.dump(list, file="dumpdata")`

Page 55. Creates a file containing an ASCII representation of the list of objects that are named for use with `data.restore` .

`data.frame(...)`

Page 51. Constructs a data frame object from vectors, matrices and other data frames.

`data.restore(file, print=F)`

Page 55. Puts data objects into the local directory that were put into a file with `data.dump` .

`debugger(data=last.dump)`

Page 98. Allows the user to browse in the function frames after an error has occurred, provided `options(error=dump.frames)` is in use.

`density(x, n=50, window="g")`

Page 136. Kernel density estimate of univariate data. Options include the choice of the window to use and the number of points at which to estimate the density.

`deparse(expr, short=F)`

Page 92. Returns a vector of character strings representing the expression.

`detach(what=2, save=T)`

> Page 38. Removes a database from the search list. The database may be specified by a number, by name, or with a logical vector.

`deviance(object, ...)`

> Page 148. Returns the deviance (twice log-likelihood for full model minus this model) or weighted residual sum of squares of the fitted model object.

`dev.ask(ask=T)`

> Page 63. Forces every graphics device to pause before starting a new plot; press the Return key to draw the next plot.

`dev.copy(device, ..., which))`

> Page 63. Starts a graphics device given by the `device` argument or uses an active device given by `which` (default, the next on the list). It copies the current graph onto the new graphics device and sets this graphics device as the current device.

`dev.cur()`

> Page 63. Returns the number and name of the graphics device currently accepting graphics commands.

`dev.list()`

> Page 63. Returns the number and names of all active devices.

`dev.off(which=dev.cur())`

> Page 63. Shuts down the device specified by the `which` argument and returns the number of the current device after it has been shut down.

`dev.print(device=postscript, ...)`

> Page 64. Starts a graphics device given by the `device` and copies the graph from the current graphics device onto this new graphics device then turns off the new graphics device.

`dev.set(which)`

> Page 63. Returns the number and name of the (newly) current graphics device. (This may not be the one you asked for if the argument `which` does not refer to an active device.)

`diag(x)`

> Page 46. Either creates a diagonal matrix or returns the diagonal of a matrix.

`diff(x, lag=1, differences=1)`

> Page 352. Returns a time series or other object which is the result of differencing the input.

`dim(x)`

> Page 20. Returns or changes the `dim` attribute, which describes the dimensions of a matrix, data frame or array.

`dimnames(x)`

> Page 43. Returns or changes the `dimnames` attribute of an array. This is a list of the same length as the dimension of the array, giving names (possibly NULL) for each dimension.

`dist(x, metric="euclidean")`

> Page 306. Returns a distance structure that represents all of the pairwise distances between objects in the data. The choices for the metric are `"euclidean"`, `"maximum"`, `"manhattan"`, and `"binary"`.

`drop1(object, ...)`

> Page 173. Returns information about the effects of dropping each term of a model.

`dump(list, fileout="dumpdata", full.precision=T)`
> Page 55. Creates an ASCII file representing a group of S objects, suitable for editing.

`dyn.load(names, undefined=0)`
> Page 116. Makes compiled FORTRAN or C routines available to be used by S functions.

`dyn.load2(names, userlibs, syslibs, size=100000)`
> Page 116. Makes compiled FORTRAN or C routines available to be used by S functions.

`eigen(x, symmetric=all(x==t(x)), only.values=F)`
> Page 49. Returns the eigenvalues and (if desired) eigenvectors of a square matrix.

`Error(terms)`
> Page 178. Used on the right-hand side of a model formula to enclose terms, separated by +, which are to be treated as random effects.

`exists(name, where, ...)`
> Page 108. Searches for an S object and returns a logical stating whether the object exists in the search path or not. The position of the object and/or its mode can be specified.

`expand.grid(...)`
> Page 69. Creates a dataframe containing all combinations of its arguments, vectors, factors or lists of these. It is most often used to make regular grids for prediction.

`expcov(r, d, alpha=0, se=1)`
> Page 391. Computes the exponential spatial covariance function.

`faces(x, ...)`
> Page 308. Represents each multivariate observation as a Chernoff face.

`factor(x, levels, labels, exclude=NA)`
> Page 22. A factor is a character vector with class attribute `"factor"` and a levels attribute which determines what character strings may be included in the vector. The function factor creates a factor object out of data and allows the levels attribute to be set.

`fft(z, inverse=F)`
> Page 360. Performs the fast Fourier transform on a vector or array of numeric or complex values.

`fitted(object)`
`fitted.values(object)`
> Page 148. Extracts the fitted values from a fitted model.

`fix(x, file=tempfile("fix"), window=F)`
> Page 93. Invokes an editor on an S object and overwrites it with the edited version.

`format(x, ...)`
> Page 56. Coerces to character strings using a common format. The function is useful for building custom output displays, and is often used in conjunction with cat to print such displays.

`gam(formula, family=gaussian, data, weights, subset)`
> Page 251. Fits a generalized additive model.

`gaucov(r, d, alpha=0, se=1)`
> Page 391. Computes a Gaussian spatial covariance function.

`get(name, where, ...)`

> Page 38. Searches for an S object and returns the object. The position of the object and/or its mode can be specified.

`glm(formula, family=gaussian, data, weights, subset)`

> Page 187. Fits a generalized linear model.

`graphics.off()`

> Page 63. Shuts down all of the active graphics devices.

`hclust(dist, method="compact", sim=)`

> Page 312. Performs hierarchical clustering on a distance or similarity structure. Choices of method are `"compact"` (also called complete linkage), `"average"`, and `"connected"` (also called single link).

`help(name="help", offline=F)`

> Page 7. Shows online help documentation, or prints it out.

`help.off()`

> Page 8. Shuts down the S-PLUS help system.

`help.start(gui)`

> Page 8. Starts the window system for help in Unix versions of S-PLUS. This system allows you to search for help files by topic and to display them.

`hist(x, nclass, breaks, plot=T, probability=F, ...)`

> Page 126. Plots a histogram.

`history(pattern=".", max=10, menu=T, reverse=T)`

> Page 59. Retrieves recent S expressions, optionally restricting the search to those expressions matching a specified pattern.

`huber(y, k=1.5, tol=1e-06)`

> Page 209. Compute Huber robust M-estimator for location with MAD scale.

`hubers(y, k=1.5, mu, s, initmu=median(y), tol=1e-06)`

> Page 209. Finds the Huber M-estimator for location with scale specified, scale with location specified or both if neither is specified. Huber's 'proposal 2' scale estimator is used.

`identify(x, y, labels=seq(along=x), n=length(x))`

> Page 72. Interactively identify points in a plot on a suitable graphics device.

`ifelse(test, yes, no)`

> Page 86. Vector version of 'if then else'.

`image(x, y, z, zlim=range(z), add=F)`

> Page 69. Creates an image of greylevels or colours that represent a third dimension.

`integrate(f, lower, upper, max.subdiv=100)`

> Approximates the integral of a real-valued function over a given interval, and estimates the absolute error in the approximation.

`interp(x, y, z, xo, yo, ncp=0, extrap=F)`

> Page 388. Interpolates the value of the third variable onto an evenly spaced grid of the first two variables, by Akima's method.

`is.na(x)`

> Page 31. Returns an object similar to the input which is filled with logicals denoting whether the corresponding element of the input is 'NA'.

`Kaver(fs, nsim, ...)`

> Page 395. Computes and averages `nsim` simulations of Ripley's K function.

`Kenvl(fs, nsim, ...)`
>   Page 395. Computes and finds the envelope (pointwise max and min) and average of `nsim` simulations of Ripley's K function.

`Kfn(pp, fs, k=100)`
>   Page 395. Computes Ripley's K function, actually $L = \sqrt{K/\pi}$.

`kmeans(x, centers, iter.max=10)`
>   Page 313. Performs K-means clustering, using the number of initial centres as the number of groups.

`ksmooth(x, y=NULL, kernel="box", bandwidth=0.5)`
>   Page 250. Estimates a probability density or performs scatterplot smoothing using kernel estimates.

`l1fit(x, y, intercept=T, print=T)`
>   Page 213. Performs an L_1 regression.

`lag(x, k=1)`
>   Page 352. Returns a time series like the input but shifted in time.

`lda(x, grouping, prior, tol)`
>   Page 316. Computes linear discriminant analysis of `x` based on the vector or factor of groups `grouping`.

`legend(x, y, legend)`
>   Page 73. Adds a legend to the current plot.

`length(x)`
>   Page 20. Returns the length of the object (such as a vector or list).

`library(section, first=F, help=NULL, lib.loc)`
>   Page 431. Attaches a section of an S library. A library section is a way of packaging a collection of S functions, datasets, documentation, and compiled code.

`lines(x, y, type="l")`
>   Page 66. Adds lines to the current plot. The type of line can be specified as well as other graphical parameters.

`list(...)`
>   Page 21. Creates a list from the specified objects.

`lm(formula, data, weights, subset, ...)`
>   Page 148. Fits a linear model.

`lm.influence(lm)`
>   Page 158. Computes statistics used in measuring the influence of the observations on the original fit of a linear model.

`lmsreg(x, y, intercept=T)`
>   Page 218. Returns a resistant regression estimate that minimizes the median of the squared residuals.

`lo(..., span=0.5, degree=1)`
>   Page 251. Allows the user to specify a loess fit in the formula for a generalized additive model.

`location.m(x, location, scale, psi.fun="bisquare", parameters)`
>   Page 208. Robust M-estimator of location.

`locator(n=500, type="n")`
>   Page 72. Returns the coordinates specified interactively on a plot. Points and/or lines may be added to the plot.

```
loess(formula, data, weights, subset, span=0.75, degree=2,
       family=c("gaussian", "symmetric"))
```
        Page 386. Fits a local regression (loess) model.

```
loglin(table, margin, start)
```
        Page 199. Estimates test statistics and parameter values for a log-linear analysis of a multidimension contingency table.

```
lowess(x, y, f=2/3, iter=3, delta=.01*range(x))
```
        Page 250. Gives a robust, local smooth of scatterplot data. Among other options is the fraction f of data smoothed at each point.

```
ltsreg(x, y, intercept=T)
```
        Page 218. Returns a resistant regression estimate that minimizes the sum of the smallest half of the squared residuals.

```
mad(y, center=median(y), constant=1.4826, na.rm=F, low=F)
```
        Page 205. Returns a robust scale estimate of the data. By default the median is taken as the centre of the data and the estimate is scaled to be a consistent estimator of the standard deviation at the Gaussian distribution.

```
mahalanobis(x, center, cov, inverted=F)
```
        Page 317. Returns a vector of the Mahalanobis distances for the rows of a data matrix.

```
matrix(data=NA, nrow, ncol, byrow=F, dimnames=NULL)
```
        Page 42. Creates a matrix.

```
mclass(m, n.clust, aux=NULL)
```
        Page 313. Uses the output from mclust to determine the classification corresponding to a given number of clusters.

```
mclust(x, method="S*", noise=F)
```
        Page 313. Performs hierarchical clustering via a wide range of clustering options, calculates a Bayesian criterion for choosing the number of clusters, and optionally allows for noise or 'outliers'.

```
median(x)
```
        Page 206. Computes a median.

```
misclass.tree(tree, detail=F)
```
        Page 344. Returns the number of misclassification errors for a classification tree.

```
missing(name)
```
        Page 94. Returns a logical value that indicates whether or not name was supplied as an argument to the function in which missing is called.

```
mode(x)
```
        Page 20. Returns or changes the type of the S object.

```
motif(options="", ...)
```
        Page 63. Creates a graphics window that conforms to the OSF/Motif graphical user interface on an X11 server.

```
mreloc(classification, x, method="S*", noise=F)
```
        Page 313. Performs iterative relocation for a given clustering criterion and classification. See also mclust .

```
ms(formula, data, start, control, trace=F)
```
        Page 239. Minimizes a sum of nonlinear functions over parameters in a data frame.

```
mstree(x, plane=T)
```
        Page 307. Returns a vector or a list which gives the minimal spanning tree and, optionally, multivariate planing information.

`names(x)`

> Page 20. Returns or changes the `"names"` attribute of an object, usually a list or a vector.

`nls(formula, data, start, control, trace=F)`

> Page 225. Fits a nonlinear regression model via least squares.

`nnet(x, y, weights, size, Wts, linout=F, entropy=F, softmax=F,`
`    skip=F, rang=0.7, decay=0, maxit=100, trace=T)`

> Page 263. Fits a single-hidden-layer neural network.

`nnet.Hess(net, x, y)`

> Page 263. Computes Hessian of the objective function at the fitted values of a `nnet` object.

`ns(x, df, knots, intercept=F)`

> Page 248. Generates a basis matrix for natural cubic splines.

`numeric(n)`

> Page 26. Allocates space for a vector of length `n`.

`objects(where=1, frame=NULL, pattern)`

> Page 37. Returns a vector of character strings which are the names of S objects in a position on the search list, or in a memory frame.

`openlook(options="", ...)`

> Page 63. Creates a graphics window that conforms to the OPEN LOOK graphical user interface on an X11 or X11/NeWS server.

`optimize(f, interval)`

> Approximates a local optimum of a continuous univariate function within a given interval.

`options( ...)`

> Page 57. Provides a means to control some of the behaviour of S such as the number of digits printed, the maximum number of characters to place on a line, and strategies for memory management.

`order(..., na.last=T)`

> Page 45. Returns an integer vector containing the permutation that will sort the input into ascending order.

`ordered(x, levels, labels, exclude=NA)`

> Page 23. Returns an object of class `"ordered"` which is an ordered factor.

`outer(X, Y, FUN="*", ...)`

> Page 46. Performs an outer product operation given two arrays (or vectors) and, optionally, a function.

`pairs(x, labels=dimnames(x)[[2]], ...)`

> Pages 67 and 301. Creates a figure which contains all of the scatterplots of one variable against another.

`par(...)`

> Page 76. Provides control over the action of the graphics device—layout, colour and so on.

`parameters(x)`

> Page 226. Returns or sets all the parameters in a parametrized data frame.

`paste(..., sep=" ", collapse=NULL)`

> Page 39. Returns a vector of character strings which is the result of pasting corresponding elements of the input vectors together. If the `collapse` argument is used, a single string is returned.

persp(x, y, z, xlab="X", ylab="Y", zlab="Z", eye)
> Page 69. Creates a perspective plot given a matrix that represents heights on an evenly spaced grid. Axes and/or a box may be placed on the plot.

plclust(tree)
> Page 312. Creates a plot of a clustering tree given a structure produced by hclust or mclust.

plot(x, ...)
> Page 65. Creates a plot on the current graphics device. A generic function which can plot many types of object.

pmax(...)
> Page 28. 'Parallel' maximum for a number of vectors.

pmin(...)
> Page 28. 'Parallel' minimum for a number of vectors.

points(x, y, type="p", ...)
> Page 66. Adds points to the current plot. The type of point can be specified as well as other graphical parameters.

poly(x, ...)
> Page 384. Returns a matrix of orthonormal polynomials, which represents a basis for polynomial regression.

polygon(x, y, density=-1, angle=45, border=T)
> Page 62. Adds a polygon with the specified vertices to the current plot.

polyroot(z)
> Page 362. Finds all roots of a polynomial with real or complex coefficients.

postscript(file="", width, height, onefile=T)
> Page 65. Allows graphics to be produced for a POSTSCRIPT printer.

post.tree(tree, digits=.Options$digits - 3, pretty=0,
          pointsize=12)
> Page 336. Generates a POSTSCRIPT presentation plot of a tree object.

ppinit(file)
> Page 395. Loads and initializes a spatial point process.

pplik(pp, R, ng=25)
> Page 396. Computes the pseudo-likelihood estimate of the parameter c of a Strauss point process.

ppreg(x, y, min.term, max.term=min.term, xpred=NULL, optlevel=2,
      bass=0, span="cv")
> Page 256. Computes projection pursuit regression, an exploratory nonlinear regression method that models y as a sum of nonparametric functions of projections of the x variables.

ppregion(xl=0, xu=1, yl=0, yu=1)
> Page 395. Sets the rectangular domain $(xl, xu) \times (yl, yu)$ for a spatial point process.

prcomp(x, retx=T)
> Page 304. Finds principal components for multivariate data, which should normally be scaled to zero mean, unit variance.

predict(object, newdata, type, se.fit=F)
> Page 149. Returns predictions from a fitted model.

```
princomp(x, covlist, scores=T, cor=F, subset=T)
```
> Page 305. Finds principal components for multivariate data. The argument `cor` determines if the scaling to a correlation matrix takes place. Another covariance matrix can be specified via `covlist`, for example the result of `cov.mve`.

```
print(x, ...)
```
> Prints the input objects.

```
prmat(obj, xl, xu, yl, yu, n)
```
> Page 388. Computes matrix of prediction in kriging over a grid.

```
prune.tree(tree, k=NULL, best=NULL, newdata)
```
> Page 342. Determines a nested sequence of subtrees of the supplied tree by recursively "snipping" off the least important splits, based upon the cost-complexity measure. If `k` is supplied, the optimal subtree for that penalty is returned, and if `best` is supplied it determines the tree size.

```
Psim(n)
```
> Page 395. Simulates a binomial spatial point process, that is a Poisson process conditioned on n points.

```
p.tree(tree, k=NULL, best=NULL, newdata)
```
> Page 345. A version of `prune.tree` that correctly computes the `"misclass"` method.

```
q()
```
> Page 5. Terminates the current S session.

```
qda(x, grouping, prior)
```
> Page 317. Computes the quadratic discriminant analysis of `x` based on the vector or factor of groups `grouping`.

```
qqline(x, ...)
```
> Page 123. Fits and plots a line through the quartiles of a normal QQ-plot.

```
qqnorm(x, ...)
```
> Page 122. A normal QQ-plot.

```
qqplot(x, y, plot=T)
```
> Page 122. QQ-plot of two data vectors (of the same length).

```
rbind(..., deparse.level=1)
```
> Page 32. Returns a matrix that is pieced together from several vectors and/or matrices.

```
read.table(file, header, sep, row.names, col.names, as.is=F)
```
> Page 33. Reads in a file in table format and creates a data frame with the same number of rows as there are lines in the file, and the same number of variables as there are fields in the file.

```
rep(x, times, length)
```
> Page 29. Replicates the input either a certain number of times or to a certain length.

```
resid(object)
residuals(object)
```
> Page 148. Extracts the residuals from a fitted model.

```
rm(..., list=NULL)
```
> Page 38. Removes objects from the working directory.

```
rreg(x, y, w=rep(1,nrow(x)), int=T, method=wt.default)
```
> Page 215. Robust regression by an M-estimator.

```
rts(x=NA, start=1, deltat=1, frequency=1)
```
> Page 349. Defines a univariate or multivariate regularly spaced time series.

`rug(x, ticksize, side=1, lwd)`
> Page 260. Adds a rug (a series of ticks indicating the data locations) to a plot.

`s(x, df=4, spar=0)`
> Page 251. Specifies a smooth function in the formula for a generalized additive model.

`sammon(d, y=cmdscale(d), k=2, niter=100, trace=T, magic=0.2,`
`       tol=1e-4)`
> Page 307. Computes Sammon's non-linear mapping, one form of non-metric multidimensional scaling.

`sample(x, size, replace=F, prob)`
> Page 124. Produces a vector of length size of objects randomly chosen from the population.

`scale(x, center=T, scale=T)`
> Page 304. Centres and then scales the columns of a matrix. By default each column in the result has mean 0 and sample standard deviation 1.

`scale.tau(y, center=median(y), tuning=1.95)`
> Page 208. Returns a robust scale estimate of the data. By default the median is taken as the centre of the data and the estimate is scaled to be a consistent estimator of the standard deviation at the Gaussian model.

`scan(file="", what=numeric(), n, sep, multi.line=F, ...)`
> Page 34. Reads data from a text file or interactively from standard input. Options are available to control how the file is read and the structure of the data object.

`screen(n, new=T)`
> Page 71. Changes the current screen in a split screen.

`search()`
> Page 36. Adds a directory or other object to the S search list or returns the current search list.

`segments(x1, y1, x2, y2)`
> Page 82. Adds line segments to the current plot.

`semat(obj, xl, xu, yl, yu, n, se)`
> Page 388. Computes a matrix of standard errors of prediction in kriging over a grid.

`seq(...)`
> Page 28. Creates a vector of evenly spaced numbers. The beginning, end, spacing and length of the sequence can be specified.

`set.seed(i)`
> Page 124. Puts the random number generator into one of $i = 1, \ldots, 1000$ reproducible states.

`sink(file, command, append=F)`
> Page 55. Directs the output to a file rather than to the terminal, or back if file is missing.

`sort(x, partial=NULL, na.last=NA)`
> Page 44. Returns a vector which is a sorted version of the input. By default missing values are deleted.

`sort.list(x, partial=NULL, na.last=T)`
> Page 45. Returns a vector of integers giving the order of the data.

`source(file, local=F)`
> Page 55. Takes input from the specified file.

`spec.ar(ar.list, n.freq, frequency, plot=F)`
>   Page 366. Computes the spectrum of a time series using the results of an autoregressive fit.

`spec.pgram(x, spans=1, taper=0.1, pad=0, detrend=T, demean=F)`
>   Page 357. Estimates the spectrum of a time series by smoothing the periodogram, and optionally plots the spectral estimate.

`spec.taper(x, p=0.1)`
>   Page 359. Tapers a time series by means of a split-cosine-bell window.

`spectrum(x, method="pgram", plot=T, ...)`
>   Page 357. Estimates and plots time-series spectra using either a smoothed periodogram or a fitted autoregression.

`sphercov(r, d, alpha=0, se=1, D=2)`
>   Page 391. Computes spherical spatial covariance function in 2 or 3 dimensions.

`spin(x, collab, highlight=rep(F, nrow(x)))`
>   Page 69. Rotates multivariate data so that the three dimensional structure of the variables chosen can be perceived.

`split(data, group)`
>   Page 377. Returns a list in which each component is a vector of values from `data` that correspond to a unique value in `group`.

`split.screen(figs, screen, erase=T)`
>   Page 71. Splits a graphics display into multiple screens.

`SSI(n, r)`
>   Page 396. Simulates a spatial sequential inhibition point process, This has n points laid down successively, each at least distance r from its predecessors.

`stars(x)`
>   Page 308. Creates star plots from a matrix of multivariate data.

`stem(x, nl, scale, twodig=F, fence=2, head=T, depth=F)`
>   Page 127. Prints a stem-and-leaf plot.

`step(object, scope, scale, direction, trace=T, keep, steps)`
>   Page 175. Performs stepwise model selection.

`stepfun(x, y, type="left")`
>   Page 72. Computes a step function from (x, y) points sorted into increasing order on x. The default `type` gives the left endpoint of the step; the alternative is `"right"`.

`stl(time.series, ...)`
>   Page 367. Decomposes a time series into frequency components of variation by a sequence of loess smoothings.

`stop(message="")`
>   Page 93. Issues a message and exits from the current function.

`storage.mode(x)`
>   Page 113. Returns or changes the type of the S object.

`Strauss(n, c=0, r)`
>   Page 395. Simulates a Strauss spatial point process with inhibition distance r and inhibition strength c.

`substitute(expr, frame)`
>   Page 92. Returns an object like `expr` but with substitutions made relative to `frame`.

`summary(object, ...)`
>   Provides a summary of an object.

```
suntools(width=800, height=632)
```
Page 63. Creates a graphics window under SunView at a Sun console.
```
supsmu(x, y, wt=rep(1,length(y)), span="cv", periodic=F, bass=0)
```
Page 250. Returns a list containing x and y components that are a smoothed version of the input data. This algorithm is designed to be fast, and by default uses cross-validation to pick the span.
```
surf.gls(np, covmod, x, y, z, nx, ...)
```
Page 391. Fits a spatial trend surface by generalized least squares.
```
surf.ls(np, x, y, z)
```
Page 384. Fits a spatial trend surface by least squares.
```
Surv(time, event)
Surv(time, time2, event)
```
Page 269. Computes a regression target for censored data for use in survival analysis.
```
survdiff(formula, data, rho=0, subset)
```
Page 273. Tests if there is a difference between two or more survival curves using the G_ρ family of tests, or for a single curve against a known alternative.
```
survfit( object, data, weights, subset, newdata, individual=F,
         conf.int=.95, se.fit=T)
```
Page 269. Computes an estimate of a survival curve or computes the predicted survivor function for a cox proportional hazards model.
```
survreg(formula, data, subset, na.action, link=c("log","identity"),
        dist=c("extreme", "logistic", "gaussian", "exponential"))
```
Page 275. Regression of censored survival data on explanatory variables.
```
sweep(A, MARGIN, STATS, FUN="-", ...)
```
Page 48. Returns an array like the input A with STATS "swept" out over the margins specified. Used to remove means, rescale to unit variances, and so on.
```
switch(EXPRESSION, ...)
```
Page 86. Branches the evaluation depending on the value of the first argument.
```
svd(x, nu=min(nrow(x),ncol(x)), nv=min(nrow(x),ncol(x)))
```
Page 49. Returns a list containing the singular value decomposition of the input.
```
symbols(x, y, circles=, squares=, rectangles=, stars=,
        thermometers=, boxplots=, add=F, inches=T)
```
Page 67. Puts a symbol on a plot at each of the specified locations. The symbols can be circles, squares, rectangles, star plots, thermometers, or boxplots.
```
synchronize(database)
```
Page 38. Forces consistency between the evaluator and the data directories, frame 0, etc. Used to write out current values, or to update the current session's view of attached databases if these have been updated.
```
t(x)
```
Page 31. Matrix transpose.
```
table(...)
```
Page 51. Returns a contingency table (array) with one dimension per argument.
```
tapply(X, INDICES, FUN, ...)
```
Page 52. Applies a function to each cell of a ragged array.
```
text(x, ...)
text.default(x, y, labels=seq(along=x))
```
Page 66. Places text at the given positions in the current plot.

`title(main, sub, xlab, ylab, axes=F)`
> Page 72. Adds titles to the current plot.

`trace(what, tracer=T, exit, at=1, print=T)`
> Page 100. Adds to the list of functions that are to be traced.

`tree(formula)`
> Page 335. Grows a tree object from a specified formula and data.

`trmat(obj, xl, xu, yl, yu, n)`
> Page 384. Computes a fitted spatial trend surface over a grid of points.

`ts.intersect(..., dframe=F)`
> Page 352. Binds several time series together into a single multivariate time series whose time domain is the intersection of the time domains of the component series.

`ts.plot(..., type="l")`
> Page 351. Plots one or more time series on the current graphics device.

`ts.union(..., dframe=F)`
> Page 352. Binds several time series together into a single multivariate time series whose time domain is the union of the time domains of the component series.

`twoway(x, trim=.5, iter=6, eps, print=F)`
> Page 211. Returns a list containing estimated row and column effects as well as a grand effect and the residuals. The default is to give estimates from a median polish.

`update(object, formula, ..., evaluate=T, class)`
> Page 148. Allows a new model to be created from an old model by providing only those arguments that need to be changed.

`uniroot(f, interval)`
> Approximates a root or zero of a continuous univariate function given an interval for which the function has opposite signs at the endpoints.

`var(x, y=x)`
> Page 28. Returns the variance of a vector or the covariance matrix of a data matrix or two vectors.

`varcomp(formula, data, method="minque0")`
> Page 167. Returns an object of class `"varcomp"` which provides estimates of variance components, coefficients and the random variables. The default estimates are MINQUE0, but REML, maximum likelihood and Winsorized estimates are alternatives.

`vcov(x, ...)`
> Page 188. Returns the variance-covariance matrix of the coefficients of a fitted model.

`warning(message="")`
> Page 93. Issues a message from a function.

`window(x, start=start(x), end=end(x))`
> Page 352. Returns a time series with a new start and/or end.

`write(x, file="data", ncolumns, append=F)`
> Page 54. Writes the contents of the data to a file in ASCII format.

`write.table(data, file="", sep=",", dimnames.write=T)`
> Page 54. Writes a data frame or matrix to a file in ASCII format.

Appendix C

S versus S-PLUS

Both the S and S-PLUS systems are continually evolving, so the description of their differences here can only refer to the versions current at the time of writing (Spring 1994). It is intended to help S users know what we use in this book which we believe to be only in S-PLUS, and to help programmers be aware of differences between the two systems.

The major differences are the functions which have been added to form S-PLUS. Major groups of such functions are

1. The classical statistical functions discussed in Chapter 5.
2. Regression functions `stepwise` and `leaps`.
3. Experimental design functions such as `design.table`, `eff.aovlist`, `se.contrast`.
4. The variance-components function `varcomp`.
5. Robust summaries such as `mad`, `location.m`, `scale.a` and `cov.mve`.
6. Robust regression functions `l1fit`, `lmsreg` and `ltsreg`.
7. The modern regression functions `ppreg`, `ace` and `avas`.
8. Survival analysis, but we prefer to use the library available to all from `statlib`.
9. The multivariate functions `kmeans` and `mclust` (but this is available from `statlib`), and functions for principal components and factor analysis.
10. Most of the time-series functions, including the ARIMA model-fitting functions but excluding seasonal decomposition.
11. A few datasets, including `geyser` and `cancer.vet`.
12. Quality-control functions (not described here).
13. The use of multiple graphical devices such as the `dev.xxx` family of functions.
14. Utilities such as `dataset.date` and `objdiff`.
15. Mathematical functions `uniroot`, `polyroot`, `nlmin` and `optimize`.

The next group of functions are those which add extensive extra functionality to existing S functions. These are related to the help system and plotting. Virtually the whole of the help system differs, including the precise format of the help files. S-PLUS has its own graphical device drivers, management of multiple devices and a system of storing plots to be sent to other devices. The internal structure of the image function is incompatible between S and S-PLUS.

Finally, there are functions with same name whose definitions differ. These are extensive. In one release of of S-PLUS, 214 functions appear in both the splus/.Functions and either the s/.Functions or stat/.Functions directories, and nine of the functions which are only in splus/.Functions are documented as S functions in the reference books.

One functional difference is that in S the statistical models functions described in Chapters 6, 7, 9 and 13 are in library statistics not in the core, so

```
library(statistics)
```

has to be declared before they can be used. Similarly, the datasets described in Chambers & Hastie (1992) (but not used in this book) are in library data.

Appendix D

Using S Libraries

A library in S is a convenient way to package together S objects for a common purpose, and to allow these to extend the system. A library section is a directory containing a .Data subdirectory, which may contain a .Help subdirectory. The directory should also contain a README file describing its contents, and may also contain object modules for dynamic loading. The details of libraries have changed with new releases of S and S-PLUS; the description here is precise only for S-PLUS 3.2.

The structure of a library section is the same as that of a working directory, but the library function makes libraries much more convenient to use. Conventionally libraries are stored in a standard place, the subdirectory library of the main S directory. Which library sections are available can be found by the library command with no arguments; for example on one of our systems this gives:

```
> library()
Library "/usr/local/newsplus/library"
The following sections are available in the library
```

SECTION	BRIEF DESCRIPTION
chron	Functions to handle dates and times.
Defunct	Some functions that are no longer supported in S-PLUS.
demo	Demo of S-PLUS. Do not attach directly. See help("demo").
examples	Functions and objects from The New S Language.
external	Handle external (large) objects.
image	Display images.
maps	Display of maps with projections.
mathematica	Interface to Mathematica system.
progdraw	Sdraw example from Programmer's Manual.
progexam	Examples from Programmer's Manual.
semantics	Functions from chapter 11 of The New S Language
classif	classification tools
date	Terry Therneau's library of date functions

haerdle	'Smoothing Techniques' by W. Haerdle
nnet	neural networks functions
spatial	spatial statistics library
survival	model-based survival package

Further information (the contents of the README file) on any section is available by

```
library(help=section_name)
```

and the library section itself is made available by

```
library(section_name)
```

This has two actions. It attaches the .Data subdirectory of the section at the end of the search list (having checked that it has not already been attached), and executes the function .First.lib if one exists within that .Data subdirectory.

Sometimes it is necessary to have functions in a library which will replace standard system functions (for example to correct bugs or to extend their functionality). This can be done by attaching the library as the second dictionary on the search list with

```
library(section_name, first=T)
```

Of course, attaching other dictionaries with attach or other libraries with first=T will push previously attached libraries down the search list.

Private libraries

So far we have only considered system-wide library sections installed under the main S directory, which usually requires privileged access to the operating system. It is also possible (under S-PLUS 3.1 and later) to have a private library, by giving library the argument lib.loc or by assigning the object lib.loc in the current session dictionary (dictionary 0). This will be vector of directory names which are searched in order for library sections before the system-wide library. For example, on another of our systems we get

```
> assign(where=0, "lib.loc", "/users/ripley/S/library")
> library()
Library "/users/ripley/S/library"
The following sections are available in the library:

SECTION          BRIEF DESCRIPTION

MASS             main library
nnet             neural nets
spatial          spatial statistics
date             Terry Therneau's library of date functions
survival         model-based survival package
```

```
Library "/packages/splus3.2/library"
The following sections are available in the library:
```

SECTION	BRIEF DESCRIPTION
chron	Functions to handle dates and times.
Defunct	Some functions that are no longer supported in S-PLUS.
demo	Demo of S-PLUS. Do not attach directly. See help("demo").
examples	Functions and objects from The New S Language.
external	Handle external (large) objects.
image	Display images.
maps	Display of maps with projections.
mathematica	Interface to Mathematica system.
progdraw	Sdraw example from Programmer's Manual.
progexam	Examples from Programmer's Manual.
semantics	Functions from chapter 11 of The New S Language.

```
    . . . .
```

Because lib.loc is local to the session, it must be assigned for each session. The .First function is often a convenient place to do so (see page 57); see its help page for other ways using .First.local or the S_FIRST environment variable.

D.1 Creating a library

The following sequence of steps can be followed to create a library section.

1. Create a directory in the library with the section name.
2. Change to that directory and create .Data and .Data/.Help subdirectories.
3. Run S and create the desired objects, which will be saved within the .Data directory. Often this is best done by

   ```
   $ S < section.q
   ```

   where section.q is a file listing the functions and containing expressions to generate the datasets. (The suffix .q is recommended to avoid any confusion with assembler code, which often has suffix .s.)
4. Use the prompt function within S to create templates of help files for the objects in the library which will be publicly accessible.
5. Edit the *.d files to create the help files.
6. Copy the *.d files to the .Data/.Help directory *without* the .d extension. For example, C-shell users may use

```
foreach help (*.d)
  cp $help .Data/.Help/$help:r
end
```

and SYS V Bourne shell users can use

```
for f in *.d
do
    cp $f .Data/.Help/'basename $f .d'
done
```

7. Create a README file in the main directory describing briefly the purpose of the library and with one-line descriptions of the public objects. (The file should be called README.TXT in Windows.)

8. Add a one-line description of the library to the README file in its parent library directory. (README.TXT in Windows.)

9. Compile and link as needed any dynamically-loadable modules within the main directory (see Section 4.7).

10. Create a .First.lib object to dynamically load any modules as required. For example, for the survival library we use:

```
.First.lib <- function(lib.loc, section)
{
    path <- paste(lib.loc, section, "survival_l.o", sep = "/")
    dyn.load(path)
    options(na.action = "na.omit")
    options(contrasts = c("contr.treatment", "contr.poly"))
    cat("default contrasts changed to contr.treatment")
    cat("default na.action changed to na.omit")
}
```

Note that the options settings are those appropriate to this survival analysis library, and may not be elsewhere, hence the warning.

11. Use help.findsum(".Data") to create an index for use by help.start.

If the library section is to be distributed, it will help to create a Makefile to perform most of these steps. One way to do so is to use

```
Splus CHAPTER arguments
```

(see its on-line help), another is to copy examples such as those in our libraries.

D.2   Sources of libraries

Many S and S-PLUS users have generously collected together their functions and datasets together into libraries and made them publically available. An archive of sources for library sections is maintained at Carnegie-Mellon University as a service to the statistical profession by Mike Meyer. To obtain details of its contents by e-mail send a message to the Internet mail address

> statlib@lib.stat.cmu.edu

with body

> send index
> send index from S

Ftp to lib.stat.cmu.edu with user statlib and gopher and World Wide Web access are also available.

Amongst the libraries available from statlib are the following which have been mentioned elsewhere in this book:

ash	1D, 2D, 3D ASH code from Scott (1992)
bootstrap.funs	Bootstrap functions from Efron & Tibshirani (1993)
classif	S functions for classification
date	Manipulation of dates by T. Therneau
dyn.load2	Dynamic loading function
fda	Flexible discriminant analysis
fig	xfig graphics driver
haerdle	S, C code and datasets from Härdle (1991)
integrate	Numerical integration code
mclust	Clustering routines (already in S-PLUS)
nesi	New Environment for Statistical Inference
postscriptfonts	Additional capabilities for postscript under Unix
robeth	ROBETH code for use with Marazzi (1993)
survival4	Survival analysis version 4 by T. Therneau
survival4.doc	Documentation for survival4
xgobi	XGobi dynamic graphics package

Each of these (except robeth and xgobi) can be requested by email by the message

> send filename from S

and all can be accessed by ftp. There are a number of other archives accessible by anonymous ftp, current details of which can be found in the file FAQ (send the message

> send FAQ from S

to statlib.) This file, the *F*requently *A*sked *Q*uestions list (with answers) from the S-news mailing list, is a mine of helpful information.

The convention is to distribute libraries as 'shar' archives; these are text files which when used as scripts for the Bourne shell sh unpack to give all the files needed for a library section. Check files such as Install for installation instructions; these usually involve editing the Makefile and typing make.

Users are encouraged to share their own efforts; Mike Meyer welcomes submissions (see the file submissions from S). Submitters should try to be aware of the potential differences between different S platforms and between S and S-PLUS.

Appendix E

Command Line Editing

Working in a windowing system such as X-windows under UNIX usually allows you to modify and re-submit previous commands by a cut-and-paste method using the mouse. This is not available when working with a simple terminal, and even with a windowing system some users prefer a keyboard-based recall and edit mechanism.

Unix versions of S-PLUS have an inbuilt command line editor that allows recall, editing and re-submission of commands using the keyboard only.

There are two conventions available for the line editor, either emacs or vi style, set according to the shell environment variable S_CLEDITOR. In a csh shell to get the emacs conventions use

```
$ setenv S_CLEDITOR emacs
```

and for the vi conventions, use vi instead of emacs. This statement would normally be included in your .login file (or equivalent) and would then be performed automatically at login time.

To make the command-line editor available during an S-PLUS session start the program with

```
$ Splus -e
```

To avoid forgetting to include the -e option a handy alias for C-shell users is to include in their .cshrc file a line like

```
alias S+ 'Splus -e'
```

The usual typographical conventions apply: ^M means *hold the* Control *down while you press the* m *key*, but Esc m means *first press the* Esc *key and then the* m *key*. Note that after Esc case is significant.

Editing Actions

The S-PLUS program keeps a history of the commands you type, including the error lines, and commands in your history may be recalled, changed if necessary, and re-submitted as new commands.

In emacs style editing is started when any control character is typed, such as ^P or ^B. Any non-control or escape characters you type while in this editing phase causes those characters to be inserted in the command you are editing, displacing any existing characters to the right of the cursor.

In vi style editing is started by pressing the Esc key. At this point the keyboard is interpreted in a similar way to what it would be in the vi editor. In particular character insertion mode is started by Esc i or Esc a, characters are typed and normal editing mode is reinstated by typing a further Esc .

In either style pressing the Return command at any time causes the command to be re-submitted.

Other useful editing actions are summarized in the Table E.1. Unfortunately it does not seem to be possible to bind the motion keys to the arrow keys.

Table E.1: Command-line editor keystroke summary

	emacs style	vi style
Vertical movement		
Start editing action		Esc
Go to the previous command	^P	k
Go to the next command	^N	j
Find the last command containing text	^R text	Esc ? text
Horizontal movement		
Go to the beginning of the command	^A	^
Go to the end of the line	^E	$
Go back one word	Esc b	b
Go forward one word	Esc f	w
Go back one character	^B	h
Go forward one character	^F	l
Editing and re-submission		
Insert text at the cursor	text	Esc i text Esc
Append text after cursor	^F text	Esc a text Esc
Delete the previous character	Delete	X
Delete the character under the cursor	^D	x
Delete the present word and 'save' it	Esc d	dw
Delete to end of command and 'save' it	^K	D
Insert (yank) the last 'saved' text here	^Y	Y
Transpose this character with the next	^T	xp
Change word to lower case	Esc l	
Change word to upper case	Esc c	
Re-submit the command to S-PLUS	Return	Return

Appendix F

Answers to Selected Exercises

Chapter 2

2.1   One way is to use

```
find.val <- function(x, val) seq(along=x)[x==val]
row.names(hills)[find.val(hills$climb, 2100)]
```

although in most cases it is easier to subscript directly by a logical vector, for example `row.names(hills)[hills$climb==2100])`.

2.2   Three solutions are:

```
res <- factor(ftv); levels(res)[-(1:2)] <- "2 or more"
res <- cut(ftv, c(-1,0,1, 10))
      levels(res) <- c("0", "1", "2 or more")
merge.levels(factor(ftv), c(1,2,3,3,3,3))
```

where the first is explained in the help page for `merge.levels`.

2.3   We used

```
mad <- function(y, mu=median(y))
            median(abs(as.vector(y)-mu))
```

where `as.vector` strips off the name attribute which `median` retains in some versions of S-PLUS.

2.4   The lazy way is to use the S-PLUS function `t.test`:

```
t.int <- function(x) t.test(x)$conf.int
sapply(split(incomes, state), t.int)
```

Doing the work ourselves is little harder:

```
t.int <- function(x)
{
    m <- mean(x)
    n <- length(x)
    lim <- qt(0.975, n-1) * sqrt(var(x)/n)
    res <- c(m - lim, m + lim)
    names(res) <- c("lower", "upper")
    res
}
sapply(split(incomes, state), t.int) # or
tapply(incomes, state, t.int)
```

2.5 Look at the functions `predict.qda` and `predict.lda`, which perform this calculation. Since for vectors `x` and `m`,

$$\|x - m\|^2 = \|x\|^2 + \|m\|^2 - 2m^T x$$

we can ignore the first term in the comparison and only compute the other two. Finally, the function

```
which.is.min <- function(x) seq(length(x))[x == min(x)]
```

will pick out the minimum if used with `apply`.

2.6 The idea is to find any indices where the strings in `out` match the names of `x` and to use their negatives as an index vector. Matching is such a common problem there is a general function, `match`, to do it.

```
x.in <- x[- match(out, names(x), nomatch=0)]
```

Note the use of `nomatch=0` to generate a zero index (and hence no action) if some string in `out` is not the name of any component in `x`.

Chapter 3

3.1 To create a barchart of `Exer` we just use `plot(Exer)`, or

```
barplot(table(Exer), names=names(table(Exer)))
```

(Try them to see the differences.) For a pie chart, we need to first tabulate the frequencies:

```
exer.freq <- table(Exer)
exer.freq
 Freq Some None
  115   98   24
```

The command `pie(exer.freq)` will now create a pie chart, but to add labels to the slices we use the `names` argument

```
pie(exer.freq, names=levels(Exer))
```

Adding a legend is accomplished by using legend with the fill argument:

```
legend(locator(1), names(exer.freq), fill=1:3)
```

For the Smoke variable a slightly different approach is needed if we wish to include the missing value in the plot.

```
smoke.freq <- summary(Smoke)
smoke.freq
 Heavy Regul Occas Never NA's
    11    17    19   189    1
```

Since the missing value represents such a small proportion of the data, we highlight it with explode=5 (because NA's is the fifth category) so it is not lost in the pie:

```
pie(smoke.freq, names=names(smoke.freq), explode=5)
legend(locator(1), names(smoke.freq), fill=1:5)
```

3.2   This is straightforward once the layout is adjusted. We just increased the sizes of the margins which are to hold text.

```
ir <- rbind(iris[,,1], iris[,,2], iris[,,3])[, 3:4]
irs <- c(rep("S", 50), rep("C", 50), rep("V", 50))
par(mar=c(7,7,7,5)) # more space on label sides
plot(ir, type="n", cex=2, lwd=2, tck=-0.02)
title("The Iris Data", cex=2)
text(ir, irs, col=c(rep(2,50), rep(3,50), rep(4, 50)))
```

On-screen the title size is limited by the displayable fonts under the openlook driver (and probably others).

3.3   Our solution was

```
x <- seq(-pi, pi, length=200)
plot(x, sin(x), type="l", axes=F, ylab="", main="sin(x)")
axis(2, pos=0, yaxs="e", las=1)
axis(1, pos = -1.1, at = pi*seq(-1, 1, 1/4), tck = -0.02,
    labels = c("-Pi", "-3Pi/4", "-Pi/2", "-Pi/4", "0",
               "Pi/4", "Pi/2", "3Pi/4", "Pi"))
```

Chapter 4

4.1   A simple function to test all potential divisors on each element in turn is

```
prime0 <- function(x)
{
    if(any(x)   <= 0 || sum(x - floor(x)) > 0)
        error("not all positive integers")
    sapply(x, prime1)
}
prime1 <- function(x)
    if(x <= 10) c(T,T,T,F,T,F,T,F,F,F)[x] else {
        !any(x %% c(2, seq(3,sqrt(x),2)) == 0)}
```

However, most of the divisions will be unnecessary, so we first screen out by small divisors:

```
prime <- function(x)
{
    if(any(x)   <= 0 || sum(x - floor(x)) > 0)
        error("not all positive integers")
    ind <- sapply(x, prime2)
    ind[ind] <- sapply(x[ind], prime1)
    ind
}
prime2 <- function(x)
 if(x <= 10) c(T,T,T,F,T,F,T,F,F,F)[x] else {
        !any(x %% c(2, seq(3,min(sqrt(x),29),2)) == 0)}
```

Sieve methods could be employed, but `prime` is probably fast enough, at less than a quarter second per random integer in $\{1, \ldots 10^{10}\}$.

4.2   First we generate some test data and a naive solution, to test the answers:

```
set.seed(777)
a <- round(runif(10000), 2)
b <- round(runif(50), 2)
res <- integer(length(b))
for (i in seq(along=b)) res[i] <- sum(a <= b[i])
res
```

Then we time this and some alternatives

```
unix.time( for (i in seq(along=b)) res[i] <- sum(a <= b[i]) )
[1] 3.80 0.13 4.00 0.00 0.00
sapply(b, function(b) sum(a <= b))
as.vector(rep(1, length(a)) %*% outer(a, b, "<="))
```

Using `outer` used 20Mb, and so was slow (32 secs).

The next ideas depend on having sorted `b`, which has negligible cost. The first idea is based on using a series of sieves, the second uses `cut`, and the last two will be familiar to students of two-sample rank tests. (They rely on the ordering of ties being preserved. We could have (did) try `rank`, but this treats ties incorrectly for this purpose and is slower.)

```
sieve <- function(a, b)
{
    l <- sort.list(b); ll <- sort.list(l)
    res <- integer(length(b))
    x <- a
    for (i in seq(along=b)) {
        ind <- x > b[l[i]]
        x <- x[ind]
        res[i] <- length(x)
    }
```

```
      (length(a) - res)[ll]
  }
  usecut <- function(a, b)
  {
      l <- sort.list(b); ll <- sort.list(l)
      breaks <-   sort(c(b[l], range(a,b)+c(-1,1)))
      as.vector(cumsum(table(cut(a, breaks))))[-(1+length(b))])[ll]
  }
  interleave <- function(a, b)
  {
      l <- sort.list(b); ll <- sort.list(l)
      x <- c(a, b[l])
      ind <- c(rep(1, length(a)), rep(0, length(b)))
      ind <- ind[order(x)]
      cumsum(ind)[ind == 0][ll]
  }
  viarank <- function(a, b)
  {
      l <- sort.list(b); ll <- sort.list(l)
      x <- sort.list(sort.list(c(a, b[l])))
      (x[(length(a)+1):(length(a)+length(b))] - 1:length(b))[ll]
  }
```

The order of merit varies with the lengths of the vectors. Timings in CPU seconds were:

length(a)	length(b)	naive	sapply	sieve	usecut	interleave	viarank
10^4	50	4.0	4.0	3.0	1.7	1.0	1.3
10^5	20	8.3	15.8	13.8	14.0	10.7	15.5
500	250	1.5	1.5	1.6	0.68	0.13	0.13

If the inequality is reversed, we can change the inequality in the first four solutions. For the last three there are difficulties with the handling of ties, most easily addressed by changing the signs of the vectors and using rev on the answer.

4.4   This is an iterative calculation, which can only be vectorized by precomputing powers of the multipliers modulus the moduli. We will use a simple and slow solution, and rely on FORTRAN for a speed-up. It would seem a good idea to follow the internal random-number generator and store a seed in the working directory.

```
  whunif <- function (n, a=0, b=1)
  {
      if(exists(".WH.seed", 1)) {
          seed <- get(".WH.seed", 1)
      } else seed <- trunc(runif(3, 1, 30269))
      x <- numeric(n)
      for(i in 1:n) {
```

```
        seed[1] <- (171*seed[1]) %% 30269
        seed[2] <- (172*seed[2]) %% 30307
        seed[3] <- (170*seed[3]) %% 30323
        x[i] <- (seed[1]/30269+seed[2]/30307+seed[3]/30323) %% 1
    }
    assign(".WH.seed", seed, where=1)
    a + (b-a) * x
}
```

This took 20 secs for 1000 random numbers.

(b) We can make use of the same ideas, but call a FORTRAN subroutine:

```
WHunif <- function (n, a=0, b=1)
{
    if(exists(".WH.seed", 1)) {
        seed <- get(".WH.seed", 1)
    } else seed <- trunc(runif(3, 1, 30269))
    storage.mode(seed) <- "integer"
    x <- .Fortran("nwh",
        as.integer(n),
        seed[1], seed[2], seed[3],
        x=single(n))$x
    assign(".WH.seed", seed, where=1)
    a + (b-a) * x
}
```

where the FORTRAN code is

```
        subroutine nwh(n, x, y, z, res)
        real res(n)
        integer x, y, z
        do 10 i = 1, n
            x = mod(171*x, 30269)
            y = mod(172*y, 30307)
            z = mod(170*z, 30323)
10          res(i) = mod(x/30269.0 + y/30307.0 + z/30323.0, 1.0)
        end
```

This took less than 0.1 sec for 1000 random numbers.

4.5   The S language has no pointers, so we have to arrange for x <- x+1 to be evaluated in the correct frame. The function

```
incr <- function(x)
    eval(substitute(x <- x+1), local=sys.parent())
```

(Lubinsky's solution) will work, but will leave a new name-value pair in the frame of the calling function (or the working directory if called from the top level). It is exactly equivalent to replacing incr(a) by a <- a + 1.

   We interpret the question to ask that the value of the actual object matching x be incremented. To do that, we have to establish where x was matched, and perform the evaluation in that frame. The possible choices are the parent frame, frame 1, frame 0 and the search list, but we can only assume that database 1 is writable, so if x was found anywhere on the search list we assign in database 1.

```
incr <- function(x)
{
    dx <- deparse(substitute(x))
    if(exists(dx, frame=sys.parent()))
        assign(dx, x+1, sys.parent())
    else if(exists(dx, frame=1)) assign(dx, x+1, 1)
    else if(exists(dx, frame=0)) assign(dx, x+1, 0)
    else assign(dx, x+1, where=1)
    invisible(NULL)
}
```

4.6   A recursive way to do this is to include v[1] with all subsets of size r−1 from v[−1] , and to generate all subsets of size r from v[−1] :

```
subsets <- function(r, n, v = 1:n)
    if(r <= 0) NULL else
    if(r >= n) v[1:n] else
    rbind(cbind(v[1], Recall(r - 1, n - 1, v[-1])),
          Recall(r, n - 1, v[-1]))
```

A follow-up exercise is to write a function that generates the subsets successively rather than simultaneously; given one subset it should produce the next in lexicographic order.

4.7   There is a lazy way to do this using the system function expand.grid:

```
expand.factors <- function(...)
{
    dfr <- expand.grid(lapply(list(...), levels))
    names(dfr) <- as.character(match.call()[-1])
    dfr
}
```

This assumes the function will be called with simple factor names as the actual arguments. A more illuminating way to do this uses a pair of nested for loops:

```
expand.factors <- function(...)
{
    lev <- lapply(list(...), levels)
    df <- NULL
    for(i in lev)
    {
        df0 <- df
        df <- NULL
        for(j in i)
            df <- rbind(df, cbind(df0, j))
    }
    df <- as.data.frame(df)
```

```
    names(df) <- as.character(match.call()[-1])
    df
}
```

A recursive solution is possible, but probably no less inelegant. As a follow-up exercise you might add some error protection and flexibility. For example, how should arguments that are not factors be handled?

Chapter 5

5.2 We used the function

```
kde2d <- function(x, y, h, n = 25, lims=c(range(x), range(y)) )
{
    nx <- length(x)
    if(length(y) != nx)
        stop("Data vectors must be the same length")
    gx <- seq(lims[1], lims[2], length = n)
    gy <- seq(lims[3], lims[4], length = n)
    if (missing(h))
        h <- c(bandwidth.nrd(x), bandwidth.nrd(y))
    h <- h/4 # for S's bandwidth scale
    ax <- outer(gx, x, "-" )/h[1]
    ay <- outer(gy, y, "-" )/h[2]
    z <- matrix(dnorm(ax), n, nx) %*%
        t(matrix(dnorm(ay),n, nx))/ (nx * h[1] * h[2])
    return(list(x = gx, y = gy, z = z))
}
```

which uses the normal reference bandwidth for each marginal distribution.

Index

Entries in this font are names of S objects. Page numbers in **bold** are to the most comprehensive treatment of the topic. Appendix B provides an annotated index to common functions, indexed here in *italics*.